AF717524

MARKOV LEARNING MODELS FOR MULTIPERSON INTERACTIONS

STANFORD MATHEMATICAL STUDIES
IN THE SOCIAL SCIENCES, V
Editors: KENNETH J. ARROW,
SAMUEL KARLIN, PATRICK SUPPES

Markov Learning Models For Multiperson Interactions

by PATRICK SUPPES
Stanford University

and RICHARD C. ATKINSON
University of California, Los Angeles

STANFORD UNIVERSITY PRESS · STANFORD, CALIFORNIA · 1960

Stanford University Press
Stanford, California

© 1960 by the Board of Trustees of the
Leland Stanford Junior University

All rights reserved

Library of Congress Catalog Card Number: 60-11632
Printed in the United States of America

TO WILLIAM K. ESTES

Preface

The aim of this book is to apply a mathematical theory of behavior to small-group experiments that closely resemble game situations. For the most part we have attempted to keep the mathematical level within the range of behavioral science. The most important prerequisite is a knowledge of discrete probability theory. The relevant portion of the theory of Markov chains has been developed in the text. Occasionally, as in the derivation of maximum-likelihood estimates of learning parameters, a familiarity with the differential calculus is assumed; however, a reader unacquainted with the calculus can simply accept these results and ignore the proofs.

Readers interested in making only a general evaluation of our research are advised to read the first six sections of Chapter 1, then skip Chapter 2, which deals primarily with statistical methods of analysis, and then to turn to the first four or five sections of each of the experimental chapters (Chapters 3–11). Chapter 12 examines three lines of future research (models for social conformity, economic markets, and continuous games) and provides a survey of our general conclusions.

We have attempted to provide a detailed theoretical analysis of much of our data, and thus certain parts of the book are not easy reading even for those who are familiar with the general area. The experimental chapters summarize the results of extensive analyses of more than a quarter of a million recorded responses by more than a thousand subjects. Several new learning models and techniques of data analysis appear here for the first time. We hope these models and statistical methods will prove generally useful in the analysis of behavioral phenomena (individual as well as social).

The research reported here was begun in the fall of 1956. Its completion was possible because of the help of many people. During 1956 to 1959, J. Merrill Carlsmith, Nancy Hirschberg, Leonard Hommel, and Franklin Krasne assisted us at Stanford University in running experiments and analyzing data; William Bogartz and Ronald Turner rendered similar services at the University of California, Los Angeles. In addition, Carlsmith and Krasne performed several of the theoretical computations. During the 1959–60 academic year William Ronald Hooper, Judith Shutt, and Eleanor Willemsen have assisted in the final round of data analysis. Evan Linick has been indispensable in programming many problems for the I.B.M. 650 Computer at Stanford; we are also indebted to Richard Hill of the Western Data Processing Center for his cooperation.

Over the course of several years Samuel Karlin, Gordon Latta, and Herbert Scarf have made a number of useful suggestions concerning mathematical aspects of the theory. Julian Blau, Rose Ginsberg, Madeleine Schlag-Rey, and Joseph Zinnes read portions of the manuscript and suggested numerous improvements. We are particularly indebted to Zinnes for his careful perusal of the section on continuous games in Chapter 12.

In the preparation of the manuscript for publication we are indebted to Louise Thursby for her patient and accurate typing and to Betty Jo Prine for the excellent drawings. Various members of the editorial department of the Stanford University Press have made many invaluable suggestions for improving the organization and exposition.

Our research has had the financial support of several foundations and government agencies: During 1956–57 partial support was obtained from the Behavioral Science Division of The Ford Foundation. Since 1958 the work at Stanford University has been generously supported by a grant from the Social Science Division of the Rockefeller Foundation. From the very beginning the research has been supported by the Group Psychology Branch of the Office of Naval Research under Contract NR 171–064 with Stanford University, and since 1958 under Contract NR 170–282 with the University of California, Los Angeles. The continued encouragement during the past four years of Luigi Petrullo and Joan Criswell of the Group Psychology Branch has been warmly appreciated.

The authors also wish to express their debt to William K. Estes and Cletus J. Burke for introducing them to mathematical learning theory. One of the authors (P.S.) became interested in the subject during a year spent with Estes at the Center for Advanced Study in the Behavioral Sciences. The other author (R.C.A.) was first exposed to learning theory as a graduate student through lectures and discussions with both Burke and Estes.

Finally, it should be noted that in every sense this book and the research preceding it were joint efforts of the two authors. We feel it is the kind of book neither of us could have written alone. Needless to say, the cooperation and patience of our wives during the writing of this book have been of enormous help.

Stanford University PATRICK SUPPES
University of California, Los Angeles RICHARD C. ATKINSON

June, 1960

Contents

MARKOV LEARNING MODELS FOR MULTIPERSON INTERACTIONS

1

The Theory

1.1. Introduction

The common part of three disciplines, all actively developed within the last two decades, forms the subject matter of this book. One of these disciplines is the study by social psychologists of individuals interacting in small groups; the second is the study of learning as a stochastic process; and the third is the study of games from the standpoint of a mathematical theory of strategy. Our primary concern is with the application of learning theory to small group experiments that closely resemble game situations.

The analysis of learning behavior as a stochastic process began only within the last decade. The first serious article in this direction was William K. Estes' "Toward a Statistical Theory of Learning" in the 1950 volume of *Psychological Review*. Shortly thereafter, Robert R. Bush and Frederick Mosteller also began publishing papers on stochastic models for learning.

Of slightly older vintage but still quite recent is the development of game theory. In spite of early work by Zermelo, Borel, and others, John von Neumann's paper of 1928 is the first real landmark in the subject. The publication by von Neumann and Oskar Morgenstern in 1944 of their treatise *Theory of Games and Economic Behavior* introduced game theory to a much wider circle.

It is pertinent to remark why, before we embarked on our empirical investigations, we thought learning theory would predict the actual behavior of individuals in game situations. To begin with, we endorse the general characterization of learning given by Bush and Mosteller in the opening pages of their book, *Stochastic Models for Learning* [1955, p. 3]: "We consider any systematic change in behavior to be learning whether or not the change is adaptive, desirable for certain purposes, or in accordance with any other such criteria. We consider learning to be 'complete' when certain kinds of stability—not necessarily stereotypy—obtain."

In our experiments we generally bring two or three subjects into an interaction or game situation without telling them everything about the game. (The degree of information given them varies from experiment to experiment.)

This restriction of information immediately turns the subjects into learners as well as players of a game. A subject's behavior naturally changes systematically as his information increases.

Readers oriented toward game theory might well ask why we restricted information so severely in most experiments that the subjects were not shown the payoff matrix. We chose this approach because statistical learning theory[1] had already been tested successfully in a number of experimental studies. Our *a priori* confidence in learning theory was based mainly on the excellent quantitative results of many of these studies. Game theory, in contrast, was not originally formulated to predict behavior, but rather to recommend it. As a theory it has been normative through and through. Yet many people who are familiar with game theory have hoped that it might describe the actual behavior of uninstructed but intelligent persons under certain stable and restricted conditions. Initially we tended to think of our experiments as a kind of competition between game theory and learning theory in their relative ability to predict behavior. However, when we turned to designing actual experiments, it seemed obvious that the only reasonable thing to do was to begin by staying close to some thoroughly investigated learning setup and not worry about fidelity to the game-theory notion of a game. So our "competition" between theoretical approaches turned into the more constructive program of seeing how far learning theory can be extended to predict behavior in situations that correspond ever more closely to real games. As we shall see in Chapters 9 and 10, problems arise when this correspondence becomes very close.

Because statistical learning theory provides the theoretical background of our work, we shall begin by making certain general remarks about the present status of this theory. Although it is of recent origin, the concepts on which it is based have been current for a long time. They are the five basic concepts of association and reinforcement psychology: the three categories of *stimulus*, *response*, and *reinforcement*, and the two processes of stimulus *sampling* and stimulus *conditioning*. A great service to this theoretical orientation was performed by such earlier behavioristic psychologists as Watson, Thorndike, Guthrie, Tolman, Hull, and Skinner when they developed these basic concepts scientifically, cleanly pruning away the tangled notions of common sense and of earlier philosophical psychology.

It is too early to evaluate the permanent significance of statistical learning theory, but it is possible to draw a historical analogy to the development of classical mechanics by Newton and his successors. As the qualitative, conceptual work of Descartes was a necessary preliminary for Newton, so the theoretical work of the behaviorists was necessary to the development of statistical learning theory. The contribution of statistical learning theory is the use of concepts of association psychology to develop a genuine quantitative theory of behavior.

[1] The phrase "statistical learning theory" is now widely used in the literature to refer to the general theory that derives from Estes' basic paper of 1950.

Earlier attempts at quantitative theory, notably Hull's, did not lead to a theory that was mathematically viable: i.e., in Hull's theory it is impossible to make non-trivial derivations leading to new quantitative predictions of behavior. Our contention is that in statistical learning theory we have a theory with the same sort of "feel" about it that theories in physics have. Non-trivial quantitative predictions can be made. Once we make an experimental identification of stimuli and reinforcing events, it is usually clear how to derive predictions about responses in a manner that is not ad hoc and is mathematically exact.

To a psychologist not familiar with game theory, it might seem more meaningful to say that we have been concerned with the application of learning theory to situations involving social interaction. From this standpoint, we have ignored concepts like friendliness, cohesiveness, group pressure, and opinion discrepancy, which have been important in the recent investigations of many social psychologists, and have attempted instead to explain the detailed features of social interaction situations in terms of conditioning concepts. The exchange of information between players in a game, for instance, can be successfully analyzed in terms of an organism's ability to discriminate between stimuli, the important point being that this information may be treated in terms of the notion of stimulus in exactly the same way that perceptual stimuli in a one-person situation are handled. The social situation, qua social, does not require the introduction of new concepts. We do not claim that our experiments on highly structured game situations justify the inference that no new fundamental concepts are required to explain any social behavior. But we think it is important to demonstrate in empirical detail and with quantitative accuracy that no new concepts are needed for a substantial class of social situations.

In the stimulus-sampling theory of learning outlined in the next section, as in many other learning theories, an experiment consists of a sequence of trials. Independent of particular theoretical assumptions, the course of events for a given subject on a trial may be roughly described as follows: (1) a set of stimuli is presented; (2) a response is made by the subject; (3) a reinforcement occurs. However, it is not easy to show empirically what we mean by stimuli, responses, and reinforcements. The high degree of invariance in the identification of responses and reinforcements in the experiments reported in subsequent chapters arises from the fact that most of our experiments were performed with very similar apparatus and under relatively homogeneous conditions.

Experimental identification of the relevant stimuli is more complex. In fact, it is characteristic of many learning studies, including some of ours, that direct identification of stimuli is not possible. For example, in a simple learning situation for which the physical stimuli are constant from trial to trial, it is not clear how to enumerate the relevant stimuli and their conditioning relations. From a methodological standpoint it seems best, for these experiments, to regard the concept of stimulus as a theoretical construct that is useful in deriving experimental predictions.

On the other hand, in discrimination experiments, identification of the stimuli is often natural, and on the basis of such an identification successful predictions of behavior may be made. Nevertheless, as will be evident in subsequent chapters (see particularly Chapters 6 and 7), identification of the stimuli does not necessarily imply direct identification of the way in which the stimuli are conditioned to responses. In our analysis, these relations of conditioning will turn out to be unobservable states of a Markov process.

As noted above, our theory of behavior is based on the concepts of stimulus, response, and reinforcement. Clearly, an experiment testing the theory may fail for either of two reasons: the theory is wrong, or the wrong experimental identification of the basic concepts has been made. Various philosophers of science have been fond of emphasizing, particularly in discussions of the general theory of relativity, that by sufficiently distorting the "natural" experimental interpretation of concepts they may save any theory from failure. This viewpoint is usually referred to as the doctrine of conventionalism. It is not appropriate here to examine how scientists apply this doctrine in practice, but in our view the issues of conventionalism are not highly relevant to a newly developing science. However beautiful the structure of a theory may be, if it does not yield stable, non-artificial experimental interpretations leading to new empirical predictions, it does not have empirical significance. We believe that the interpretations of our stimulus-sampling theory of learning do have such a stable character.

The organization of this book is as follows: The present chapter describes the theory underlying our experiments. Chapter 2 is concerned with various methods, statistical and otherwise, that we used in analyzing the data. Because the fundamental theory is probabilistic in character, the conceptual separation between the first two chapters is not absolutely clear. Roughly speaking, we proceed as follows: In this chapter we derive theoretical quantities that do not depend on observed quantities, for example, asymptotic mean probabilities of response and their associated variances; in Chapter 2 we derive quantities that do depend on observed data, for example, the maximum-likelihood estimate of the learning parameter.

Chapters 3–11 are devoted to a detailed presentation of the experiments. Chapter 3 is concerned with some simple zero-sum, two-person games, and Chapter 4 with some non-zero-sum, two-person games. Chapter 5 considers a symmetric non-zero-sum, two-person game. Chapters 6 and 7 study the effects of communication between the two players in terms of what learning theorists call stimulus discrimination. Chapter 8 deals with a three-person, simple majority game. Chapter 9 describes some experiments in which the subjects were told various things about the payoff matrix. Chapter 10 analyzes the effects of monetary payoffs and contains a detailed consideration of noncontingent reinforcement. Chapter 11 studies a one-person situation designed for the measurement of utility on the basis of paired comparison data. This study has been included because of the many close historical and conceptual connections between game theory and utility. As in the other chapters, the primary theoretical analysis is derived from statistical learning theory.

In the final chapter, Chapter 12, we summarize our main conclusions and consider in outline various extensions of the theory. These extensions range from the Asch-type situation of social conformity to experimental simulation of an economic market.

1.2. Stimulus-Sampling Theory of Learning

The basic theory used in our experiments is a modification of stimulus-sampling theory as first formulated by Estes [1950] and Estes and Burke [1953].[2] The exact way in which our theory deviates from theirs is indicated later, but certainly there is no deviation in basic ideas.

We begin our discussion of the general theory by formulating in a non-technical manner its fundamental axioms or assumptions. An exact mathematical formulation is to be found in Estes and Suppes [1959*b*].

The first group of axioms deals with the conditioning of sampled stimuli, the second group with the sampling of stimuli, and the third with responses.

Conditioning Axioms

C1. *On every trial each stimulus element is conditioned to exactly one response.*

C2. *If a stimulus element is sampled on a trial, it becomes conditioned with probability θ to the response (if any) that is reinforced on that trial; if it is already conditioned to that response, it remains so.*

C3. *If no reinforcement occurs on a trial, there is no change in conditioning on that trial.*

C4. *Stimulus elements that are not sampled on a given trial do not change their conditioning on that trial.*

C5. *The probability θ that a sampled stimulus element will be conditioned to a reinforced response is independent of the trial number and the preceding pattern of events.*

Sampling Axioms

S1. *Exactly one stimulus element is sampled on each trial.*

S2. *Given the set of stimulus elements available for sampling on a trial, the probability of sampling a given element is independent of the trial number and the preceding pattern of events.*

Response Axiom

R1. *On any trial that response is made to which the sampled stimulus element is conditioned.*

[2] The phrase "stimulus-sampling theory" has a more restricted meaning than "statistical learning theory"; the latter also encompasses the kind of linear models discussed in Sec. 1.5.

There are a number of remarks to be made about these axioms, including a comparison with the Estes-Burke theory and modifications of it that have been proposed by other people.

To begin with, the axioms assume that there is a fixed number of responses and reinforcements, and a fixed set of stimulus elements for any specific experimental situation. (The formulation in Estes and Suppes [1959*b*] makes this obvious.) In all the experiments considered in this book the number of stimulus elements is assumed to be small. Because of the small number of stimulus elements, we are able to consider explicitly the appropriate Markov process derivable from the theory. We return to this point in detail later.

Readers familiar with the Estes-Burke stimulus-sampling theory will recognize the basic modification we have incorporated in C1–C5; namely, the conversion of the conditioning process itself into a probabilistic process. For Estes and Burke, the sampling of a stimulus element results in its becoming conditioned or connected to the reinforced response with probability 1, that is, deterministically. The experiments we consider lend themselves naturally to the assumption that exactly one stimulus element is presented on each trial and that the element is sampled with probability 1. If the conditioning process were then assumed to be deterministic, we would have a theory that predicts responses exactly if the sequence of stimulus presentations is known. This is, of course, too strong a theory. The conditioning model used here is a rather drastic simplification of the actual state of affairs, and it is precisely the probabilistic element of the theory that provides our predictions with the right degree of definiteness.

If, as is natural for our discrimination experiments, we suppose that exactly one stimulus element is presented to the subject on each trial, we may assume that the element is then sampled with probability θ, and thus keep the Estes-Burke deterministic theory of conditioning. In this case, an additional assumption must be made concerning the probability of a response when no sample is drawn. Various assumptions may be introduced to cover the empty sample trials, but most of them seem to be either awkward to work with or ad hoc in character. It may be mentioned in this connection that Estes and Burke initially introduced their deterministic conditioning assumptions for situations in which it was natural to assume that a large number of stimulus elements were present. We have concentrated on a different type of situation, and this probably accounts for our changed emphasis: conditioning is probabilistic, sampling is required. For certain experimental situations, their theory and ours yield different numerical values for observable quantities, but in the cases we have looked at so far the differences are too small to make a direct test feasible. Consequently it seems best to regard our conditioning assumptions as an obvious and not very deep generalization of the Estes-Burke theory. Further generalization may be obtained by introducing dependencies among the probabilities that a stimulus element will be conditioned.

Other kinds of generalization are possible. For example, Restle ([1955*a*], [1955*b*]) postulates two processes, conditioning and adaptation: relevant stimuli are conditioned; irrelevant ones are adapted out. Atkinson introduces the notion that trace stimuli [1957] and observing responses [1958] determine the stimulus elements to be sampled. LaBerge [1959*a*] weakens Axiom C1 and assumes that initially some stimulus elements are neutral in the sense that they are conditioned to no response. Independent of the merits of these various generalizations, they, like the initial work of Estes and Burke, are aimed at models with a large number of stimulus elements. For reasons that will be evident before the end of this chapter, such models are very unwieldy for analyzing our rather complicated experiments.

On the other hand, generalizations that give a more complete account of motivation are highly pertinent to our work. Two tentative models of this sort are described later.

Axioms C3 and S2 have not usually been explicitly formulated by statistical learning theorists, but they are necessary for strict derivation of the appropriate Markov process representing the course of learning. Axioms of this character are often called *independence-of-path assumptions.*

The theory formulated by our axioms would be more flexible and general if (1) Axiom S1 were replaced by the postulate that a fixed number of stimuli are sampled on each trial or that stimuli are sampled with independent probabilities, and (2) Axiom R1 were then changed to read: The probability of a response is the proportion of sampled stimulus elements conditioned to that response. However, for the set of experiments reported in this book it is scarcely possible experimentally to distinguish between S1 and R1 and their generalizations.

To avoid any confusion about our use of "stimulus," "stimulus element," and "element," we emphasize that "stimulus" and "stimulus element" are used as synonyms and that "element" almost always means "stimulus element." We make this point because some learning theorists tend to consider stimuli as objects manipulated by the experimenters and stimulus elements as different but related objects sampled by the subject. Perplexing epistemological questions may be raised about these different kinds of objects and their relations to each other, but we do not want to raise them here. We have found for our purposes that the following distinction is sufficient. The experimenter may on a given trial *present* a set of stimuli from which the subject *samples* a subset. In general the sampled subset will not be identical with the whole set presented. This notion of presentation set is worked out in formal detail in Estes and Suppes [1959*b*].

In the next three sections we consider detailed applications of the theory formulated in the above axioms. We first introduce some of the concepts and notation that we use throughout the rest of the book, beginning with the basic sample space and the main random variables we define on this sample space. Many statistical learning theorists shy away from using random variables explicitly, but we find their use indispensable, for the notion of a random variable is exactly what gives precision to the probability assertions we want to make. We remind the reader that a random variable is a (measurable) function defined on the sample space, and that the over-all probability measure on the sample space induces a probability distribution on the values of the random variable.

Our sample space X for a given experiment is just the set of all possible outcomes x of the experiment. It is helpful to think of x as a given subject's behavior in a particular realization of the experiment (or, in our situations, sometimes to think of x as the behavior of a particular pair or triple of subjects). For brevity we shall in the sequel speak of subject x when we really mean "behavior x of the subject."

For simplicity, let us first consider notation for an experiment involving only single subjects, not pairs or triples. A subject is given a sequence of trials. On each trial he makes either one of two responses, A_1 or A_2. Using boldface letters for random variables, we may thus define the *response random variable*:

$$(1.2.1) \qquad \mathbf{A}_n(x) = \begin{cases} 1 & \text{if subject } x \text{ makes response } A_1 \text{ on trial } n\,, \\ 0 & \text{if subject } x \text{ makes response } A_2 \text{ on trial } n\,. \end{cases}$$

After x's response, the correct response is appropriately indicated to him. Indication of the correct response constitutes reinforcement. On each trial, exactly one of two reinforcing events, either E_1 or E_2, occurs. The occurrence of E_i means that A_i (for $i = 1, 2$) was the correct response. Thus we may define the *reinforcement random variable*:

$$(1.2.2) \qquad \mathbf{E}_n(x) = \begin{cases} 1 & \text{if on trial } n, \text{ reinforcement } E_1 \text{ occurred for subject } x\,, \\ 2 & \text{if on trial } n, \text{ reinforcement } E_2 \text{ occurred for subject } x\,. \end{cases}$$

In experiments that permit non-reinforcement on some trials, such a trial is called an E_0 trial, and the value of $\mathbf{E}_n$ is 0. The asymmetry between the values of $\mathbf{A}_n$ and $\mathbf{E}_n$ is justified by the fact that we want to sum the $\mathbf{A}_n$'s but not the $\mathbf{E}_n$'s. When no summation is involved, we prefer to use the values 1 and 2; the usage readily generalizes to more than two responses.

The A and E notation is standard in the literature. Since the additional random variables that we need have not been explicitly used, we introduce some new notation.

The enumeration of which stimulus elements are conditioned to which responses may be represented by a *state-of-conditioning random variable* $\mathbf{C}_n$. We use the term "state" because the possible values of this random variable on a given trial correspond to the possible states of the fundamental Markov process we introduce in the next section. We could give a general definition of the state-of-conditioning random variable, but it is simpler not to because the definition depends on the number of stimulus elements in the experiment. For application in the next section, where only a single stimulus element s_1 is assumed, we define $\boldsymbol{C}_n$ as follows:

$$(1.2.3) \qquad \boldsymbol{C}_n(x) = \begin{cases} 1 & \text{if on trial } n \text{, element } s_1 \text{ is conditioned to } A_1 \text{ for subject } x \text{,} \\ 2 & \text{if on trial } n \text{, element } s_1 \text{ is conditioned to } A_2 \text{ for subject } x \text{.} \end{cases}$$

Note that for an experiment with a single stimulus element, it follows from Axioms S1 and R1 that[3]

$$(1.2.4) \qquad P(\boldsymbol{A}_n = 1 \mid \boldsymbol{C}_n = 1) = 1 \,, \qquad P(\boldsymbol{A}_n = 1 \mid \boldsymbol{C}_n = 2) = 0 \,.$$

We also need a *random variable* $\boldsymbol{F}_n$ *for effectiveness of conditioning*. The value of this random variable depends on whether the sampled stimulus element becomes conditioned to the reinforced response. If it does, the conditioning situation is effective; otherwise not. In view of the Sampling Axiom S1, which says that exactly one element is sampled on each trial, the definition of $\boldsymbol{F}_n$ is simple:

$$(1.2.5) \qquad \boldsymbol{F}_n(x) = \begin{cases} 1 & \text{if on trial } n \text{, the stimulus element sampled is effectively conditioned,} \\ 0 & \text{otherwise .} \end{cases}$$

It follows immediately from Axiom C2 that

$$(1.2.6) \qquad P(\boldsymbol{F}_n = 1) = \theta \,, \qquad P(\boldsymbol{F}_n = 0) = 1 - \theta \,.$$

We shall use F_1 and F_0 informally to indicate effective and ineffective conditioning, respectively. This usage is similar to our use of A_1, A_2, E_1, and E_2.

In most experimental studies response probabilities are computed by averaging over a block of trials as well as subjects. For this purpose, we introduce the random variable $\bar{\boldsymbol{A}}_N$ for the sum of responses:

$$(1.2.7) \qquad \bar{\boldsymbol{A}}_N(x) = \sum_{n=1}^{N} \boldsymbol{A}_n(x) \,.$$

Although this definition indicates that we sum from trial 1 of the experiment,

[3] Notation like $P(\boldsymbol{A}_n = 1)$ is standard. The explicit definition in terms of the sample space is $P(\boldsymbol{A}_n = 1) = P(\{x : \boldsymbol{A}_n(x) = 1\})$, where $\{x : \boldsymbol{A}_n(x) = 1\}$ is the set of all experimental outcomes x which have response A_1 on trial n.

ordinarily we do not. When the summation begins on trial m, and it is necessary to be explicit, the following notation will be used:[4]

$$\bar{A}_{m,N}(x) = \sum_{n=m+1}^{m+N} A_n(x) . \tag{1.2.8}$$

Obviously a troublesome problem is what to do about Definitions (1.2.1), (1.2.2), (1.2.3), (1.2.5), (1.2.7), and (1.2.8) when we turn from experiments with individual subjects to experiments with pairs of subjects. The simplest systematic device is to add a superscript (1) or (2) to designate the first or second member of the pair. The practical difficulty is that this notation is rather cumbersome. Often we shall call one member of the pair Player A and the other Player B. We then define

$$\begin{aligned} \boldsymbol{A}_n &= \boldsymbol{A}_n^{(1)} , & \alpha_n &= P(\boldsymbol{A}_n = 1) , \\ \boldsymbol{B}_n &= \boldsymbol{A}_n^{(2)} , & \beta_n &= P(\boldsymbol{B}_n = 1) , \\ & & \gamma_n &= P(\boldsymbol{A}_n = 1 , \boldsymbol{B}_n = 1) . \end{aligned} \tag{1.2.9}$$

Note that γ_n is the probability of a joint event, i.e., an A_1 response by Player A and a B_1 response by Player B. In a similar fashion we define sums of random variable and Cesáro mean probabilities:

$$\begin{aligned} \bar{\boldsymbol{A}}_N &= \sum_{n=1}^{N} \boldsymbol{A}_n , & \bar{\alpha}_N &= \frac{1}{N} \sum_{n=1}^{N} \alpha_n , \\ \bar{\boldsymbol{B}}_N &= \sum_{n=1}^{N} \boldsymbol{B}_n , & \bar{\beta}_N &= \frac{1}{N} \sum_{n=1}^{N} \beta_n , \\ & & \bar{\gamma}_N &= \frac{1}{N} \sum_{n=1}^{N} \gamma_n . \end{aligned} \tag{1.2.10}$$

Summation starting at trial m rather than at trial 1 is defined in an analogous fashion. Finally, if the appropriate limits exist, we define

$$\begin{aligned} \alpha &= \lim_{n\to\infty} \alpha_n , & \beta &= \lim_{n\to\infty} \beta_n , & \gamma &= \lim_{n\to\infty} \gamma_n , \\ \bar{\alpha} &= \lim_{N\to\infty} \bar{\alpha}_N ; & \bar{\beta} &= \lim_{N\to\infty} \bar{\beta}_N ; & \bar{\gamma} &= \lim_{N\to\infty} \bar{\gamma}_N . \end{aligned} \tag{1.2.11}$$

Because the other random variables are not often referred to explicitly in subsequent chapters we do not introduce an abbreviated notation for their use in the two-person experiments.

1.3. Simple Example: Markov Model for Noncontingent Case

We shall now illustrate the method of deriving, from the learning axioms stated in the last section, the Markov learning process for a given experimental situation. The sequence of events on a trial is:

Stimulus sampled → Response made → Reinforcement occurs → Conditioning of sampled stimulus.

---

[4] In the experimental literature there has been a fair amount of confusion between the random variables $\boldsymbol{A}_n$ and $\bar{\boldsymbol{A}}_{m,N}$.

We consider what is from a theoretical standpoint one of the simplest cases: noncontingent reinforcement. This case is defined by the condition that the probability of E_1 on any trial is constant and independent of the subject's responses. It is customary in the literature to call this probability π. Thus,

$$P(\boldsymbol{E}_n = 1) = \pi\,, \qquad P(\boldsymbol{E}_n = 2) = 1 - \pi\,. \tag{1.3.1}$$

We assume further that the set S of stimulus elements contains exactly one element, which we label s_1. The definition of the random variable $\boldsymbol{C}_n$ for the state of conditioning is thus given by (1.2.3). It is possible to interpret s_1 as the ready signal for a trial, but a physical identification of s_1 is not necessary. Moreover, if two or more stimulus elements are postulated, rather than one, there is no obvious physical interpretation of the stimuli.

What we may prove from our axioms is that the sequence of random variables $\boldsymbol{C}_1, \boldsymbol{C}_2, \boldsymbol{C}_3, \cdots, \boldsymbol{C}_n, \cdots$ is a Markov chain.[5] This means, among other things, that given a knowledge of the conditioning on trial n, the conditional probability

$$P(\boldsymbol{C}_{n+1} = j \mid \boldsymbol{C}_n = i) \tag{1.3.2}$$

is unchanged by knowledge of the conditioning on any trials preceding n. This fact is characteristic of Markov *processes*. The process is a Markov *chain* when the transition probabilities (1.3.2) are independent of n, that is, constant over trials. When we have a chain, the transition probabilities may be represented by a matrix $[p_{ij}]$; obviously the process is completely characterized by this matrix and the initial probabilities of a response. For explicitness we write

$$p_{ij} = P(\boldsymbol{C}_{n+1} = j \mid \boldsymbol{C}_n = i)\,, \tag{1.3.3}$$

that is, p_{ij} is the probability that $\boldsymbol{C}_{n+1} = j$, given that $\boldsymbol{C}_n = i$. In the usual language of Markov processes, the values i and j of the random variable $\boldsymbol{C}_n$ are the states of the process. When there is but one stimulus element and two responses, there are only two states in the process, 1 and 2 [see (1.2.3)].

We now use the axioms of the preceding section and the particular assumptions for the noncontingent case to derive the transition matrix $[p_{ij}]$. In making such a derivation it is convenient to represent by a tree the various possible events that may happen on a trial. Each sequence of branches emanating from a point must represent a mutually exclusive and exhaustive set of possibilities. Thus, suppose that at the end of trial n subject x is in state 1. At the beginning of trial $n+1$ subject x will make response A_1; then either E_1 or E_2 will occur, E_1 with probability π, and E_2 with probability $1-\pi$. We thus have the partial tree given in Fig. 1.1.

To complete the tree, we need to indicate the possibilities at a and b. At a, the stimulus s_1 will become conditioned to A_1 with probability θ (event F_1) and remain unchanged with probability $1-\theta$ (event F_0). But since x is already in

[5] For an exact definition of Markov chains, see Feller [1957] or Kemeny and Snell [1960]. For more complicated schedules of reinforcement this particular sequence of random variables may not be a Markov chain.

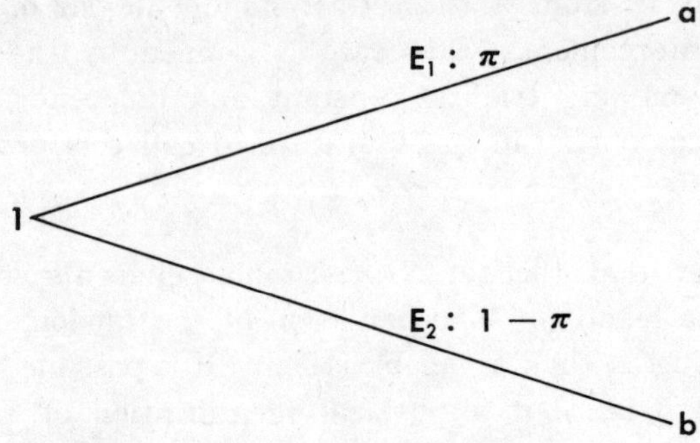

FIGURE 1.1

state 1, this means there is only one possibility at a: stay in state 1. At b the situation is different. With probability θ the stimulus s_1 will become conditioned to A_2 (event F_1), since the occurrence of E_2 reinforced A_2; that is, x will go from state 1 to state 2. And with probability $1-\theta$ event F_0 occurs, so that x will stay in state 1. We then have the complete tree given in Fig. 1.2.

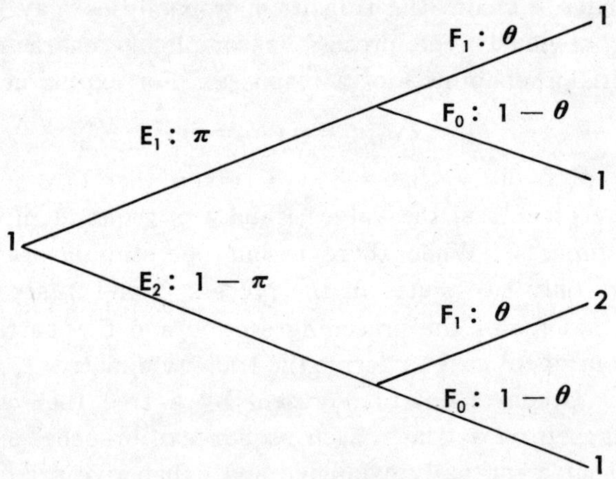

FIGURE 1.2

Assuming now that x is in state 2 at the end of trial n, we derive, by exactly the same argument, the tree given in Fig. 1.3 (clearly we always need exactly as many trees as there are states in the process to compute the transition matrix $[p_{ij}]$).

Each path of a tree, from its beginning point to a terminal point, represents a possible outcome on a given trial. The probability of each path is obtained

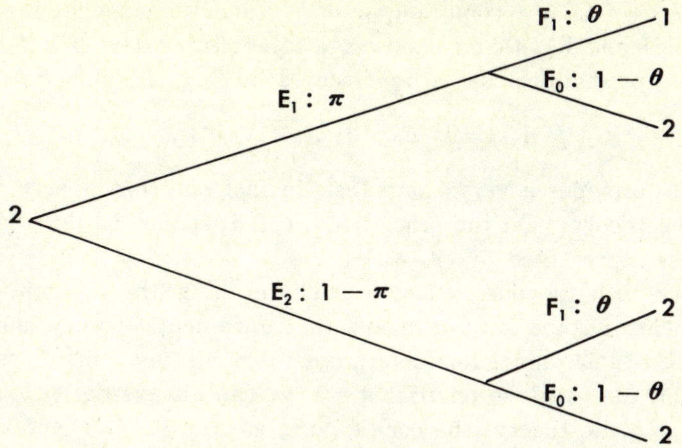

FIGURE 1.3

by multiplying the appropriate conditional probabilities. Thus for the tree of Fig. 1.2, the probability of the top path of the four may be represented by

$$\begin{aligned} P(\boldsymbol{E}_n = 1,\ \boldsymbol{F}_n = 1) &= P(\boldsymbol{F}_n = 1 \mid \boldsymbol{E}_n = 1)P(\boldsymbol{E}_n = 1) \\ &= \theta\pi\ . \end{aligned}$$

Of the four paths of Fig. 1.2, three lead from state 1 to state 1. Thus

$$\begin{aligned} p_{11} &= P(\boldsymbol{C}_{n+1} = 1 \mid \boldsymbol{C}_n = 1) \\ &= \theta\pi + (1-\theta)\pi + (1-\theta)(1-\pi) = \theta\pi + (1-\theta)\ . \end{aligned}$$

Similarly, $p_{12} = \theta(1-\pi)$. Notice, of course, that $p_{11} + p_{12} = 1$. By similar computations for Fig. 1.3, we obtain

$$\begin{aligned} p_{21} &= \theta\pi\ , \\ p_{22} &= (1-\theta)\pi + \theta(1-\pi) + (1-\theta)(1-\pi) = 1 - \theta\pi\ . \end{aligned}$$

Combining the above results, we have as the transition matrix for the noncontingent case with one stimulus element the following:

$$\text{(1.3.4)} \qquad \begin{array}{c|cc} & 1 & 2 \\ \hline 1 & \theta\pi + (1-\theta) & \theta(1-\pi) \\ 2 & \theta\pi & 1-\theta\pi \end{array}$$

Before examining some of the predictions that may be derived for the Markov chain represented by (1.3.4), let us consider the relation of this process to the learning axioms of Sec. 1.2. The central problem may be illustrated by an example. Suppose that subject x makes an A_1 response on trial n, and that reinforcing event E_1 then occurs. If there is a single stimulus element, it must be conditioned to A_1 in order for A_1 to occur on trial n. But if E_1 then occurs, according to Axiom C2 the conditioning of s_1 cannot change and we predict, using Axioms S1 and R1, that response A_1 will occur on trial $n+1$ with probability 1. This prediction may be represented by

$$P(\boldsymbol{A}_{n+1} = 1 \mid \boldsymbol{E}_n = 1,\ \boldsymbol{A}_n = 1) = 1\,. \tag{1.3.5}$$

Equation (1.3.5) provides a very sharp test, in fact one that is sure to lead to rejection of the theory. On the other hand, (1.3.5) cannot be derived from the Markov process represented by (1.3.4).

The difficulty with the one-element model (meaning the assumption of one stimulus element) is that for this model the fundamental theory laid down by Axioms C1–C5, S1, S2, and R1 is deterministic in all but minor respects. In particular, from the response on trial $n+1$ we can always derive exactly what was, according to the theory, the conditioning on trial n. In effect, the random variables $\boldsymbol{A}_n$, $\boldsymbol{E}_n$, $\boldsymbol{C}_n$, and $\boldsymbol{F}_n$ are all observable, and the values of the dependent variables $\boldsymbol{A}_n$, $\boldsymbol{C}_n$, and $\boldsymbol{F}_n$ can be predicted almost deterministically from experimental protocols for individual subjects.[6]

On the other hand, the Markov process defined by (1.3.4) leads only to probabilistic predictions for the values of $\boldsymbol{A}_n$ and $\boldsymbol{C}_n$, and to no predictions about $\boldsymbol{E}_n$ and $\boldsymbol{F}_n$. The assumed distributions of $\boldsymbol{E}_n$ and $\boldsymbol{F}_n$ are used in the derivation of (1.3.4) but are not directly observed. Contrary to the opinion of many psychologists, it is possible to compare the fit of (1.3.4) to the protocol of an individual subject. In fact, a standard goodness-of-fit test is available (see Sec. 2.4). Naturally a goodness-of-fit test can also be made for a sample of protocols drawn from a population of subjects. Thus the Markov model defined by (1.3.4) is thoroughly testable, but it is in no respect deterministic.

Granted that a deterministic theory of learning seems too much to ask for at this stage of development, it does not follow that the fundamental theory embodied in the axioms of Sec. 1.2 should be abandoned, or perhaps regarded as a make-believe theory from which realistic stochastic processes like (1.3.4) may be derived. For the fundamental theory is sharply deterministic only when we assume there is no more than one stimulus element. Notice that the axioms of Sec. 1.2 say nothing about the *number* of stimulus elements. When we assume there are two stimulus elements, say s_1 and s_2 in the noncontingent case, the random variables $\boldsymbol{C}_n$ and $\boldsymbol{F}_n$ are not observable, and few deterministic predictions can be made.

The one-element model has so many special features that it will be useful to discuss the two-element model in some detail. Let us introduce $\boldsymbol{S}_n$ as the *sampling random variable*, defined for the two-element model as follows:

$$\boldsymbol{S}_n(x) = \begin{cases} 1 & \text{if } s_1 \text{ is sampled on trial } n\,, \\ 2 & \text{if } s_2 \text{ is sampled on trial } n\,. \end{cases} \tag{1.3.6}$$

[6] If $\boldsymbol{A}_n$ and $\boldsymbol{E}_n$ have the same value, the value of $\boldsymbol{F}_n$ cannot be observed, but it does not matter in this case. If the values of $\boldsymbol{A}_n$ and $\boldsymbol{E}_n$ differ on a trial, the value of $\boldsymbol{F}_n$ is determined uniquely by the value of $\boldsymbol{A}_{n+1}$; hence it is observable since $\boldsymbol{A}_{n+1}$ is observable.

The trial sequence beginning with C_n and ending with C_{n+1} may then be represented thus:

$$C_n \to S_n \to A_n \to E_n \to F_n \to C_{n+1} \,. \tag{1.3.7}$$

We also need to define the random variable C_n, representing the state of conditioning, for the two-element model. For expository purposes we make the value of C_n simply the *set* of stimulus elements conditioned to the A_1 response. Here and subsequently 0 is used to designate the empty set as well as the number zero.[7] Thus

$$C_n = \begin{cases} \{s_1, s_2\} & \text{if } s_1 \text{ and } s_2 \text{ are both conditioned to } A_1, \\ \{s_1\} & \text{if } s_1 \text{ is conditioned to } A_1 \text{ and } s_2 \text{ is conditioned to } A_2, \\ \{s_2\} & \text{if } s_2 \text{ is conditioned to } A_1 \text{ and } s_1 \text{ is conditioned to } A_2, \\ 0 & \text{if neither element is conditioned to } A_1. \end{cases} \tag{1.3.8}$$

As (1.3.8) indicates, the two-element model leads to a four-state Markov process. In deriving this process we give below only the first two trees, i.e., those for which we are at the beginning of the trial either in state $\{s_1, s_2\}$ or in state $\{s_1\}$. The trees for the other two states are in all essentials the same. Note that we have a set of branches for each of the intermediate steps in (1.3.7); i.e., the successive branches correspond to the sequence

$$S_n \to A_n \to E_n \to F_n \,. \tag{1.3.9}$$

But since by Axioms C1 and S1 the value of A_n is uniquely determined by the values of C_n and S_n, that is, the response is uniquely determined by the state of conditioning and the single sampled stimulus element, we may reduce (1.3.9) to

$$S_n \to E_n \to F_n \,. \tag{1.3.10}$$

Each of the random variables in (1.3.10) has two possible values, which means that there are eight possible paths in each of the four trees corresponding to the four possible states of conditioning. However, some reduction in the number of paths may be made by observing that if the sampled stimulus is conditioned to the response that is reinforced, the effectiveness of conditioning is irrelevant. For example,

$$\begin{aligned} &P(C_{n+1} = \{s_1\} \mid C_n = \{s_1\},\ S_n = 1,\ E_n = 1,\ F_n = 1) \\ &\qquad = P(C_{n+1} = \{s_1\} \mid C_n = \{s_1\},\ S_n = 1,\ E_n = 1,\ F_n = 0) = 1 \,. \end{aligned} \tag{1.3.11}$$

We take account of this irrelevancy to reduce the number of paths. Thus there are not eight but six paths in the tree of any state.

Given (1.3.10) and conditional probabilities like (1.3.11), it is straightforward to derive the trees shown in Figs. 1.4 and 1.5. The tree for the state of conditioning represented by the empty set is symmetrical to Fig. 1.4, and the tree for $\{s_2\}$ is similar to Fig. 1.5.

One assignment of probabilities in both figures which is not justified by any of the preceding discussion is that of the equiprobabilities of 1/2 to the sampling

[7] We use the familiar brace notation for sets: thus $\{s_1, s_2\}$ is the set of two stimulus elements s_1 and s_2.

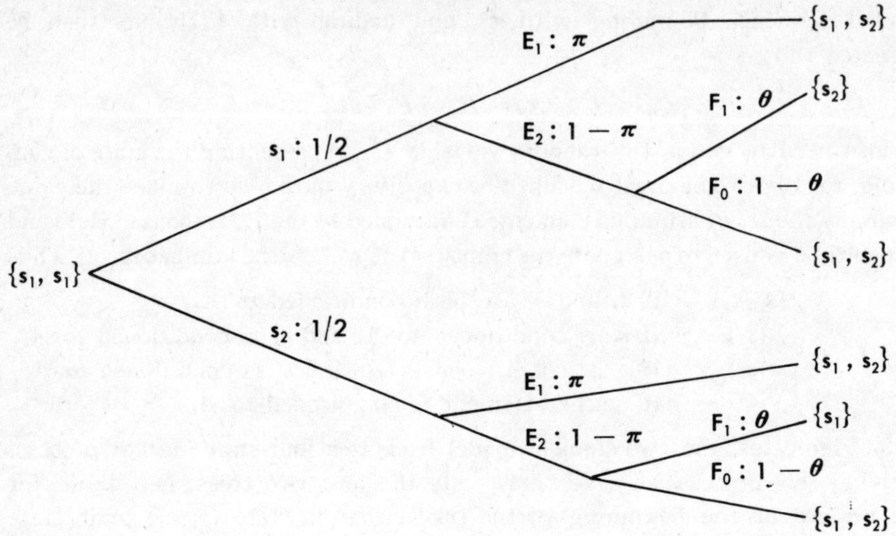

FIGURE 1.4

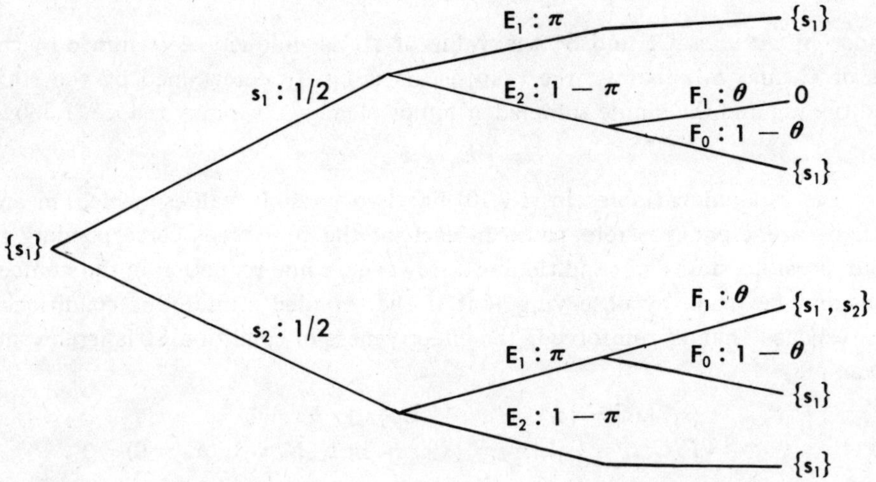

FIGURE 1.5

of s_1 and of s_2. Axiom S1 requires simply that exactly one stimulus element be sampled, and the special experimental conditions of the noncontingent case do not indicate the probabilities with which s_1 and s_2 will be sampled. We make this additional assumption in its simplest form, but it is clear how the analysis could be carried through on the supposition that s_1 is sampled with probability ω and s_2 with probability $1 - \omega$. When we come to discrimination experiments, in which one player discriminates between known responses of the other player, we shall see that the probabilities corresponding to 1/2 for s_1 and s_2 fall out in a natural manner from the theory (see Chapters 6 and 7).

From Figs. 1.4 and 1.5, and the two additional trees not displayed, the transition matrix obtained for the two-element noncontingent model is as follows:[8]

(1.3.12)

	0	$\{s_1\}$	$\{s_2\}$	$\{s_1, s_2\}$
0	$1 - \theta\pi$	$\theta\pi/2$	$\theta\pi/2$	0
$\{s_1\}$	$\theta(1 - \pi)/2$	$1 - \theta/2$	0	$\theta\pi/2$
$\{s_2\}$	$\theta(1 - \pi)/2$	0	$1 - \theta/2$	$\theta\pi/2$
$\{s_1, s_2\}$	0	$\theta(1 - \pi)/2$	$\theta(1 - \pi)/2$	$1 - \theta(1 - \pi)$

Detailed analysis is not required to see that the two-element model does not suffer from the defects of the one-element model. The probabilistic character of the Markov chain represented by (1.3.12) is also part of the elementary process given by Figs. 1.4 and 1.5. Knowing which response and which reinforcing event occurred on trial n does not permit us to make a deterministic prediction about the response on trial $n + 1$. It should be noted that the states of conditioning are not observable; consequently deterministic predictions are generally not possible in the two-element model.

On the other hand, statistical analysis of the degree to which the Markov chain of (1.3.12) fits empirical data is more difficult and less satisfactory than the corresponding analysis for the Markov chain represented by (1.3.4). We return to this problem in the next chapter. If a two-element model were adopted for each subject, all our two-person models would have at least sixteen states in the appropriate Markov chain. It is mainly to avoid the burden of this increased complexity that we have generally restricted ourselves to the one-element models.

We now turn to the derivation of asymptotic probabilities of response for the Markov chains (1.3.4) and (1.3.12). That is, we want to find the quantity

$$\lim_{n\to\infty} P(A_n = 1) \tag{1.3.13}$$

if the appropriate limit exists. In most experiments estimates of (1.3.13) are obtained by averaging over subjects *and* a final block of trials. Thus, we could as well ask for the Cesáro mean asymptotic probability

$$\lim_{N\to\infty} \frac{1}{N} \sum_{n=m+1}^{m+N} P(A_n = 1)\,. \tag{1.3.14}$$

It is well known that when both of these limits exist, they are identical, although (1.3.14) may exist and (1.3.13) not. It is also well known that for any finite-state Markov chain the limit (1.3.14) exists.

[8] In Sec. 2.3 we shall examine the possibility of collapsing states $\{s_1\}$ and $\{s_2\}$ into a single state to yield a three-state process, in which the states are designated 0, 1, 2, depending simply on the *number* of stimulus elements conditioned to A_1.

Some notation less awkward than (1.3.13) is useful. If (p_{ij}) is the transition matrix, then $p_{ij}^{(n)}$ is the probability of being in state j at trial $r+n$, given that at trial r we were in state i. We define this quantity recursively:

$$p_{ij}^{(1)} = p_{ij} , \qquad p_{ij}^{(n+1)} = \sum_{\nu} p_{i\nu} p_{\nu j}^{(n)} . \tag{1.3.15}$$

Moreover, if the appropriate limit exists and is independent of i, we set

$$u_j = \lim_{n\to\infty} p_{ij}^{(n)} . \tag{1.3.16}$$

In particular, in the one-element model discussed above,

$$u_1 = \lim_{n\to\infty} P(\boldsymbol{A}_n = 1) .$$

The limiting quantities u_j exist for any finite-state Markov chain that is irreducible and aperiodic. A Markov chain is *irreducible* if there is no closed proper subset of states, that is, no proper subset of states such that once within this set the probability of leaving it is 0. For example, the chain whose transition matrix is

	1	2	3
1	$\frac{1}{2}$	$\frac{1}{2}$	0
2	$\frac{1}{4}$	$\frac{3}{4}$	0
3	0	$\frac{1}{2}$	$\frac{1}{2}$

is reducible, because the set $\{1, 2\}$ of states is a proper closed subset. A Markov chain is *aperiodic* if there is no fixed period for return to any state; in other words, a chain is periodic if a return to some initial state j is impossible except in t, $2t$, $3t$, $\cdots$ trials, for $t > 1$. Thus the chain whose matrix is

	1	2	3
1	0	1	0
2	0	0	1
3	1	0	0

has period $t = 3$ for return to each state.

With two exceptions, all the Markov chains considered in this book are irreducible and aperiodic. Moreover, since each has only a finite number of states, the limiting quantities u_j exist in all cases and are independent of the initial state on trial 1. If there are r states, we call the vector $\boldsymbol{u} = (u_1, u_2, \cdots, u_r)$ the *stationary probability* vector of the chain. It may be shown (Feller [1957], Fréchet [1938]) that the components of this vector are the solutions of the r linear equations

$$u_j = \sum_{\nu} u_{\nu} p_{\nu j} \qquad (j = 1, \cdots, r) \tag{1.3.17}$$

such that $\sum u_j = 1$.

Thus to find the asymptotic probabilities u_j of a state, we need find only the solutions of the r linear equations (1.3.17). The intuitive basis of this system of equations seems clear. Consider a two-state process. Then the probability p_{n+1} of being in state 1 is just

$$p_{n+1} = p_{11}p_n + p_{21}(1 - p_n),$$

but at asymptote $p_{n+1} = p_n = u_1$ and $1 - p_n = u_2$, whence

$$u_1 = p_{11}u_1 + p_{21}u_2,$$

which is the first of the two equations of the system (1.3.17) when $r = 2$.

We now find the asymptotic probabilities u_j for the one-element and two-element noncontingent models. Recalling (1.3.4), we have the following transition matrix for the one-element model:

	1	2
1	$\theta\pi + (1-\theta)$	$\theta(1-\pi)$
2	$\theta\pi$	$1 - \theta\pi$

The two equations given by (1.3.17) are

$$\begin{aligned} u_1 &= [\theta\pi + (1-\theta)]u_1 + \theta\pi u_2, \\ u_2 &= \theta(1-\pi)u_1 + (1-\theta\pi)u_2, \end{aligned} \tag{1.3.18}$$

and the normalizing assumption is

$$u_1 + u_2 = 1. \tag{1.3.19}$$

Using (1.3.19) and the first equation of (1.3.18), we obtain

$$u_1 = [\theta\pi + (1-\theta)]u_1 + \theta\pi(1 - u_1),$$

and solving for u_1,

$$u_1 = \pi. \tag{1.3.20}$$

For the two-element model, the five equations (including the normalizing assumption) are, on the basis of (1.3.12),

$$\begin{aligned} u_1 &= (1-\theta\pi)u_1 + \tfrac{1}{2}\theta(1-\pi)u_2 + \tfrac{1}{2}\theta(1-\pi)u_3, \\ u_2 &= \tfrac{1}{2}\theta\pi u_1 + (1-\tfrac{1}{2}\theta)u_2 + \tfrac{1}{2}\theta(1-\pi)u_4, \\ u_3 &= \tfrac{1}{2}\theta\pi u_1 + (1-\tfrac{1}{2}\theta)u_3 + \tfrac{1}{2}\theta(1-\pi)u_4, \\ u_4 &= \tfrac{1}{2}\theta\pi u_2 + \tfrac{1}{2}\theta\pi u_3 + (1-\theta+\theta\pi)u_4, \\ u_1 &+ u_2 + u_3 + u_4 = 1. \end{aligned} \tag{1.3.21}$$

To indicate that systems of equations like (1.3.21) may often be solved by a little insight rather than by applying routine methods that guarantee an answer, we proceed to solve this system. Note initially that the first three equations simplify to

$$\pi u_1 = \tfrac{1}{2}(1-\pi)(u_2 + u_3), \tag{1.3.22}$$

$$u_2 = \pi u_1 + (1-\pi)u_4, \tag{1.3.23}$$

(1.3.24) $$u_3 = \pi u_1 + (1 - \pi)u_4 .$$

We conclude from (1.3.23) and (1.3.24) that

(1.3.25) $$u_3 = u_2 ,$$

whence we may express u_1 and u_4 in terms of u_2, using (1.3.22) and (1.3.23):

(1.3.26) $$u_1 = \frac{1 - \pi}{\pi} u_2 ,$$

(1.3.27) $$u_4 = \frac{\pi}{1 - \pi} u_2 .$$

Substituting from (1.3.25), (1.3.26), and (1.3.27) and into the fifth equation of (1.3.21), we get

(1.3.28) $$\left(\frac{1 - \pi}{\pi} + 1 + 1 + \frac{\pi}{1 - \pi}\right) u_2 = 1 .$$

Simplifying (1.3.28), we obtain

$$[(1 - \pi)^2 + 2\pi(1 - \pi) + \pi^2]u_2 = \pi(1 - \pi) ;$$

but since the coefficient of u_2 is $(1 - \pi + \pi)^2 = 1$, we infer that $u_2 = \pi(1 - \pi)$, and by appropriate substitution,

$$u_1 = (1 - \pi)^2 , \qquad u_3 = \pi(1 - \pi) , \qquad u_4 = \pi^2 .$$

Since the states of conditioning cannot be observed in the two-element model, *a fortiori* the asymptotic probabilities of these states cannot. However, for any trial n we may relate the probabilities of states of conditioning to the probability of an A_1 response, as follows:

(1.3.29) $$P(\boldsymbol{A}_n = 1) = P_n(\{s_1 , s_2\}) + \tfrac{1}{2}P_n(\{s_1\}) + \tfrac{1}{2}P_n(\{s_2\}) .$$

Note why the coefficient $\frac{1}{2}$ occurs. If the subject is in state $\{s_1\}$, then if s_1 is sampled on trial n he will make response A_1, but if s_2 is sampled he will make response A_2. And by assumption the probability of sampling s_1 is $\frac{1}{2}$. From (1.3.29) and the asymptotic probabilities of the states $\{s_1 , s_2\}$, $\{s_1\}$, and $\{s_2\}$, the following asymptotic probability of an A_1 response is obtained:

$$\lim_{n \to \infty} P(\boldsymbol{A}_n = 1) = u_4 + \tfrac{1}{2}u_2 + \tfrac{1}{2}u_3 = \pi^2 + \tfrac{1}{2}\pi(1 - \pi) + \tfrac{1}{2}\pi(1 - \pi) = \pi .$$

This result agrees with the asymptotic probability of an A_1 response in the one-element model. Moreover, it is not difficult to show that this asymptotic probability of response will be obtained on the assumption of any number r of stimulus elements and any sampling probability distribution $\omega_1 , \cdots, \omega_r$.

In the next chapter we shall be concerned with the general question of goodness-of-fit of a model to observed data, but the fundamental character of this general question should not obscure the importance of certain particular quantities such as the asymptotic probabilities of response. It is generally agreed that the first and grossest test of a model in the area of statistical learning theory is its ability to predict observed asymptotic response probabilities. From this gross test we may move in either of two directions: toward further asymptotic results or toward analysis of the rate of learning. Let us begin with the first.

It is often charged that statistical learning theory predicts only the average behavior of a group of subjects and not the behavior of individual subjects. This misconception rests on some rather widespread misunderstandings of probabilistic theories of behavior. We may illustrate what we mean by considering the variance of the sum of the random variables $\boldsymbol{A}_{m+1}, \cdots, \boldsymbol{A}_{m+n}$, at asymptote. Following (1.2.8), this quantity is defined as

$$\bar{\boldsymbol{A}}_{m,N} = \sum_{n=m+1}^{m+N} \boldsymbol{A}_n \, . \tag{1.3.30}$$

In a given experiment consisting of, say, 240 trials for each subject, we may compute $\bar{\boldsymbol{A}}_{m,N}(x)$ for each subject x over the last 100 trials; that is, we compute $\bar{\boldsymbol{A}}_{140,100}$. By averaging over subjects we obtain the expectation $E(\bar{\boldsymbol{A}}_{m,N})$, which is predicted to be $N\pi$. But from a given sample of subjects drawn from a homogeneous population, we may also compute the variance of $\bar{\boldsymbol{A}}_{m,N}$, which can then be compared with the predicted variance. Moreover, the same variance is predicted for the corresponding number of blocks of N trials for a given subject at asymptote. In either case we have a probabilistic prediction about individual subjects. For example, if $N = 10$ and we consider the last 100 trials for a given subject at asymptote, then we may directly compare the variance of the 10 blocks of 10 trials each with the theoretically computed variance of $\bar{\boldsymbol{A}}_{m,N}$.

We now turn to the derivation of this variance. For notational simplicity we drop the m subscript, particularly since at asymptote it does not matter at which trial we begin.

Theorem. *In the one-element noncontingent model the variance of* $\bar{\boldsymbol{A}}_N$ *at asymptote is*

$$\operatorname{var}(\bar{\boldsymbol{A}}_N) = N\pi(1-\pi)\frac{2-\theta}{\theta} - \frac{2\pi(1-\pi)(1-\theta)[1-(1-\theta)^N]}{\theta^2} \, . \tag{1.3.31}$$

Proof. By the classical theorem for the sum of random variables (see Feller [1957], p. 216),

$$\operatorname{var}(\bar{\boldsymbol{A}}_N) = \sum_{n=1}^{N} \operatorname{var}(\boldsymbol{A}_n) + 2 \sum_{1 \leq j < k \leq N} \operatorname{cov}(\boldsymbol{A}_j \boldsymbol{A}_k) \, . \tag{1.3.32}$$

(Note that for simplicity we sum from $n = 1$; this does not affect the result at asymptote.)

Now since at asymptote $E(\boldsymbol{A}_n) = \pi$, the variance is $\operatorname{var}(\boldsymbol{A}_n) = \pi(1-\pi)$, and

$$\operatorname{cov}(\boldsymbol{A}_j \boldsymbol{A}_k) = E(\boldsymbol{A}_j \boldsymbol{A}_k) - E(\boldsymbol{A}_j)E(\boldsymbol{A}_k) = E(\boldsymbol{A}_j \boldsymbol{A}_k) - \pi^2 \, .$$

Furthermore

$$E(\boldsymbol{A}_j \boldsymbol{A}_k) = E(\boldsymbol{A}_k \mid \boldsymbol{A}_j)E(\boldsymbol{A}_j) = \pi E(\boldsymbol{A}_k \mid \boldsymbol{A}_j) = \pi p_{11}^{(k-j)} \, .$$

Now for the transition matrix of the one-element model we may prove by induction that

$$p_{11}^{(n)} = \pi + (1-\pi)(1-\theta)^n \, , \tag{1.3.33}$$

whence, combining the above results, we get

$$\begin{aligned}\operatorname{cov}(\boldsymbol{A}_j\boldsymbol{A}_k) &= \pi[\pi + (1-\pi)(1-\theta)^{k-j}] - \pi^2 \\ &= \pi(1-\pi)(1-\theta)^{k-j} \, .\end{aligned}$$

Finally, we need the following summation:

$$\begin{aligned}\sum_{1\leqq j<k\leqq N} (1-\theta)^{k-j} &= (1-\theta)^{2-1} + [(1-\theta)^{3-2} + (1-\theta)^{3-1}] \\ &\quad + [(1-\theta)^{4-3} + (1-\theta)^{4-2} + (1-\theta)^{4-1}] \\ &\quad + \cdots + [(1-\theta) + \cdots + (1-\theta)^{N-1} \\ &= \frac{1-(1-\theta)^2}{\theta} + \cdots + \frac{1-(1-\theta)^N}{\theta} - (N-1) \\ &= \left[\frac{N-1}{\theta} - (N-1)\right] - \frac{1}{\theta}[(1-\theta)^2 + \cdots + (1-\theta)^N] \\ &= \frac{N-1}{\theta}(1-\theta) - \left(\frac{1-\theta}{\theta^2}\right)[1-\theta-(1-\theta)^N] \\ &= \frac{N}{\theta}(1-\theta) - \left(\frac{1-\theta}{\theta^2}\right)[1-(1-\theta)^N] \, .\end{aligned}$$

Using this result and substituting in (1.3.32), we then have

$$\begin{aligned}\operatorname{var}(\bar{\boldsymbol{A}}_N) &= N\pi(1-\pi) + 2\pi(1-\pi)\frac{N}{\theta}(1-\theta) - 2\pi(1-\pi)\left(\frac{1-\theta}{\theta^2}\right)[1-(1-\theta)^N] \\ &= N\pi(1-\pi)\frac{2-\theta}{\theta} - \frac{2\pi(1-\pi)(1-\theta)[1-(1-\theta)^N]}{\theta^2} \, ,\end{aligned}$$

which completes the proof.

Later, (1.3.31) will be compared with some observed data. Many further interesting asymptotic quantities can be presented, but we defer additional computations until we consider some data for the noncontingent case in Chapter 10. Other indications of how the theory may be applied to the individual subject's protocols are given in Chapter 2.

We conclude this section with a discussion of the rate of learning for the one-element noncontingent model. The learning rate may be represented by the absolute probability $u_1^{(n)}$ of an A_1 response on trial n, where

$$u_1^{(n)} = \sum_{i=1}^{2} u_i^{(1)} p_{i1}^{(n-1)} \, . \tag{1.3.34}$$

Here $u_i^{(1)}$ is the initial probability of response A_i, that is, the probability of this response on trial 1.[9]

[9] For explicitness, note that $u_1^{(1)} = P(\boldsymbol{A}_1 = 1)$ and $u_1^{(n)} = P(\boldsymbol{A}_n = 1)$.

Similar to (1.3.33), we have $p_{21}^{(n)} = \pi - \pi(1-\theta)^n$. The absolute probability $u_1^{(n)}$ is then

(1.3.35)
$$\begin{aligned} u_1^{(n)} &= u_1^{(1)}[\pi + (1-\pi)(1-\theta)^{n-1}] + u_2^{(1)}[\pi - \pi(1-\theta)^{n-1}] \\ &= \pi - (\pi - u_1^{(1)})(1-\theta)^{n-1} . \end{aligned}$$

For ($\pi > u_1^{(1)}$ and $0 < \theta < 1$), it is clear from (1.3.35) that $u_1^{(n)}$ is a strictly increasing function of n.

At this point we stop our theoretical analysis of the noncontingent case, but we return to it in Secs. 2.5 and 2.7 and again in Chapter 10.

1.4 Markov Model for Zero-Sum, Two-Person Games

In the preceding section stimulus-sampling learning theory was applied to the simplest one-person experimental case. We now want to apply it to one of the simpler two-person cases. The general experimental situation may be described as follows. On a given trial each of the two players (i.e., subjects) independently makes one of two responses. As indicated in Sec. 1.2, the players are designated as A and B, with A making response A_1 or A_2 and B making response B_1 or B_2. The probability that a given player's response will be reinforced on a particular trial depends on the actual responses made on the trial. It is by virtue of this dependence or contingency that the game aspect of the experiment arises. For example, if Player A makes response A_1 and Player B response B_1, then there is a probability a_1 that responses A_1 and B_2 are reinforced, and a probability $1-a_1$ that responses A_2 and B_1 are reinforced.[10] Thus with probability a_1 it turns out that A made a correct response and B an incorrect one; with probability $1-a_1$ the situation is reversed. The zero-sum character of the game is due to the fact that whichever one of the four response pairs (A_1B_1), (A_1B_2), (A_2B_1), (A_2B_2) actually occurs, exactly one player wins in the sense of having made a correct response. Thus the reinforcement probabilities may be described by the following payoff matrix:

(1.4.1)

	B_1	B_2
A_1	a_1	a_2
A_2	a_3	a_4

In thinking of each trial as the play of a simple 2×2 game, it is to be noted that the payoff is not a matter of being correct or incorrect, but the probability a_i of being correct.[11] The experimenter's selection of actual reinforcements on the basis of the a_i's corresponds to a chance move by the referee in a game.

[10] Using the E_i notation introduced in Sec. 1.2, there is a probability a_1 that an E_1 event occurs for Player A and an E_2 event for Player B, and a probability $1-a_1$ that an E_2 event occurs for Player A and an E_1 event for Player B.

[11] Technically we then have a constant-sum game, which is strategically equivalent to the zero-sum game obtained by subtracting $\frac{1}{2}$ from each a_i.

The choice of an appropriate strategy is made in terms of the expected values of these chance moves.

Restricting ourselves to the assumption of one stimulus element for each player, we find that the stipulation of (1.4.1) completely determines the derivation of the Markov process from the axioms of Sec. 1.2. We assume, of course, that both players satisfy the axioms. On each trial Player A's single stimulus element is conditioned to response A_1 or A_2, and Player B's single element is conditioned to response B_1 or B_2. There are, consequently, four possible states of conditioning, which will be represented by the four ordered pairs (1, 1), (1, 2), (2, 1), (2, 2). The first member of each pair indicates the conditioning of Player A's stimulus element, and the second member the conditioning of Player B's element.

In view of the detailed derivations given in the preceding section, we restrict ourselves here to deriving only one of the four trees for the Markov process. Let us assume that Player A's element is conditioned to A_1 and Player B's to B_2, i.e., that we are in the state (1, 2). We then have the tree given in Fig. 1.6. We want to make several remarks about this tree. First, in order to avoid

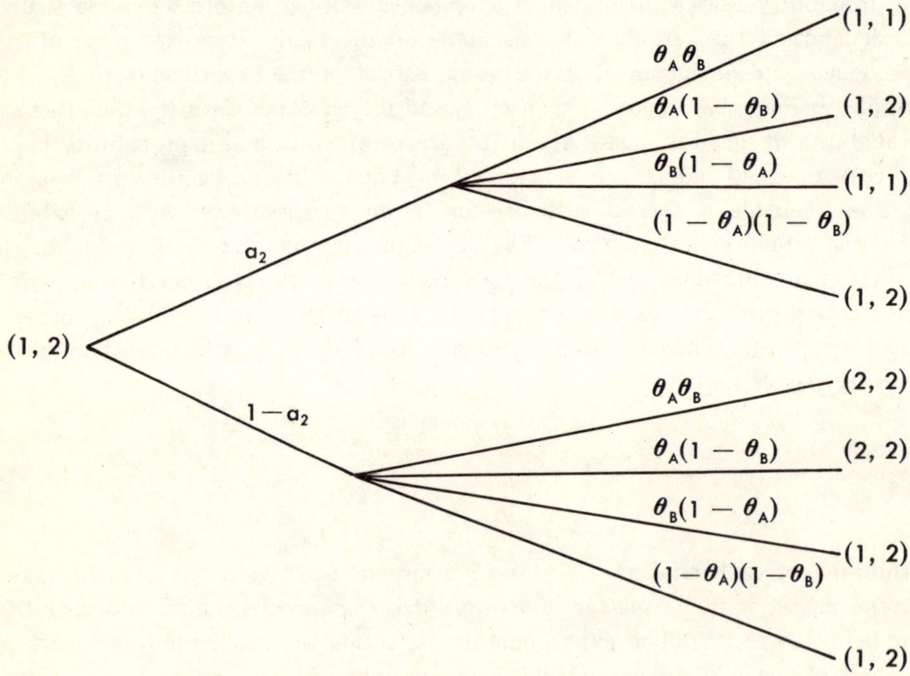

FIGURE 1.6

a further multiplication of notation, the actual reinforcement and conditioning events have not been indicated, since these events are unequivocally specified by their probabilities. For example, a_2 on the upper main branch indicates that

Player A made a correct response and Player B an incorrect one; that is, both players received an E_1 reinforcing event. The probability $\theta_A\theta_B$ at the top indicates that conditioning was effective for both players. However, the effectiveness of conditioning is irrelevant when the response actually made is reinforced (a point already discussed in Sec. 1.3); and therefore, if we consider only relevant conditioning, this tree may be reduced as shown in Fig. 1.7

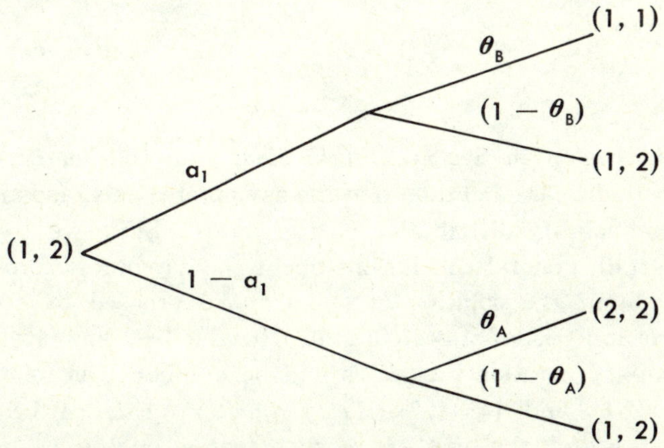

FIGURE 1.7

It should be clear from both figures that θ_A is the probability of effective conditioning for Player A, and θ_B the corresponding probability for B. That is, we have as a generalization of (1.2.6)

$$P(F_n^{(A)} = 1) = \theta_A\,, \qquad P(F_n^{(B)} = 1) = \theta_B\,. \tag{1.4.2}$$

We can see from Figs. 1.6 and 1.7 that it is impossible in a single trial to go from the state (1, 2) to the state (2, 1). Analysis of the other three trees leads to similar conclusions: it is not possible in a single trial to go from (2, 1) to (1, 2), from (1, 1) to (2, 2), or from (2, 2) to (1, 1). This means that the anti-diagonal of the transition matrix must be identically zero. With this result we bring into the Markov process itself the kind of overly strong deterministic predictions that arise in the one-element stimulus model for the noncontingent case, but not in the Markov process for that case, as represented by (1.3.4). In this respect, then, the situation is worse in the case under present discussion, for the Markov process in the response random variables yields deterministic predictions. As any experimenter would expect and as we shall see in Chapter 3, these predictions are not borne out by actual data. As in the noncontingent case, these difficulties may be met by passing from a one-element to a multi-element stimulus model for each player. The multi-element model will be discussed in later chapters.

Figure 1.7 and the other three trees corresponding to it yield the following transition matrix:

$$
\text{(1.4.3)} \quad
\begin{array}{c|cccc}
 & (1,1) & (1,2) & (2,1) & (2,2) \\
\hline
(1,1) & \begin{array}{l} a_1(\theta_A - \theta_B) \\ + (1-\theta_A) \end{array} & a_1\theta_B & (1-a_1)\theta_A & 0 \\
(1,2) & a_2\theta_B & \begin{array}{l} a_2(\theta_A - \theta_B) \\ + (1-\theta_A) \end{array} & 0 & (1-a_2)\theta_A \\
(2,1) & (1-a_3)\theta_A & 0 & \begin{array}{l} a_3(\theta_A - \theta_B) \\ + (1-\theta_A) \end{array} & a_3\theta_B \\
(2,2) & 0 & (1-a_4)\theta_A & a_4\theta_B & \begin{array}{l} a_4(\theta_A - \theta_B) \\ + (1-\theta_A) \end{array}
\end{array}
$$

From the discussion in Sec. 1.3, it is clear that this matrix represents an irreducible, aperiodic case and thus that the asymptotes exist and are independent of the initial probability distribution on the states. Since we are interested in establishing certain general conclusions about the asymptotic probabilities, it is necessary to obtain the solution of (1.4.3). We proceed by solving for the asymptotic probabilities of any four-state, irreducible, and aperiodic Markov chain. To simplify notation the states will be numbered as follows: $1 = (1, 1)$, $2 = (1, 2)$, $3 = (2, 1)$, and $4 = (2, 2)$. Let $[p_{ij}]$ $(i, j = 1, 2, 3, 4)$ be the transition matrix. Then we seek the numbers u_j such that $u_j = \sum_\nu u_\nu p_{\nu j}$, and $\sum u_j = 1$. The general solution is given by $u_j = D_j/D$ $(j = 1, \cdots, 4)$, where

$$
\begin{aligned}
D_1 &= [p_{21}p_{32}p_{43} + p_{31}p_{42}p_{23} + p_{41}(p_{22}-1)(p_{33}-1)] \\
&\qquad - [p_{41}p_{32}p_{23} + p_{21}p_{42}(p_{33}-1) + p_{31}(p_{22}-1)p_{43}]\,, \\
D_2 &= -[(p_{11}-1)p_{32}p_{43} + p_{31}p_{42}p_{13} + p_{41}p_{12}(p_{33}-1)] \\
&\qquad + [p_{41}p_{32}p_{13} + (p_{11}-1)p_{42}(p_{33}-1) + p_{31}p_{12}p_{43}]\,, \\
D_3 &= [(p_{11}-1)(p_{22}-1)p_{43} + p_{21}p_{42}p_{13} + p_{41}p_{12}p_{23}] \\
&\qquad - [p_{41}(p_{22}-1)p_{13} + (p_{11}-1)p_{42}p_{23} + p_{21}p_{12}p_{43}]\,, \\
D_4 &= -[(p_{11}-1)(p_{22}-1)(p_{33}-1) + p_{21}p_{32}p_{13} + p_{31}p_{12}p_{23}] \\
&\qquad + [p_{31}(p_{22}-1)p_{13} + (p_{11}-1)p_{32}p_{23} + p_{21}p_{12}(p_{33}-1)]\,, \\
D &= \sum D_j\,.
\end{aligned}
$$

When the anti-diagonal of $[p_{ij}]$ is identically zero, as is the case with (1.4.3), the first four equations above simplify to

$$
\begin{aligned}
D_1 &= -[p_{21}p_{42}(p_{33}-1) + p_{31}(p_{22}-1)p_{43}]\,, \\
D_2 &= -p_{31}p_{42}p_{13} + [(p_{11}-1)p_{42}(p_{33}-1) + p_{31}p_{12}p_{43}]\,, \\
D_3 &= [(p_{11}-1)(p_{22}-1)p_{43} + p_{21}p_{42}p_{13}] - p_{21}p_{12}p_{43}\,, \\
D_4 &= -(p_{11}-1)(p_{22}-1)(p_{33}-1) + [p_{31}(p_{22}-1)p_{13} + p_{21}p_{12}(p_{33}-1)]\,.
\end{aligned}
$$

Applying these equations to (1.4.3) and simplifying, we obtain

$$
\begin{aligned}
D_1 &= [a_2(1-a_3)(1-a_4) + (1-a_2)(1-a_3)a_4]\theta_A^2\theta_B \\
&\qquad + [a_2a_3(1-a_4) + a_2(1-a_3)a_4]\theta_A\theta_B^2 , \\
D_2 &= [a_1(1-a_3)(1-a_4) + (1-a_1)a_3(1-a_4)]\theta_A^2\theta_B \\
&\qquad + [a_1a_3(1-a_4) + a_1(1-a_3)a_4]\theta_A\theta_B^2 , \\
D_3 &= [(1-a_1)(1-a_2)a_4 + (1-a_1)a_2(1-a_4)]\theta_A^2\theta_B \\
&\qquad + [(1-a_1)a_2a_4 + a_1(1-a_2)a_4]\theta_A\theta_B^2 , \\
D_4 &= [(1-a_1)(1-a_2)a_3 + a_1(1-a_2)(1-a_3)]\theta_A^2\theta_B \\
&\qquad + [(1-a_1)a_2a_3 + a_1(1-a_2)a_3]\theta_A\theta_B^2 .
\end{aligned}
\tag{1.4.4}
$$

Since D is the sum of the D_j's and since $u_j = D_j/D$, we may infer from (1.4.4) that the asymptotic probability u_j of each state is a function only of the ratio θ_A/θ_B and the experimenter-determined values a_i. To see this, first note that

$$
u_j = \frac{c_j\theta_A^2\theta_B + d_j\theta_A\theta_B^2}{c\theta_A^2\theta_B + d\theta_A\theta_B^2} , \tag{1.4.5}
$$

where the coefficients c_j, d_j, c, and d are functions of the a_i's given by equations (1.4.4); namely,

$$
\begin{aligned}
c_1 &= a_2(1-a_3)(1-a_4) + (1-a_2)(1-a_3)a_4 , \\
c_2 &= a_1(1-a_3)(1-a_4) + (1-a_1)a_3(1-a_4) , \\
c_3 &= (1-a_1)(1-a_2)a_4 + (1-a_1)a_2(1-a_4) , \\
c_4 &= (1-a_1)(1-a_2)a_3 + a_1(1-a_2)(1-a_3) ; \\
d_1 &= a_2a_3(1-a_4) + a_2(1-a_3)a_4 , \\
d_2 &= a_1a_3(1-a_4) + a_1(1-a_3)a_4 , \\
d_3 &= (1-a_1)a_2a_4 + a_1(1-a_2)a_4 , \\
d_4 &= (1-a_1)a_2a_3 + a_1(1-a_2)a_3 ; \\
c &= c_1 + c_2 + c_3 + c_4 ; \\
d &= d_1 + d_2 + d_3 + d_4 .
\end{aligned}
\tag{1.4.6}
$$

Dividing the numerator and denominator of (1.4.5) by $\theta_A\theta_B^2$, we obtain the desired result:

$$
u_j = \frac{c_j(\theta_A/\theta_B) + d_j}{c(\theta_A/\theta_B) + d} . \tag{1.4.7}
$$

With (1.4.6) and (1.4.7) at hand it is a matter of elementary arithmetic to compute the asymptotic probabilities u_j for any fixed ratio θ_A/θ_B of the learning parameters. Moreover, for some experimental situations it is reasonable to expect that the rate of learning will be approximately the same for both players, and therefore to assume that $\theta_A = \theta_B$. With this additional assumption, a parameter-free prediction of asymptotic response probabilities can be made independent of, and prior to, any analysis of experimental data. The results of such predictions are reported in Chapter 3.

From (1.4.7) we can also draw some interesting conclusions about the relationship of the asymptotic response probabilities u_j to the ratio θ_A/θ_B. Setting $\varepsilon = \theta_A/\theta_B$ and differentiating (1.4.7) with respect to ε, we obtain

$$\frac{d}{d\varepsilon}u_j = \frac{c_j d - c d_j}{(c\varepsilon + d)^2} . \tag{1.4.8}$$

If $c_j d \neq c d_j$, then u_j has no maximum for ε in the open interval $(0, \infty)$, which is the permissible range of values for the ratio θ_A/θ_B. In fact, since the sign of the derivative is independent of ε, we know that u_j is either a monotonically decreasing or a monotonically increasing function of ε: strictly decreasing if $c_j d < c d_j$, and strictly increasing if $c_j d > c d_j$. Obviously both cases must obtain, since $u_1 + u_2 + u_3 + u_4 = 1$. Moreover, because of the monotonicity of u_j in ε, it is easy to compute the bounds of u_j from (1.4.7). Thus,

$$\lim_{\varepsilon\to\infty} u_j = \frac{c_j}{c} , \qquad \lim_{\varepsilon\to 0} u_j = \frac{d_j}{d} . \tag{1.4.9}$$

If u_j is an increasing function with respect to ε, then $d_j/d < c_j/c$, and its values lie in the open interval $(d_j/d, c_j/c)$; if decreasing, in the interval $(c_j/c, d_j/d)$. Numerical values of these intervals are given in Chapter 3 for the sets of experimental parameters a_i actually used.

In connection with the asymptotic response probabilities of Players A and B, we want to show that between $\alpha = \lim_{n\to\infty} P(\boldsymbol{A}_n = 1)$ and $\beta = \lim_{n\to\infty} P(\boldsymbol{B}_n = 1)$ there obtains a certain linear relation that is independent of the ratio θ_A/θ_B.[12] (In Sec. 1.5 we show that this same linear relation obtains in the non-Markov linear model.) To begin with, note that

$$\alpha = u_1 + u_2 , \tag{1.4.10}$$

since $P(\boldsymbol{A}_n = 1) = P(\boldsymbol{A}_n = 1, \boldsymbol{B}_n = 1) + P(\boldsymbol{A}_n = 1, \boldsymbol{B}_n = 0)$; and correspondingly that

$$\beta = u_1 + u_3 . \tag{1.4.11}$$

From (1.4.7), (1.4.10), and (1.4.11) it follows that

$$(c\rho + d)\alpha = (c_1 + c_2)\rho + d_1 + d_2 ,$$
$$(c\rho + d)\beta = (c_1 + c_3)\rho + d_1 + d_3 .$$

By elementary operations we may eliminate ρ from these two equations and obtain

$$(c_1 + c_3 - c\beta)d\alpha - (c_1 + c_3 - c\beta)(d_1 + d_2)$$
$$= (c_1 + c_2 - c\alpha)d\beta - (c_1 + c_2 - c\alpha)(d_1 + d_3) .$$

The quadratic term $cd\alpha\beta$ cancels out and we have

$$(c_1 d + c_3 d + c d_1 + c d_3)\alpha = (c_1 d + c_2 d + c d_1 + c d_2)\beta$$
$$- (c_1 + c_2)(d_1 + d_3) + (c_1 + c_3)(d_1 + d_2) .$$

[12] Note that γ as defined in (1.2.11) is simply u_1.

Substituting (1.4.6) into this equation and simplifying, we obtain

$$[(a_3 + a_4 - a_1 - a_2) + (a_1a_2 - a_3a_4)]\alpha = (a_1a_3 - a_2a_4)\beta + \tfrac{1}{2}(a_3 + a_4 - a_1 - a_2) + a_4(a_2 - a_3) . \tag{1.4.12}$$

It would be pleasant to state that we have proved a variance theorem for the sum of A_1 and B_1 responses, corresponding to the variance theorem proved for the noncontingent case in the preceding section [see equation (1.3.31)]. The analytical difficulties of proving such a theorem are not insurmountable, but the effort required to obtain the explicit higher transition probabilities $p_{ij}^{(n)}$ for (1.4.3) seemed to us considerably greater than the value of the result. The compromise we settled on was to prove some relatively simple theorems about variances for arbitrary 4×4 irreducible, aperiodic Markov chains, and then proceed from these by numerical computation. The theorems are given in this section; the computations are reported in Chapters 3 and 4. The generalization of the theorems to $n \times n$ chains is obvious.

We first define the following random variables:

$$X_n(x) = \begin{cases} 1 & \text{if on trial } n \text{ state 1 occurs ,} \\ 0 & \text{otherwise .} \end{cases} \tag{1.4.13}$$

Thus with reference to (1.4.3), if $X_n = 1$, then state (1, 1) occurred on trial n, that is, responses A_1 and B_1 were made. Note that here the proper interpretation of the sample space point x is as a *pair* of subjects, players A and B; not as a single subject.

$$Y_n(x) = \begin{cases} 1 & \text{if on trial } n \text{ state 2 occurs ,} \\ 0 & \text{otherwise .} \end{cases} \tag{1.4.14}$$

Again with reference to (1.4.3), state 2 corresponds to (1, 2), that is, to the pair of responses A_1 and B_2.

$$Z_n(x) = \begin{cases} 1 & \text{if on trial } n \text{ state 3 occurs ,} \\ 0 & \text{otherwise .} \end{cases} \tag{1.4.15}$$

With reference to (1.4.3), state 3 corresponds to the state (2, 1), that is, to the pair of responses A_2 and B_1.

By analogy with (1.4.10) and (1.4.11) we have the following two functional equations for A_n and B_n in terms of X_n, Y_n, and Z_n:

$$A_n = X_n + Y_n , \qquad B_n = X_n + Z_n . \tag{1.4.16}$$

For the sum of any of these five random variables over a block of N trials we use the notation already introduced for A_n and B_n, namely $\bar{X}_N$, $\bar{Y}_N$, $\bar{Z}_N$, $\bar{A}_N$, and $\bar{B}_N$. Also, from previous notation the asymptotic probabilities of X_n, Y_n, Z_n, A_n, and B_n are u_1, u_2, u_3, α, and β, respectively.

We conclude this section with the following theorem:

THEOREM. *At asymptote we have the following variances for the sum of N random variables:*

$$\text{var}(\bar{X}_N) = Nu_1(1 - Nu_1) + 2u_1 \sum_{j=1}^{N-1} (N-j)p_{11}^{(j)} , \tag{1.4.17}$$

$$\text{var}(\bar{Y}_N) = Nu_2(1 - Nu_2) + 2u_2 \sum_{j=1}^{N-1} (N-j)p_{22}^{(j)} , \tag{1.4.18}$$

$$\text{var}(\bar{Z}_N) = Nu_3(1 - Nu_3) + 2u_3 \sum_{j=1}^{N-1} (N-j)p_{33}^{(j)} , \tag{1.4.19}$$

$$\begin{aligned} \text{var}(\bar{A}_N) = \text{var}(\bar{X}_N) + \text{var}(\bar{Y}_N) + 2u_1 \sum_{j=1}^{N-1} (N-j)p_{12}^{(j)} \\ + 2u_2 \sum_{j=1}^{N-1} (N-j)p_{21}^{(j)} - 2N^2 u_1 u_2 , \end{aligned} \tag{1.4.20}$$

$$\begin{aligned} \text{var}(\bar{B}_N) = \text{var}(\bar{X}_N) + \text{var}(\bar{Z}_N) + 2u_1 \sum_{j=1}^{N-1} (N-j)p_{13}^{(j)} \\ + 2u_3 \sum_{j=1}^{N-1} (N-j)p_{31}^{(j)} - 2N^2 u_1 u_3 . \end{aligned} \tag{1.4.21}$$

PROOF. We prove only (1.4.17) and (1.4.20); proofs of the other three cases are similar.

By the fundamental theorem for the variance of a sum of random variables (see Feller [1957], p. 216), at asymptote

$$\text{var}(\bar{X}_N) = N\,\text{var}(X_n) + 2 \sum_{1 \leq j < k \leq N} \text{cov}(X_j X_k) . \tag{1.4.22}$$

Now at asymptote

$$\text{var}(X_n) = u_1(1 - u_1) , \tag{1.4.23}$$

and

$$\text{cov}(X_j X_k) = E(X_j X_k) - E(X_j)E(X_k) = E(X_j X_k) - u_1^2 . \tag{1.4.24}$$

Moreover,

$$E(X_j X_k) = E(X_k \mid X_j)E(X_j) = u_1 E(X_k \mid X_j) = u_1 p_{11}^{(k-j)} \qquad (\text{for } j < k) . \tag{1.4.25}$$

We need to evaluate one sum:

$$\begin{aligned} \sum_{1 \leq j < k \leq N} p_{11}^{(k-j)} &= p_{11}^{(2-1)} + [p_{11}^{(3-2)} + p_{11}^{(3-1)}] \\ &\quad + [p_{11}^{(4-3)} + p_{11}^{(4-2)} + p_{11}^{(4-1)}] + \cdots \\ &\quad + [p_{11}^{(N-[N-1])} + \cdots + p_{11}^{(N-1)}] \\ &= (N-1)p_{11}^{(1)} + (N-2)p_{11}^{(2)} + \cdots + p_{11}^{(N-1)} \\ &= \sum_{j=1}^{N-1} (N-j)p_{11}^{(j)} . \end{aligned} \tag{1.4.26}$$

Substituting (1.4.25) into (1.4.24) and applying the summation result (1.4.26), we infer that

$$\begin{aligned} 2 \sum_{1 \leq j < k \leq N} \text{cov}(X_j X_k) &= 2u_1 \sum_{j=1}^{N-1} (N-j)p_{11}^{(j)} - 2 \sum_{1 \leq j < k \leq N} u_1^2 \\ &= 2u_1 \sum_{j=1}^{N-1} (N-j)p_{11}^{(j)} - N(N-1)u_1^2 . \end{aligned} \tag{1.4.27}$$

Finally, combining the last term on the right of (1.4.27) and multiplying the right-hand side of (1.4.23) by N, we have

$$Nu_1(1-u_1) - N(N-1)u_1^2 = Nu_1(1-Nu_1)\,. \tag{1.4.28}$$

The term on the right of (1.4.28) and the first term on the right of (1.4.27) yield the desired result for var $(\bar{X}_N)$.

We turn now to (1.4.20). From (1.4.16) we have immediately

$$\operatorname{var}(\bar{A}_N) = \operatorname{var}(\bar{X}_N + \bar{Y}_N) = \operatorname{var}(\bar{X}_N) + \operatorname{var}(\bar{Y}^N) + 2\operatorname{cov}(\bar{X}_N\bar{Y}_N)\,. \tag{1.4.29}$$

Now as before

$$\operatorname{cov}(\bar{X}_N\bar{Y}_N) = E(\bar{X}_N\bar{Y}_N) - E(\bar{X}_N)E(\bar{Y}_N) = E(\bar{X}_N\bar{Y}_N) - N^2u_1u_2\,. \tag{1.4.30}$$

We note that the product $\bar{X}_N\bar{Y}_N$ is a product of sums and that $E(X_nY_n) = 0$ for $1 \leqq n \leqq N$, whence

$$\begin{aligned} E(\bar{X}_N\bar{Y}_N) &= E\Big(X_1\cdot\sum_{n\neq 1} Y_n\Big) + E\Big(X_2\cdot\sum_{n\neq 2} Y_n\Big) + \cdots + E\Big(X_N\cdot\sum_{n\neq N} Y_n\Big) \\ &= u_1\sum_{j=2}^{N} p_{12}^{(j-1)} + \Big[u_2p_{21}^{(1)} + u_1\sum_{j=3}^{N} p_{12}^{(j-2)}\Big] \\ &\qquad + \Big[u_2\sum_{j=1}^{2} p_{21}^{(j)} + u_1\sum_{j=4}^{N} p_{12}^{(j-3)}\Big] + \cdots \\ &\qquad + \Big[u_2\sum_{j=1}^{N-2} p_{21}^{(j)} + u_1p_{12}^{(1)}\Big] + u_2\sum_{j=1}^{N-1} p_{21}^{(j)} \\ &= u_1[(N-1)p_{12}^{(1)} + (N-2)p_{12}^{(2)} + \cdots + p_{12}^{(N-1)}] \\ &\qquad + u_2[(N-1)p_{21}^{(1)} + (N-2)p_{21}^{(2)} + \cdots + p_{21}^{(N-1)}] \\ &= u_1\sum_{j=1}^{N-1}(N-j)p_{12}^{(j)} + u_2\sum_{j=1}^{N-1}(N-j)p_{21}^{(j)}\,. \end{aligned} \tag{1.4.31}$$

Substituting the results of (1.4.30) and (1.4.31) into (1.4.29) yields the desired result, which completes the proof.

Generally, in comparing these variances with empirical data we shall divide them by N^2 to normalize to the relative frequencies of responses in N trials.

Models for more complicated experimental situations are presented in later chapters. And in Chapter 3 the multi-stimulus-element model is discussed for the simple zero-sum game situation that has been the focus of this section.

*1.5. Alternative Linear Model[13]

For those experiments in which the available stimuli are the same on all trials we can use a model that dispenses with the concept of stimuli. In such a "pure" reinforcement model there is only one assumption: that the probability of a response on a given trial is a linear function of the probability of that response on the previous trial. A one-person experiment may be represented

[13] Starred sections may be omitted without loss of continuity by readers interested only in the main lines of development.

simply as a sequence $(\boldsymbol{A}_1, \boldsymbol{E}_1, \boldsymbol{A}_2, \boldsymbol{E}_2, \cdots, \boldsymbol{A}_n, \boldsymbol{E}_n, \cdots)$ of the response and reinforcement random variables defined by (1.2.1) and (1.2.2). Any sequence of values of these random variables represents a possible experimental outcome. (For analysis of an experiment in which more than two responses or reinforcements are possible, (1.2.1) and (1.2.2) need to be modified so that the value of $\boldsymbol{A}_n$ is a number j representing the response on trial n, and the value of $\boldsymbol{E}_n$ is a number k representing the reinforcing event on trial n. However, this modification is not necessary in this section.)

The linear theory is formulated for the probability of a response on trial $n+1$, *given* the entire preceding sequence of responses and reinforcements.[14] For this preceding sequence we use the notation x_n. Thus, x_n is a sequence of length $2n$ with 0's and 1's in the odd positions indicating responses A_1 and A_2, and 1's and 2's in the even positions indicating reinforcing events E_1 and E_2. The axioms of the linear theory are as follows:

Axiom L1. *If* $\boldsymbol{E}_n = 1$ *and* $P(x_n) > 0$, *then*

$$P(\boldsymbol{A}_{n+1} = 1 \mid x_n) = (1-\theta)P(\boldsymbol{A}_n = 1 \mid x_{n-1}) + \theta \,.$$

Axiom L2. *If* $\boldsymbol{E}_n = 2$ and $P(x_n) > 0$, *then*

$$P(\boldsymbol{A}_{n+1} = 1 \mid x_n) = (1-\theta)P(\boldsymbol{A}_n = 1 \mid x_{n-1}) \,.$$

Here, as usual, θ is to be thought of as the learning parameter.

For the noncontingent case of Sec. 1.3 we can derive from L1 and L2 the same asymptotic mean result as for the Markov model, i.e.,

$$\lim_{n\to\infty} P(\boldsymbol{A}_n = 1) = \pi \,. \tag{1.5.1}$$

On the other hand, the expression for the variance of the Cesáro sum $\bar{\boldsymbol{A}}_N$ given by (1.3.31) is different for the linear model. We shall not derive it here, but in Estes and Suppes [1959*a*] it is shown to be

$$\operatorname{var}(\bar{\boldsymbol{A}}_N) = \frac{\pi(1-\pi)}{(2-\theta)\theta}\{N\theta(4-3\theta) - 2(1-\theta)[1-(1-\theta)^N]\} \,. \tag{1.5.2}$$

The following interesting result obtains:

Theorem. *In the noncontingent case, for every* $N \geqq 2$ *and for every* θ *and* π *in the open interval* $(0, 1)$, *the variance of* $\boldsymbol{A}_N$ *at asymptote is less in the linear model than in the one-element Markov model.*

Proof. We seek the conditions on N and θ under which the following inequality holds [$\pi(1-\pi)/\theta$ has been canceled from both (1.3.31) and (1.5.2)]:

$$\frac{N\theta(4-3\theta)}{2-\theta} - \frac{2(1-\theta)[1-(1-\theta)^N]}{2-\theta} < N(2-\theta) - \frac{2(1-\theta)[1-(1-\theta)^N]}{\theta} \,.$$

[14] In the language of stochastic processes, this means that we have replaced the Markov chains of earlier sections by chains of infinite order.

This simplifies to $1-(1-\theta)_N < N\theta$. Now

$$1-(1-\theta)^N = \theta[1+(1-\theta)+(1-\theta)^2+\cdots+(1-\theta)^{N-1}],$$

whence, canceling θ, which by hypothesis is not 0, we have

$$1+(1-\theta)+(1-\theta)^2+\cdots+(1-\theta)^{N-1} < N\,.$$

Clearly this strict inequality holds under the conditions of the hypothesis of the theorem, and the proof is thereby completed.

To extend the linear model to two-person situations, we assume that both subjects satisfy Axioms L1 and L2. For the study of higher moments, which will not be considered here (see Lamperti and Suppes [1959]), we also need the assumption that, given the sequence of past responses and reinforcements, the probabilities of responses of the two players on trial n are statistically independent. It is shown in Estes and Suppes [1959*a*] that the following recursive equations hold for the two-person, zero-sum situation defined in Sec. 1.4 [we use here the notation of (1.2.9)]:

$$\begin{aligned}\alpha_{n+1} &= [1-\theta_A(2-a_2-a_4)]\alpha_n+\theta_A(a_4-a_3)\beta_n+\theta_A(1-a_4)\\ &\qquad +\theta_A(a_1+a_3-a_2-a_4)\gamma_n\,;\\ \beta_{n+1} &= [1-\theta_B(a_3+a_4)]\beta_n+\theta_B(a_2-a_4)\alpha_n+\theta_B a_4\\ &\qquad +\theta_B(a_3+a_4-a_1-a_2)\gamma_n\,.\end{aligned} \tag{1.5.3}$$

It is further shown in Lamperti and Suppes [1959] that the limits α, β, and γ exist, whence we obtain from (1.5.3) two linear relations, which are independent of θ_A and θ_B:

$$\begin{aligned}(2-a_2-a_4)\alpha &= (a_4-a_3)\beta+(a_1+a_3-a_2-a_4)\gamma+(1-a_4)\,;\\ (a_3+a_4)\beta &= (a_2-a_4)\alpha+(a_3+a_4-a_1-a_2)\gamma+a_4\,.\end{aligned} \tag{1.5.4}$$

By eliminating γ from these two equations we obtain the linear relation (1.4.12) in α and β. Unfortunately, this relationship represents one of the few quantitative results that can be directly computed when this model is applied to multiperson situations. Our relative neglect of the linear model in the sequel is due mainly to its mathematical intractability in comparison with the Markov models already discussed.

1.6. Comparisons with Game Theory

As we remarked in the first section, it is possible to view game theory as a descriptive, empirical theory of behavior, but in fact this does not seem to be a very promising approach. Our a posteriori reason is that for our experiments it did not make good predictions. Actually, however, there are several general reasons why one should not be surprised by the poor predictive success of game theory. In the classical sense of psychological theories, game theory is not a behavior theory. It does not provide an analysis of how the organism receives cues or stimuli from its environment and then adjusts its behavior accordingly.

In other words, it does not provide, even in schematic form, a formulation of the elementary process that would lead an organism to select the appropriate game-theoretic choice of a strategy.

From a general methodological standpoint, game theory, like classical economics, is concerned with the ideal behavior of a rational man. This concern with the rational man is the basis of another strong bond between classical economics and game theory: both are very much theories of equilibrium. In neither discipline, moreover, does the derivation of an equilibrium condition depend on any assumptions about the particular dynamic process by which the equilibrium point is reached. This static character no doubt accounts for the uneasiness with which many psychologists view the concept of utility. The economist and game theorist take the utility function of the individual consumer or player as given, whereas a psychologist immediately tends to inquire where the utility function came from, and to seek the environmental factors controlling the process by which a particular set of values or utilities is acquired.

Granted that game theory is an equilibrium theory, it may still be argued that it has predictive pertinence to our experiments. For it can be maintained that when individuals have reached an asymptotic level of behavior (in learning-theory terms), they will be in equilibrium with their environment (including the other players in a game situation), and hence that the optimality concepts of game theory may well apply to their patterns of choice. For example, even when subjects are not shown the payoff matrix, after many trials they may have learned enough about the prospects of winning and losing to approximate an optimal game strategy.

We now define three concepts of optimality in game theory, which later will be used for comparison with learning-theory predictions.[15]

In the first place, when it is applicable, the appealing *sure-thing* principle may be used to select an optimal strategy. A strategy satisfies the sure-thing principle if no matter what your opponent does, you are at least as well off with this strategy in comparison to any other available to you, and possibly better off. For example, consider the following 2×2 matrix of a two-person, zero-sum game.

	B_1	B_2
A_1	5	7
A_2	2	-1

On each play of the game, Player A chooses row A_1 or A_2 and Player B chooses column B_1 or B_2. If A_1 and B_1 are chosen, then A receives \$5 and B loses this amount; if A_2 and B_2 are chosen, A loses \$1 and B receives this amount; and so on. Obviously, Player A's choice of A_1 satisfies the sure-thing principle,

[15] For a detailed presentation of game theory, see McKinsey [1952], Luce and Raiffa [1957], or Blackwell and Girshick [1954].

for no matter what B does, A is better off with A_1 than with A_2. However, the weakness of the sure-thing principle is exemplified by B's situation. Neither B_1 nor B_2 satisfies the sure-thing principle, and so, in his choice of a strategy, B must seek another principle of optimality.

Intuitively, it seems clear that B should always select B_1, since he will lose only \$5 rather than \$7 when A chooses A_1. The optimality principle indicating the selection of B_1 is von Neumann's famous minimax principle [1944]. The idea is that B should minimize his maximum loss. If he chooses B_2, his maximum loss is \$7; if he chooses B_1, it is \$5. So he should choose B_1.

When the sure-thing principle applies, it agrees with the minimax principle. Thus, A minimizes his maximum loss when he picks A_1 (actually, he maximizes his minimum gain, which amounts to the same thing). If A picks A_1, his minimum gain is \$5; if he picks A_2, his minimum "gain" is the loss of \$1. Thus to maximize his minimum gain he should pick A_1.

Let (a_{ij}), with $i = 1, \cdots, n$ and $j = 1, \cdots, m$, be the payoff matrix of an $n \times m$ zero-sum game. If

$$\max_i \min_j a_{ij} = \min_j \max_i a_{ij} = v ,$$

we call v the value of the game, and any strategies i^* and j^* such that

$$\min_j a_{i^*j} = \max_i \min_j a_{ij} \qquad \text{and} \qquad \max_i a_{ij^*} = \min_j \max_i a_{ij}$$

are *pure* minimax strategies (i^* for A and j^* for B). A *pure* strategy is one that selects a given row or column with probability 1.

Unfortunately, pure minimax strategies do not always exist. For instance, there are none for the payoff matrix

(1.6.1)

	B_1	B_2
A_1	-1	3
A_2	0	-2

because $\max_i \min_j a_{ij} = -1$ but $\min_j \max_i a_{ij} = 0$. The implication here is that in repeated plays of this game a fixed choice of A_1 or A_2 by A, or a fixed choice of B_1 or B_2 by B, will not be optimal against an intelligent opponent.

The insufficiency of pure strategies may be remedied by introducing probability *mixtures* of pure strategies. For instance, Player A might choose A_1 with probability 1/3 and A_2 with probability 2/3. A probability mixture is called a *mixed strategy*. Such a strategy for A may be designated $\xi = (\xi_1, \xi_2)$, where ξ_i is the probability of choosing A_i $(i = 1, 2)$.[16] Similarly, mixed strategies for B are designated $\eta = (\eta_1, \eta_2)$. Von Neumann's fundamental theorem states that *mixed* minimax strategies exist for any zero-sum, two-person game with a finite number

[16] If n strategies are available to A, then $\xi = (\xi_1, \cdots, \xi_n)$ is an n-dimensional vector such that $\xi_i \geqq 0$ $(i = 1, \cdots, n)$ and $\sum \xi_i = 1$.

of strategies available to each player. In other words, there are probability mixtures ξ^* and η^* such that

$$\max_{\xi} \sum_{i,j} a_{ij}\xi_i\eta_j^* = \min_{\eta} \sum_{i,j} a_{ij}\xi_i^*\eta_j = v\,, \tag{1.6.2}$$

and v is called the *value* of the game.[17] What (1.6.2) shows is that Player A may assure himself of winning at least v by playing ξ^* and Player B may assure himself of losing at most v by playing η^*.

We shall not go into the general methods of finding the value of a game and its minimax strategies, but we can illustrate the simple technique for 2×2 games with payoff matrix (a_{ij}). For simplicity, let $x = \xi_1$ and $y = \eta_1$. Then it may be shown that it is sufficient to consider x and y separately against the use of pure strategies by the other player. Thus we seek numbers x, y, and v such that

$$\begin{aligned} &xa_{11} + (1-x)a_{21} \geqq v\,, \qquad && ya_{11} + (1-y)a_{12} \leqq v\,, \\ &xa_{12} + (1-x)a_{22} \geqq v\,; && ya_{21} + (1-y)a_{22} \leqq v\,; \end{aligned} \tag{1.6.3}$$

$$0 \leqq x, y \leqq 1\,.$$

In any numerical case, the solution of these inequalities is a simple matter. For (1.6.1), we obtain $x = 1/3$, $y = 5/6$, $v = -1/3$. (In example (1.6.1), x and y are unique. This is not always the case; as a trivial instance, if $a_{ij} = 0$ for every i and j, any ξ and any η is a mixed minimax strategy.)

The choice of optimal strategies for non-zero-sum, two-person games is a complicated matter which as yet does not have a satisfactory solution. (A two-person game is non-zero-sum when what one player receives at the end of a play is not the negative of what the other player receives.) A natural division of non-zero-sum games is into *cooperative* and *non-cooperative* games. In a co-operative game the players are permitted to communicate and bargain before selecting a strategy; in a non-cooperative game no such communication and bargaining is permitted.

Subsequent chapters devoted to non-zero-sum games are entirely concerned with the non-cooperative type. Probably the best concept of optimality yet proposed for such games is Nash's notion of an equilibrium point ([1950], [1951]). An *equilibrium point* is a set of strategies, one for each player, such that if all the players but one follow their given strategies, the remaining player cannot do better by following any strategy other than one belonging to the equilibrium point. It was shown by Nash that every non-zero-sum, n-person game in which each player has a finite number of strategies has an equilibrium point among its mixed strategies. Techniques for finding an equilibrium point will be considered in Chapter 4.

---

[17] Note that $\sum_{i,j} a_{ij}\xi_i\eta_j$ is just the expectation of A's gain with respect to the two independent probability mixtures ξ and η.

*1.7. Hypothesis Models

Many behavioral scientists believe that subjects' response patterns in the type of choice experiments we are considering can best be explained in terms of possible hypotheses entertained by the subject (see, for example, Bruner, Goodnow, and Austin [1956]). We show in this section that a broad class of hypothesis models may be formulated within the framework of our fundamental postulates. For this class of models, the choice between stimulus-sampling language and hypothesis language to describe the theory becomes a matter of taste.

The formal identity of the two approaches may be illustrated first by considering in some detail a hypothesis model for the case in which reinforcement is noncontingent. For simplicity, let us begin with four hypotheses:

(1.7.1)

h_1: an E_1 reinforcement is followed by an E_1;
h_1': an E_1 reinforcement is followed by an E_2;
h_2: an E_2 reinforcement is followed by an E_2;
h_2': an E_2 reinforcement is followed by an E_1.

The extension to a broader class of hypotheses is straightforward and will be considered later. Given the above four hypotheses and the fact that h_1 and h_1 (and h_2 and h_2') are incompatible, a *strategy* for the subject consists of believing, or acting as if he believed, one of the following four pairs of hypotheses: $\{h_1, h_2\}$, $\{h_1, h_2'\}$, $\{h_1', h_2\}$, $\{h_1', h_2'\}$. That is, a strategy is a set of mutually compatible hypotheses that prescribe a unique action or response on each trial, given information about the preceding reinforcements and responses. Thus, strategy $\{h_1, h_2\}$ indicates that an A_1 response will be made if, on the preceding trial, an E_1 reinforcement occurred, and an A_2 response will be made if it is preceded by an E_2 reinforcement.

One of the most important requirements of the model is to specify the mechanism for learning. A strategy, we may postulate, is held to as long as it is correct. What happens when it has been shown to be incorrect? One feasible postulate is that it is then dropped from the strategy and replaced by the correct relevant hypothesis with probability θ. (In the present setup, exactly one applicable hypothesis is correct on each trial.) By adopting independence-of-path assumptions similar to those given in Sec. 1.2, we are led to a Markov chain whose states are the four possible strategies.

A typical tree is given in Fig. 1.8. The first two branches indicate the reinforcement that occurred on the preceding trial, and thus tell us which of the two hypotheses in the strategy $\{h_1', h_2\}$ applies. Some readers may be unhappy with the introduction of the probability θ for changing a wrong hypothesis; they may feel that it should be changed with probability 1. Unfortunately, by this latter assumption the entire Markov chain would be uniquely determined

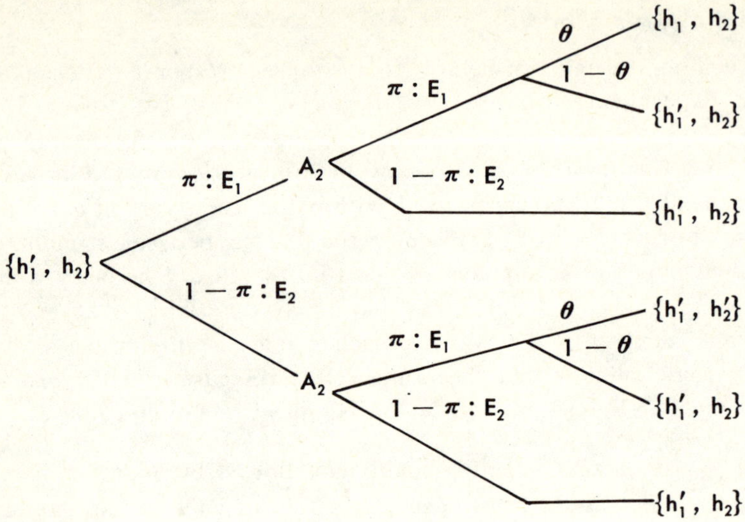

FIGURE 1.8

prior to the experiment, and even a cursory inspection of the data from noncontingent experiments indicates that such a model would provide an extremely poor fit. The possibility of estimating the parameter θ from the data increases the probability of a good fit and does not change the conceptual scheme of the hypothesis model in any substantive way.

With Fig. 1.8 in front of us, the procedure for deriving the transition matrix associated with the hypothesis model is clear. However, our concern at the moment is to demonstrate that the hypothesis model may be just as naturally interpreted as a stimulus-sampling model, and that it is derivable, with one additional assumption, from the basic postulates of Sec. 1.2. We identify the stimuli that may be sampled by the subject as the two possible reinforcing events on the preceding trial. Which one the subject samples is, of course, determined by which event, E_1 or E_2, actually occurred on the preceding trial. Let s_1 represent a preceding E_1 event, and s_2 a preceding E_2 event. Following the analysis of Sec. 1.3, the four possible states of conditioning are $\{s_1, s_2\}$, $\{s_1\}$, $\{s_2\}$, and 0, where the state is defined by the set of stimuli conditioned to the A_1 response. With the probability θ now playing the role of the conditioning parameter, we derive, in terms of these four states of conditioning, a Markov chain that is identical to the one derived from the hypothesis model. The formal correspondence between strategies and states of conditioning should be obvious:

(1.7.2)
$$\{h_1, h_2\} \to \{s_1\}, \qquad \{h'_1, h_2\} \to 0,$$
$$\{h_1, h'_2\} \to \{s_1, s_2\}, \qquad \{h'_1, h'_2\} \to \{s_2\}.$$

For instance, strategy $\{h_1, h_2\}$ says: (1) to follow an E_1 reinforcement with an

A_1 response, because the hypothesis h_1 is that an E_1 follows an E_1; and (2) to follow an E_2 reinforcement with an A_2 response, because the hypothesis h_2 is that an E_2 follows an E_2. The state of conditioning $\{s_1\}$ leads to exactly the same behavior. For if an E_1 occurs on the preceding trial, then, by virtue of our assumption about the identification of stimuli, the subject samples s_1 and makes response A_1 because s_1 is conditioned to A_1. Similarly, if an E_2 occurs on the preceding trial, the subject samples stimulus s_2 and makes response A_2 because s_2 is conditioned to A_2 in the state of conditioning $\{s_1\}$.

When the stimuli that are available for sampling vary from trial to trial, as they do in the model just described, the subject may learn to discriminate between stimuli and respond differently; that is, conditioning may be systematically different for the two stimuli. Let us examine this question for the present model, thereby also analyzing in greater detail the formally identical hypothesis model. (At the beginning of Chapter 6 we discuss discrimination in some detail.)

From Fig. 1.8 and the discussion of the two-element model in Sec. 1.3, it is easily seen that the transition matrix of our discrimination model for the noncontingent case is as follows:

(1.7.3)

	0	$\{s_1\}$	$\{s_2\}$	$\{s_1, s_2\}$
0	$1-\theta\pi$	$\theta\pi^2$	$\theta\pi(1-\pi)$	0
$\{s_1\}$	$\theta\pi(1-\pi)$	$1-2\theta\pi(1-\pi)$	0	$\theta\pi(1-\pi)$
$\{s_2\}$	$\theta(1-\pi)^2$	0	$1-\theta[\pi^2+(1-\pi)^2]$	$\theta\pi^2$
$\{s_1, s_2\}$	0	$\theta(1-\pi)^2$	$\theta\pi(1-\pi)$	$1-\theta(1-\pi)$

By the methods of Sec. 1.3 it can be shown that the asymptotic probabilities u_i for this matrix are exactly the same as for the matrix (1.3.12), which represents the two-element model when equiprobable sampling of elements is assumed. Thus $u_1 = (1-\pi)^2$, $u_2 = u_3 = \pi(1-\pi)$, $u_4 = \pi^2$, and, following (1.3.29),

$$\begin{aligned}\lim_{n\to\infty} P(\boldsymbol{A}_n = 1) &= u_4 + \pi u_2 + (1-\pi)u_3 \\ &= \pi^2 + \pi^2(1-\pi) + \pi(1-\pi)^2 = \pi\,. \end{aligned} \tag{1.7.4}$$

On the other hand, since the transition matrices (1.3.12) and (1.7.3) are not identical, the two models must differ in some of their behavioral predictions. We shall consider these differences in Chapter 10. However, it is worth noting that both models agree in not requiring that $P(\boldsymbol{A}_{n+1} = 1 \mid \boldsymbol{E}_n = 1, \boldsymbol{A}_n = 1) = 1$. As remarked in Sec. 1.3, this requirement is an undesirable feature of the one-element model.

Unfortunately, even the simple hypothesis model just considered—together with its stimulus-sampling counterpart—is too complicated for detailed use in connection with most of the experiments reported in subsequent chapters. More complex hypotheses and their stimulus-sampling analogues quickly become impossible to analyze. For example, consider the analysis for the noncontingent situation in which the hypotheses concern the previous three pairs of reinforcements and responses. In this case there are 2^6 different conditions, each requiring a hypothesis. Since each hypothesis predicts that the given sequence of three responses and three reinforcements will be followed by an E_1 or an E_2, there are 2^{2^6} possible strategies, any one of which may be adopted by the subject. This calculation alone indicates how rapidly hypothesis models become analytically and computationally unmanageable, even for the simplest schedules of reinforcement.

Of course, the stimulus-sampling analogue is beset by the same difficulty. Each sequence of three reinforcements and three responses constitutes a stimulus that is conditioned to the A_1 or A_2 response. So, just as there are 2^{2^6} strategies, there are 2^{2^6} states of conditioning, which leads to a Markov chain that is quite difficult to analyze. A reasonable simplification is to treat several sequences of reinforcements and responses as equivalent for the subject and, therefore, as conditioned to the same response. For instance, suppose we wish to consider a hypothesis model in terms of the last m reinforcements. There are then 2^{2^m} strategies or states of conditioning, but this number may be reduced to $m+1$ by considering only the number of E_1 reinforcements in the last m trials. There are then $m+1$ states of conditioning. Some results on a similar model are reported in Chapter 10. It is apparent that for many reinforcement schedules such a simple reduction to the number of past reinforcements of a given kind will not begin to approximate the conditioning processes occurring in a subject.

Some investigators who are interested in hypothesis-type models (for instance, Newell, Shaw, and Simon [1958], [1959]), have urged the desirability of a completely deterministic theory. For them, the parameter θ would be an unfortunate appendage. In our view, it is impossible at the present level of observation and theory to construct a purely deterministic model of choice or response behavior. We hold that whether models are conceptualized in terms of stimulus sampling or hypothesis testing, probabilistic parameters must be introduced from the beginning if there is to be any hope of making accurate quantitative predictions about actual behavior for a variety of reinforcement schedules.

On the other hand, the probabilistic theory of stimulus sampling considered in this book is often challenged by the claim that it cannot correctly predict, and thus explain, responses to a deterministic reinforcement schedule, for example the schedule of simple alternation: $E_1, E_2, E_1, E_2, E_1, E_2, \cdots$. We all know that in appropriate circumstances human beings can learn such sequences, i.e., make responses that accurately predict the next reinforcing event. Galanter and Smith [1958] have in fact shown that subjects can master some very complex patterns quite rapidly.

In claiming that stochastic models of the Estes-Burke and Bush-Mosteller types do not explain the learning of the simple alternation sequence, reference is usually made to the linear model described in Sec. 1.5. Critics usually point out that in terms of the kind of analysis used for the noncontingent case, the linear model predicts an average probability of $\frac{1}{2}$ for responding A_1 on any trial. Moreover, according to this model the probability of making an A_1 response after an E_1 reinforcement is higher than before the reinforcement, which is contrary to the observed data; for the subject quickly learns to follow an E_1 reinforcement with an A_2 response.

If we approach the learning of the simple alternation schedule in terms of stimulus conditioning and not simply in terms of a response-reinforcement model like the linear one, more satisfactory results may be obtained. In the stimulus-conditioning approach, just as in the model that is represented by the transition matrix (1.7.3), the two stimuli correspond to the two possible preceding reinforcements, E_1 and E_2. However, if we try to work with the same four states of conditioning as previously, difficulties arise. These difficulties become obvious if one attempts to draw the tree corresponding to Fig. 1.8. If, for example, on trial n the subject is in state $\{s_1, s_2\}$, we cannot indicate the probability of sampling s_1 or s_2. This probability will be either 0 or 1, depending on the particular place in the alternating-reinforcement schedule. To fix this probability we must also indicate in the conditioning state which stimulus is sampled, i.e., which reinforcement occurred on the last trial. To accommodate this complication we shall use the notation $(i; j, k)$ for the states of the Markov chain. The numbers i, j, and k are either 1 or 2. The number i indicates which reinforcement occurred on the preceding trial (and thus which stimulus is sampled), and j indicates the conditioning of the s_1 stimulus (thus $j = 2$ means that E_1 as a stimulus is conditioned to the A_2 response), and k indicates the conditioning of the E_2 reinforcement as a stimulus. Following the analysis laid down in previous examples, the appropriate trees can be constructed. As an example, consider the state $(2; 1, 2)$:

$$(2; 1, 2) \text{ --- } s_2 = E_2 \text{ --- } A_2 \text{ --- } E_1 \begin{cases} \xrightarrow{\ \theta\ } (1; 1, 1) \\ \xrightarrow{\ 1-\theta\ } (1; 1, 2) \end{cases}$$

Note that much of the tree is deterministic in character. The state itself specifies that s_2 is sampled, which must be followed by an A_2 response; for, in the state $(2; 1, 2)$, an s_2 stimulus is conditioned to an A_2 response. Then an E_1 reinforcement must occur, since an E_2 reinforcement occurred on the previous trial. Probability enters only with regard to the effectiveness of conditioning. The transition matrix for the eight possible states has the following attractively simple form:

(1.7.5)

	(1; 1, 1)	(1; 1, 2)	(1; 2, 1)	(1; 2, 2)	(2; 1, 1)	(2; 2, 1)	(2; 1, 2)	(2; 2, 2)
(1; 1, 1)					$(1-\theta)$	θ		
(1; 1, 2)							$(1-\theta)$	θ
(1; 2, 1)						1		
(1; 2, 2)								1
(2; 1, 1)	1							
(2; 2, 1)			1					
(2; 1, 2)	θ	$(1-\theta)$						
(2; 2, 2)			θ	$(1-\theta)$				

The above matrix differs from most of the others in this book in that it is not irreducible. The two states (1; 2, 1) and (2; 2, 1) form a proper subset such that once within this set, the probability of leaving it is 0; moreover, the probability of entering this set is 1. Of course, once entry has occurred, the subject will make the correct response on every trial. An approach identical to the present one may be used to derive a model for learning any deterministic sequence of reinforcements.

To conclude, we repeat that hypothesis models of the kind formulated here may actually be derived as stimulus-discrimination models within the basic framework of stimulus-sampling theory. In this sense, the elusive notions of strategy and hypothesis can be reduced to the more primitive concepts of stimulus sampling and conditioning.

1.8. Survey of Recent Literature

It is not feasible to survey the extensive literature of experimental and social psychology that is relevant to the contents of this book. Therefore, we have restricted ourselves to those studies that approach social behavior in terms of some specific mathematical model and have in addition referred to some of the better known attempts to apply reinforcement theory at the qualitative level to social situations. Certain studies are referred to in more or less detail in later chapters.

One of the first detailed attempts to account for social behavior in terms of a reinforcement theory analysis was provided by Miller and Dollard [1941]. They demonstrated experimentally that imitation of one organism by another can be controlled and manipulated by rewards and punishments. Further, they have shown that imitation is evoked by the stimulus of another person's behavior, and that imitative behavior generalizes to other stimulus situations. Since their original publication, many experiments have been reported in which imitative behavior is viewed in terms of stimulus-response concepts. More recently, several studies have been designed to facilitate analysis of the data by means of stochastic models of learning; in particular, the work of Schwartz [1953],

Kanareff and Lanzetta [1958], and Neimark and Rosenberg [1959]. We shall describe only the first study, since the other two are similar. On each trial of Schwartz's experiment, two children were asked to guess whether the experimenter was going to say "a" or "b." One child was told to go first on all trials; it was explained that this procedure was being used to save time. According to the design, the first child to respond was correct eight out of ten times. The second child had the option of imitating the first child or not; if he did imitate, he would be right eight out of ten times. It was found that the frequency of imitation increased during a sequence of trials when the first child was correct but appreciably decreased after the first child was wrong. Bush and Mosteller [1955, Ch. 12] applied a stochastic learning model to the data obtained by Schwartz. Their analysis indicates that imitation problems can be handled mathematically in much the same manner as other learning problems.

Group decision processes also have been studied within the framework of stochastic learning models. Hays and Bush [1954] carried out an experiment in which a three-man group was asked to predict on each trial whether an E_1 or E_2 event would occur; the actual probability of an E_1 was .75. Two models were proposed for analyzing the data. The first, called the "group-actor model," assumed that the three-man group behaved as an individual organism and predicted that the asymptotic frequency of predicting E_1 would be .75. The second model, called the "voting model," assumed that each individual acted independently according to a simple stochastic learning model, but that the group decision was made by majority rule. The analysis of these data did not provide a clear choice between the two models, but it was established that if the subjects behaved independently, they learned more slowly than most individuals in two-choice learning experiments.

A closely related type of experimental situation, involving social persuasion and conformity, also has been analyzed in terms of a variety of mathematical models. We mention particularly Cervin [1957], [1958], [1959], Anderson and Hovland [1957], Cohen [1958], Flament [1958], [1959], and Suppes and Krasne [1959]. The studies of Cohen, Flament, and Suppes and Krasne deal with the Asch [1956] type of social conformity experiment, which is discussed in Chapter 12.

Several other mathematical models for small-group behavior have been proposed. Christie, Luce, and Macy [1952] drew upon the concepts of cybernetics and electrical engineering to formulate a model for communication and learning in a task-oriented group. Their work is among the most interesting yet done. However, we do not analyze it here because a task-oriented group, as defined by these authors, is one in which all individuals in the group are actively working together to achieve a common goal. None of the multiperson experimental situations we consider have this character, and the models of Christie, Luce, and Macy are too specific to generalize in any direct fashion to our experiments.

Simon [1952] translated into mathematical form a set of hypotheses first set forth by Homans [1950]. Homans' hypotheses concerned the relation, over time, of three attributes of a group: the amount of activity engaged in together; the amount of interaction between the members; and the friendliness between the members. Simon took some of Homans' propositions, stated them in mathematical form as postulates of a dynamic system, and then drew deductions from them. Later Simon and Guetzkow [1955] undertook a more complicated task along the same lines, namely, to formulate mathematically the hypotheses set forth by Festinger [1950] concerning opinion deviance and variables related to it: pressure to communicate, pressure toward uniformity, the receptivity of communications, and group cohesion. The model that Simon and Guetzkow developed on the basis of Festinger's hypotheses is more complex than the Simon-Homans model. Both of these developments have proved to be interesting in their own right; but unfortunately no detailed attempt has been made to compare deductions with observed results—undoubtedly because the authors have not prescribed rules of correspondence between theoretical concepts and observable variables that would permit an empirical evaluation of the models. There is, in fact, a real question about the possibility of arriving at objective methods by which such concepts as group friendliness or cohesiveness may be measured in an interval or ratio scale. Without such measurements the theory might just as well be left in its original qualitative form.

Several fairly elaborate probabilistic models for small-group behavior have been proposed by Bales and his associates (see, for example, Bales, Strodtbeck, and Mills [1951], and Stephan and Mischler [1952]). However, no serious investigation of their mathematical properties has been undertaken. An interesting review of these models is provided by Coleman [1954].

More recently a series of studies have been conducted by Atkinson and Suppes [1958], [1959], Burke [1959], [1960], and Estes [1957*b*] for two-person interaction situations in which the reinforcement schedules are prescribed by the responses of the pair of subjects. The purpose of these studies has been to apply statistical learning theory to a certain class of social situations. We shall not review the methodology of these studies, since they are closely allied to the experiments reported in Chapters 3 and 4 of this book.

Related to this work is a two-person interaction study by Ratoosh [1958] (see also Scodel, Ratoosh, and Minas [1959]). Pairs of subjects were shown the following payoff matrix:

		Player B	
		I	II
Player A	I	(3, 3)	(0, 5)
	II	(5, 0)	(1, 1)

In this game the players choose either Response I or II, and the entries in the cells of the matrix indicate the payoff (in cents) to each player; the payoff to

Player A is indicated first. The procedure was repeated over a series of trials, and it was found that nearly all subjects showed a strong preference for Response II, even though some of their verbal comments indicated that they realized such behavior was not in their long-term best interests. The research reported in Chapters 5, 9, and 10 of this book relates to Ratoosh's experimental routine.

Another highly interesting set of results has been reported by Sidowski [1957], [1959], and Sidowski, Wyckoff, and Tabory [1956]. In these studies an attempt was made to re-evaluate the essential features of a social situation, as viewed within the framework of conditioning theory, and to investigate the simplest setup that could be considered truly social within this framework. The orientation of the research conducted by these investigators is well represented by the work of Sidowski [1957]. In this study the influence of reinforcement and punishment in a minimal social situation was investigated. Two subjects, isolated from each other, were provided with two push buttons, by means of which each could give the other a shock or a score. The subjects were divided into an Informed Group and an Uninformed Group. Each subject in the Informed Group was told that the subject paired with him controlled his shocks and scores; subjects in the Uninformed Group were not told they were interacting in any way. The two major groups were each subdivided into three subgroups in which the subjects could give each other Shock Only, Score Only, and Shock and Score, respectively. The results indicated that learning occurred in the Score Only and the Shock and Score conditions, but not in the Shock Only conditions. Learning, therefore, was demonstrated in those groups in which the subject had an opportunity to receive reward. Learning under these conditions occurred regardless of whether or not the subject was told that he was interacting with another subject; in fact, there were no significant differences attributable to this distinction.

Flood [1954*a*], [1954*b*] has developed several mathematically sophisticated models for interaction and for "game-learning-theory," as he calls it. However, a discussion of these models would lead us too far afield from learning. For the same reason, we only make reference to work by Foa and Zacks [1959] on a stochastic facet theory of social interaction in the dyad.

Kalisch, Milnor, Nash, and Nering [1954] report some experimental n-person games, and Stone [1958] an experiment in bargaining games. The experimental situations on which these studies were based differ considerably from ours, but the theory used to analyze the data is essentially the relevant portions of game theory, which we have already discussed in Sec. 1.6.

In conclusion, we cannot fail to mention the work of Rashevsky on social behavior. Probably the best reference is his book *Mathematical Biology of Social Behavior*, published in 1951. For our purposes, the model of greatest interest deals with imitative behavior. The equations that form the basis of the model are derived from Rashevsky's theory of the central nervous system, with an added assumption about stimuli generated by other individuals. Roughly speaking, he assumes that each individual who is responding acts as a stimulus on other individuals, so that the stimulus which determines behavior at any given instant of time is the cumulative effect of everyone's behavior. The experiments reported in Chapters 6, 7, and 12 of our book relate to this conceptual scheme. Unfortunately, the Rashevsky model is not well enough defined to provide an analysis of the data.

1.9. Summary

This chapter develops the learning theory that we use in analyzing multiperson-interaction experiments that closely resemble game situations. The fundamental postulates of the theory are formulated and analyzed in terms of the basic categories of *stimulus*, *response*, and *reinforcement*, and of the processes of stimulus *sampling* and *conditioning*. Next we show how the theory may be used to derive a Markov model for the simple case of noncontingent reinforcement. In the first model, it is assumed that only one stimulus is available for sampling. After noting the serious empirical difficulties confronting this assumption, we consider the model that assumes multiple stimulus elements. The general method of deriving quantitative results is emphasized, since we use the same approach throughout the book for more complicated situations.

In Sec. 1.4 we derive the model for a two-person, zero-sum game and compute the asymptotic response probabilities for both players. We also consider some useful expressions for computing the variances of response probabilities and mean learning curves. Here we conclude the main theoretical developments and discuss in the rest of the chapter related matters that are not essential to understanding the subsequent experimental chapters. In Sec. 1.5 a linear model that may serve as an alternative to stimulus-sampling Markov models is presented. In Sec. 1.6 the most important concepts of game theory are developed in an elementary fashion to facilitate later comparison with experimental results. In Sec. 1.7 models that are initially conceived in terms of hypotheses and strategies, rather than in terms of stimuli and conditioning, are derived within our basic theoretical framework. Finally in Sec. 1.8 we survey the recent literature that is particularly relevant to the subject matter of the book.

2

Methods of Analysis

2.1. Introduction

The aim of this chapter is to present those methods of data analysis that apply specifically to our models. Although some of these methods are standard, they may not be familiar to all readers. The emphasis is on maximum-likelihood methods and on what we shall term "pseudomaximum-likelihood methods."

In Sec. 2.2 we consider maximum-likelihood estimates for the learning parameters first in the case of noncontingent reinforcement, and then for the two-person interaction situation described in Sec. 1.4. Necessarily, as we shall see, the methods described require the assumption of exactly one element in the stimulus set of each subject.

In Sec. 2.3 we extend the methods of parameter estimation to models based on the assumption of more than one stimulus element. The methods developed formally resemble the maximum-likelihood methods of the preceding section but in fact do not have the maximum-likelihood property. In this context we also consider a method for estimating jointly the number N of stimuli and the value θ of the learning parameter.

In Sec. 2.4 we describe various χ^2 tests for testing the following three hypotheses: (1) that the transition probabilities of a first-order chain are constant; (2) that (in case the transition probabilities are constant) they are specific numbers; and (3) that the process is a uth-order (rather than rth-order) Markov chain. (For the derivation of these tests, the reader is referred to the recent statistical literature on Markov chains.)

In Sec. 2.5 we give some methods that are peculiar to the models for making what is "almost" a maximum-likelihood estimate of the learning parameter. These methods are of perhaps more theoretical than practical importance. They hinge upon first making a least-squares estimate of the not directly observed transition numbers m_{ij} from state i to state j in the transition matrix for the multi-element model.

Finally, in Sec. 2.6 we develop a generalized conditioning model in which the single parameter θ is replaced by several conditioning parameters. Simple maximum-likelihood methods of estimation for these new parameters are derived.

2.2. Maximum-Likelihood Estimates

When the states of a Markov chain are observable and the transition probabilities depend on a real parameter, say θ, then it is often not difficult to obtain the maximum-likelihood estimate $\hat{\theta}$ of θ.

Keeping in mind that we are mainly interested in Markov chains that consist of a sequence of response random variables, let $a_1, a_2, \cdots, a_n$ represent a finite sequence of values of the random variables from trial 1 to trial n. Let s be the number of subjects. Then the maximum-likelihood estimate of the learning parameter θ is the number $\hat{\theta}$ (if it exists) such that for all θ'

$$\prod_{\sigma=1}^{s} f^{(\sigma)}(a_1, a_2, \cdots, a_n; \hat{\theta}) \geqq \prod_{\sigma=1}^{s} f^{(\sigma)}(a_1, a_2, \cdots, a_n; \theta') . \tag{2.2.1}$$

Here $f^{(\sigma)}(a_1, a_2, \cdots, a_n; \hat{\theta})$ is the probability of the sequence of responses $a_1, a_2, \cdots, a_n$ for subject σ when the learning parameter is $\hat{\theta}$. It is important to understand that the notation $a_1, a_2, \cdots, a_n$ for each σ does not imply that every subject has the same sequence of responses. It would be more explicit, but also more cumbersome, to write $a_1^{(\sigma)}, a_2^{(\sigma)}, \cdots, a_n^{(\sigma)}$. The use of the product notation in (2.2.1) expresses the assumption that the probabilities of response sequences of different subjects are statistically independent.

Consider now for a single subject $f(a_1, a_2, \cdots, a_n; \theta)$. By virtue of the fundamental Markov property of the process, we have

$$f(a_n \mid a_{n-1}; \theta) f(a_{n-1} \mid a_{n-2}; \theta) \cdots f(a_2 \mid a_1; \theta) f(a_1; \theta) = f(a_n, a_{n-1}, \cdots, a_1; \theta) . \tag{2.2.2}$$

Thus, summing over trials and subjects, we want to maximize

$$\prod_{\sigma=1}^{s} \prod_{m=2}^{n} f^{(\sigma)}(a_m \mid a_{m-1}; \theta) f^{(\sigma)}(a_1; \theta) . \tag{2.2.3}$$

Let N be the number of states in the process; $p_{ij}(\theta)$ the probability of going from state i to state j with parameter value θ; n_{ij} the observed number of transitions from state i to state j aggregated over trials and subjects (thus the n_{ij} are tabulated from experimental data); $p_i(\theta)$ the probability of being in state i on trial 1; and n_i the number of subjects in state i on trial 1 (i.e., here the number of subjects who make response i on trial 1). Substituting this notation in (2.2.3), we may then replace (2.2.1) by[1]

$$\prod_{i,j=1}^{N} p_i^{n_i}(\hat{\theta}) p_{ij}^{n_{ij}}(\hat{\theta}) \geqq \prod_{i,j=1}^{N} p_i^{n_i}(\theta') p_{ij}^{n_{ij}}(\theta') . \tag{2.2.4}$$

[1] Suppose, for instance, that there are two subjects with sequences of responses 12121 and 11222 for the first five trials, where 1 indicates response A_1 and 2 indicates response A_2. Then clearly $n_1 = 2$, $n_2 = 0$, because both subjects made the A_1 response on the first trial, and, as may easily be checked, $n_{11} = 1$, $n_{12} = 3$, $n_{21} = 2$, and $n_{22} = 2$. The use of the n_{ij} as exponents in (2.2.4) comes from the simple equation

$$p_1^2 p_{11} p_{12}^3 p_{21}^2 p_{22}^2 = p_1 p_{12} p_{21} p_{12} p_{21} p_1 p_{11} p_{12} p_{22} p_{22} .$$

That is, to find $\hat{\theta}$ we want to maximize

$$\prod_{i,j=1}^{N} p_i^{n_i}(\theta) p_{ij}^{n_{ij}}(\theta)$$

with respect to θ, but the value of θ that maximizes this expression also maximizes its logarithm, which in many cases is easier to work with. Consequently we shall usually seek to maximize

$$L(\theta) = \sum_i \left[n_i \log p_i(\theta) + \sum_j n_{ij} \log p_{ij}(\theta) \right]. \tag{2.2.5}$$

Ordinarily the function $L(\theta)$ will have a local maximum, and so we can find θ as an appropriate solution of

$$\frac{dL(\theta)}{d\theta} = \sum_i \left[\frac{n_i p_i'(\theta)}{p_i(\theta)} + \sum_j \frac{n_{ij} p_{ij}'(\theta)}{p_{ij}(\theta)} \right] = 0 , \tag{2.2.6}$$

where p' is the derivative with respect to θ of p.

We now apply the general results (2.2.5) and (2.2.6) to the one-element model for the noncontingent case discussed in Chapter 1. The transition matrix (p_{ij}) is given by (1.3.4); moreover, if we begin with trial 1, $p_i(\theta)$ is independent of θ, and so we obtain from (2.2.6)

$$\frac{dL(\theta)}{d\theta} = -\frac{n_{11}(1-\pi)}{1-\theta(1-\pi)} + \frac{n_{12}}{\theta} + \frac{n_{21}}{\theta} - \frac{n_{22}\pi}{1-\theta\pi} = 0 . \tag{2.2.7}$$

This simplifies to the quadratic equation given in the following theorem.

THEOREM. *The maximum-likelihood estimate of θ for the one-element model in the noncontingent case is the smaller root of the equation*

$$\begin{aligned}(n_{11} + n_{12} + n_{21} + n_{22})\pi(1-\pi)\theta^2 \\ - [n_{11}(1-\pi) + n_{12} + n_{21} + n_{22}\pi]\theta + (n_{12} + n_{21}) = 0 .\end{aligned} \tag{2.2.8}$$

Numerical solutions of this equation for various experimental data are presented in subsequent chapters. From Descartes' rule of signs, it follows that there are no negative roots of (2.2.8), but occasionally the smaller root will be greater than 1. The question naturally arises of what interpretation to place on $\hat{\theta} > 1$. Clearly such a value of $\hat{\theta}$ cannot be interpreted as a probability, even though the stochastic character of the matrix (1.3.4) may not itself be disturbed by $\hat{\theta} > 1$. It seems to us that the best compromise is to think of such an estimate as approximating a true value of θ which, as a probability, is very close to 1. However, it has been our experience that when $\hat{\theta} > 1$ the over-all fit of the model is usually rather poor. We shall return to this point in later chapters. The proof that the smaller root is the appropriate one is straightforward and will be omitted. Moreover, when the data fit the model exactly, $\theta = 2$ is always the larger root.[2]

[2] The proof most easily proceeds by expressing the n_{ij} in terms of N, π, and θ, where N is the total number of transitions, i.e., $N = \sum n_{ij}$. Independent of the model, $n_{12} = n_{21}$ almost exactly, because the number of entries into a row must equal the number of exits from the row, with the exception of initial and terminal trials. On this assumption we have $n_{11} = [\pi - \theta\pi(1-\pi)]N$, $n_{12} = n_{21} = \theta\pi(1-\pi)N$, $n_{22} = [(1-\pi) - \theta\pi(1-\pi)]N$.

It is to be noticed that in terms of observable quantities we cannot make a similar maximum-likelihood estimate of θ for the two-element noncontingent model whose transition matrix is given by (1.3.12). The difficulty is that the states of the Markov chain are unobservable, and that the sequence of response random variables, which is a sequence of observables, is not a Markov chain, but a chain of infinite order (i.e., a chain in which the probability of response on a given trial depends on all preceding responses).

The quadratic equation (2.2.8) suggests that the maximum-likelihood estimate of θ would be the root of a polynomial of higher degree in the zero-sum, two-person situation described by matrix (1.4.3). (For the present, we shall assume $\theta_A = \theta_B$ in (1.4.3) and thus consider the estimation of a single parameter.) Surprisingly enough, this is not the case. In fact, the estimate has a much simpler form.

THEOREM. *If* $0 < a_i < 1$, *then the maximum-likelihood estimate of* θ *for the one-element model in the zero-sum, two-person case is*

$$\hat{\theta} = \frac{n_{12} + n_{13} + n_{21} + n_{24} + n_{31} + n_{34} + n_{42} + n_{43}}{n_{11} + n_{22} + n_{33} + n_{44} + n_{12} + n_{13} + n_{21} + n_{24} + n_{31} + n_{34} + n_{42} + n_{43}} . \tag{2.2.9}$$

PROOF. As in the noncontingent case, $p_i(\theta)$ is independent of θ, and we have from (1.4.3)

$$\frac{dL(\theta)}{d\theta} = -\frac{n_{11}}{1-\theta} + \frac{n_{12}a_1}{a_1\theta} + \frac{n_{13}(1-a_1)}{(1-a_1)\theta} + \frac{n_{21}a_2}{a_2\theta} - \frac{n_{22}}{1-\theta} + \frac{n_{24}(1-a_2)}{(1-a_2)\theta}$$
$$+ \frac{n_{31}(1-a_3)}{(1-a_3)\theta} - \frac{n_{33}}{1-\theta} + \frac{n_{34}a_3}{a_3\theta} + \frac{n_{42}(1-a_4)}{(1-a_4)\theta} + \frac{n_{43}a_4}{a_4\theta} - \frac{n_{44}}{1-\theta} = 0 .$$

Simplifying, we obtain

$$\frac{n_{12} + n_{13} + n_{21} + n_{24} + n_{31} + n_{34} + n_{42} + n_{43}}{\theta} - \frac{n_{11} + n_{22} + n_{33} + n_{44}}{1-\theta} = 0 .$$

We solve this linear equation to obtain the desired result. If a_i is 0 or 1, this estimate is changed in the obvious way.

The simultaneous estimating of θ_A and θ_B in the zero-sum, two-person situation is more complicated. Specifically

$$\frac{\partial L(\theta_A, \theta_B)}{\partial \theta_A} = \frac{n_{11}(a_1 - 1)}{a_1(\theta_A - \theta_B) + 1 - \theta_A} + \frac{n_{13}}{\theta_A} + \frac{n_{22}(a_2 - 1)}{a_2(\theta_A - \theta_B) + 1 - \theta_A}$$
$$+ \frac{n_{24}}{\theta_A} + \frac{n_{31}}{\theta_A} + \frac{n_{33}(a_3 - 1)}{a_3(\theta_A - \theta_B) + 1 - \theta_A}$$
$$+ \frac{n_{42}}{\theta_A} + \frac{n_{44}(a_4 - 1)}{a_4(\theta_A - \theta_B) + 1 - \theta_A} = 0 ; \tag{2.2.10}$$
$$\frac{\partial L(\theta_A, \theta_B)}{\partial \theta_B} = -\frac{n_{11}a_1}{a_1(\theta_A - \theta_B) + 1 - \theta_A} + \frac{n_{12}}{\theta_B} + \frac{n_{21}}{\theta_B}$$
$$- \frac{n_{22}a_2}{a_2(\theta_A - \theta_B) + 1 - \theta_A} - \frac{n_{33}a_3}{a_3(\theta_A - \theta_B) + 1 - \theta_A}$$
$$+ \frac{n_{34}}{\theta_B} + \frac{n_{43}}{\theta_B} - \frac{n_{44}a_4}{a_4(\theta_A - \theta_B) + 1 - \theta_A} = 0 .$$

The maximum-likelihood estimates of θ_A and θ_B are determined by the simultaneous solution of the above equations.

Maximum-likelihood estimates for conditioning parameters in other two- and three-person interaction situations are presented in later chapters.

2.3. Pseudomaximum-Likelihood Estimates

When more than one stimulus element is assumed, the problems of estimating the parameter θ are considerably more difficult. As remarked in Chapter 1, the states of the Markov chain are the possible states of conditioning of the stimuli, and these states are unobservable. The observable sequence of response random variables $\langle \boldsymbol{A}_1, \boldsymbol{A}_2, \cdots, \boldsymbol{A}_n, \cdots \rangle$ is itself a chain of infinite order; that is, the probability distribution on trial n of $\boldsymbol{A}_n$ depends, not just on the response on trial $n-1$, but on the responses made on all preceding trials.

As in the case of the one-element model, (2.2.1) defines the maximum-likelihood estimate $\hat{\theta}$, but (2.2.2) no longer holds. We nonetheless take (2.2.3) as the expression to maximize, even though the resulting estimate is not the maximum-likelihood one. The justification for this approach is that it represents a crude but computationally practical approximation of the maximum-likelihood estimate. If the transition numbers n_{ijk} and the accompanying theoretical expressions $f(a_m \mid a_{m-1}a_{m-2}; \theta)$ are used, a still better estimate is obtained at the cost of considerably more work. Similarly, use of the transition numbers n_{ijkl} yields a still better approximation. Let us call (2.2.3) the first-order pseudomaximum-likelihood estimate θ^*, and the estimate that uses two preceding trials the second-order estimate, and so on. It is not difficult to show that for all the models considered in this book the n-order pseudomaximum-likelihood estimates rapidly converge to the maximum-likelihood estimate $\hat{\theta}$. This result follows directly from showing that the chains of infinite order constituted by the response random variables satisfy the conditions for convergence given in Lamperti and Suppes [1959].

As a simple example, we compute the (first-order) pseudomaximum-likelihood estimate θ^* for the two-element noncontingent model discussed in Sec. 1.3. To begin with, we need the probabilities

$$\lim_{n\to\infty} P(\boldsymbol{A}_{n+1} = i \mid \boldsymbol{A}_n = j) \qquad (i, j = 1, 2)\,. \tag{2.3.1}$$

To abbreviate notation in derivations we replace the random variable notation $\boldsymbol{A}_{n+1} = i$ by the event notation $A_{i,n+1}$, etc; that is, $A_{i,n+1}$ is the event of response i on trial $n+1$. Similarly, $E_{k,n}$ is the event of reinforcing event E_k on trial n.

Moreover, for purposes of subsequent generalization and to illustrate some useful technical simplifications, we shall consider the two-element noncontingent model in terms of a Markov chain whose states are the *number* of stimulus elements conditioned to the A_1 response. Thus there will be three states, 0, 1, 2, rather than the four states $0, \{s_1\}, \{s_2\}, \{s_1, s_2\}$ of the process defined by

(1.3.12). From the discussion in Sec. 1.3 it should be obvious how to construct the trees of the Markov chain in these three states; we simply give here the transition matrix.

$$\text{(2.3.2)} \qquad \begin{array}{c|ccc} & 0 & 1 & 2 \\ \hline 0 & 1-\theta\pi & \theta\pi & 0 \\ 1 & \frac{1}{2}\theta(1-\pi) & 1-\frac{1}{2}\theta & \frac{1}{2}\theta\pi \\ 2 & 0 & \theta(1-\pi) & 1-\theta(1-\pi) \end{array}$$

We now use (2.3.2) to compute the quantities (2.3.1). Let $C_{i,n}$ be the event that i stimulus elements are conditioned to A_1 on trial n. We first note that

$$\text{(2.3.3)} \qquad \begin{aligned} P(A_{1,n+1} \mid A_{1,n}) &= \sum_{i=0}^{2} P(A_{1,n+1}C_{i,n+1} \mid A_{1,n}) \\ &= P(A_{1,n+1} \mid C_{1,n+1})P(C_{1,n+1} \mid A_{1,n}) \\ &\qquad + P(A_{1,n+1} \mid C_{2,n+1})P(C_{2,n+1} \mid A_{1,n}) , \end{aligned}$$

because

$$P(A_{1,n+1} \mid C_{i,n+1}A_{1,n}) = P(A_{1,n+1} \mid C_{i,n+1})$$

(a consequence of the "independence-of-path" axioms) and

$$P(A_{1,n+1} \mid C_{0,n+1}) = 0 .$$

Also, from the sampling and response axioms, we have

$$\text{(2.3.4)} \qquad P(A_{1,n+1} \mid C_{1,n+1}) = \tfrac{1}{2} \quad \text{and} \quad P(A_{1,n+1} \mid C_{2,n+1}) = 1 ;$$

thus (2.3.3) simplifies to

$$\text{(2.3.5)} \qquad P(A_{1,n+1} \mid A_{1,n}) = \tfrac{1}{2}P(C_{1,n+1} \mid A_{1,n}) + P(C_{2,n+1} \mid A_{1,n}) .$$

Our problem now is to compute the two quantities on the right-hand side of (2.3.5). We first observe that

$$\text{(2.3.6)} \qquad \begin{aligned} P(C_{1,n+1} \mid A_{1,n}) &= \sum_{i=0}^{2} \frac{P(C_{1,n+1}A_{1,n}C_{i,n})}{P(A_{1,n})} \\ &= \frac{P(C_{1,n+1} \mid A_{1,n}C_{1,n})P(A_{1,n} \mid C_{1,n})P(C_{1,n})}{P(A_{1,n})} \\ &\qquad + \frac{P(C_{1,n+1} \mid A_{1,n}C_{2,n})P(A_{1,n} \mid C_{2,n})P(C_{2,n})}{P(A_{1,n})} , \end{aligned}$$

where again we have eliminated the state C_0 because $P(A_{1,n} \mid C_{0,n}) = 0$. Now

$$\text{(2.3.7)} \qquad \begin{aligned} P(C_{1,n+1} \mid A_{1,n}C_{1,n}) &= P(C_{1,n+1} \mid E_{1,n}A_{1,n}C_{1,n})P(E_{1,n}) \\ &\qquad + P(C_{1,n+1} \mid E_{2,n}A_{1,n}C_{1,n})P(E_{2,n}) \\ &= \pi + (1-\theta)(1-\pi) , \end{aligned}$$

since in order for $A_{1,n}$ to occur, the sampled element must be conditioned to A_1. It follows that if E_1 occurs, regardless of the effectiveness of conditioning, the new conditioning state is the same. However, if E_2 occurs, the conditioning stays the same only with probability $(1-\theta)$. Now for the second term on the right-hand side of (2.3.6), we can pass from $C_{2,n}$ to $C_{1,n+1}$ only if an E_2 reinforcement occurs to reduce the number of stimulus elements conditioned to A_1. Thus

(2.3.8) $$P(C_{1,n+1} \mid A_{1,n}C_{2,n}) = \theta(1 - \pi) .$$

It easily follows from the asymptotic results in Sec. 1.3 that

(2.3.9) $$\lim C_{0,n} = (1 - \pi)^2 , \qquad \lim C_{1,n} = 2\pi(1 - \pi) , \qquad \lim C_{2,n} = \pi^2 ,$$

and of course $\lim P(A_{1,n}) = \pi$. We thus infer from (2.3.4) and (2.3.6–9) that

(2.3.10) $$\begin{aligned}\lim P(C_{1,n+1} \mid A_{1,n}) &= [\pi + (1 - \theta)(1 - \pi)]\frac{1}{2}\,\frac{2\pi(1 - \pi)}{\pi} + \frac{\theta(1 - \pi)\pi^2}{\pi} \\ &= \pi(1 - \pi) + (1 - \theta)(1 - \pi)^2 + \theta\pi(1 - \pi) .\end{aligned}$$

We next compute $\lim P(C_{2,n+1} \mid A_{1,n})$. It is clear that given the A_1 response on trial n, the sampled stimulus element on trial n must be conditioned to A_1; and thus to have event $C_{2,n+1}$ it is necessary to have $C_{2,n}$. That is, given $A_{1,n}$, conditioning state $C_{2,n+1}$ can be reached only from $C_{2,n}$. Thus

(2.3.11) $$\begin{aligned}\lim P(C_{2,n+1} \mid A_{1,n}) &= \lim \frac{P(C_{2,n+1} \mid A_{1,n}C_{2,n})P(A_{1,n} \mid C_{2,n})P(C_{2,n})}{P(A_{1,n})} \\ &= \frac{[\pi + (1 - \theta)(1 - \pi)]\pi^2}{\pi} \\ &= \pi - \theta\pi(1 - \pi) .\end{aligned}$$

Going back now to (2.3.3) and applying (2.3.4), (2.3.10), and (2.3.11), we have

(2.3.12) $$\begin{aligned}\lim P(A_{1,n+1} \mid A_{1,n}) &= \tfrac{1}{2}[\pi(1 - \pi) + (1 - \theta)(1 - \pi)^2 + \theta\pi(1 - \pi)] \\ &\qquad + \pi - \theta\pi(1 - \pi) \\ &= \pi + \frac{(1 - \theta)(1 - \pi)}{2} ,\end{aligned}$$

which is the desired result. From considerations of symmetry, or by direct computation, it is easily established that

(2.3.13) $$\lim P(A_{2,n+1} \mid A_{2,n}) = 1 - \pi + \frac{(1 - \theta)\pi}{2} ,$$

and thus

(2.3.14) $$\begin{aligned}\lim P(A_{1,n+1} \mid A_{2,n}) &= \pi - \frac{(1 - \theta)\pi}{2} , \\ \lim P(A_{2,n+1} \mid A_{1,n}) &= 1 - \pi - \frac{(1 - \theta)(1 - \pi)}{2} .\end{aligned}$$

It then follows from (2.3.12–14) that the first-order pseudomaximum-likelihood estimate θ^* is a solution of the equation

(2.3.15) $$\frac{dL^*(\theta)}{d\theta} = -\frac{n_{11}(1 - \pi)}{1 + \pi - \theta(1 - \pi)} + \frac{n_{12}}{1 + \theta} + \frac{n_{21}}{1 + \theta} - \frac{n_{22}\pi}{2 - (1 + \theta)\pi} = 0 ,$$

which simplifies to the quadratic equation of the following theorem.

THEOREM. *The first-order pseudomaximum-likelihood estimate θ^* for the two-element model in the noncontingent case is the smaller root of the equation*

$$(2.3.16)\quad \begin{aligned}&(n_{11}+n_{12}+n_{21}+n_{22})\pi(1-\pi)\theta^2\\ &-2[n_{11}(1-\pi)^2+(n_{12}+n_{21})(\pi^2-\pi+1)+n_{22}\pi^2]\theta\\ &-n_{11}(1-\pi)(2-\pi)+(n_{12}+n_{21})(1+\pi)(2-\pi)-n_{22}\pi(1+\pi)=0\,.\end{aligned}$$

(If the model fits the data exactly, the larger root is always $\theta = 3$.)

The statistical properties of this method of estimation need investigation. However, it can be shown that the method is consistent for all the models considered in this book.

Equations (2.3.12) and (2.3.13) are special cases for $N = 2$ of more general results for arbitrary N. In order to use pseudomaximum-likelihood methods to estimate jointly N and θ, it is necessary to derive the general result.

THEOREM. *For the N-element model (with equiprobable sampling of stimulus elements), in the noncontingent case*

$$(2.3.17)\quad \lim P(A_{1,n+1} \mid A_{1,n}) = \pi + \frac{(1-\theta)(1-\pi)}{N}$$

and

$$(2.3.18)\quad \lim P(A_{2,n+1} \mid A_{2,n}) = 1 - \pi + \frac{(1-\theta)\pi}{N}\,.$$

PROOF. We simply generalize the method of attack used for $N = 2$. Corresponding to (2.3.3), we have

$$(2.3.19)\quad \begin{aligned}P(A_{1,n+1} \mid A_{1,n}) &= \sum_{i=1}^{N}\sum_{j=1}^{N} P(A_{1,n+1}C_{i,n+1}A_{1,n}C_{j,n})/P(A_{1,n})\\ &= \sum_i\sum_j P(A_{1,n+1} \mid C_{i,n+1})P(C_{i,n+1} \mid A_{1,n}C_{j,n})\\ &\qquad \cdot P(A_{1,n+1} \mid C_{j,n})P(C_{j,n})/P(A_{1,n})\,.\end{aligned}$$

By analogy with (2.3.4),

$$(2.3.20)\quad P(A_{1,n} \mid C_{j,n}) = j/N\,,$$

and corresponding to (2.3.7), (2.3.8), and (2.3.11), we have

$$(2.3.21)\quad P(C_{i,n+1} \mid A_{1,n}C_{j,n}) = \begin{cases}0\,, & i > j\\ \pi + (1-\theta)(1-\pi)\,, & i = j\\ \theta(1-\pi)\,, & i = j-1\\ 0\,, & i < j-1\,.\end{cases}$$

The results in Sec. 1.3 for $N - 2$ easily generalize to show that

$$(2.3.22)\quad \lim P(C_{i,n}) = \binom{N}{i}\pi^i(1-\pi)^{N-i}\,,$$

where $\binom{N}{i} = N!/[i!(N-i)!]$, which is the binomial coefficient. Substituting (2.3.20) and (2.3.21) in (2.3.19), we obtain

(2.3.23)
$$P(A_{1,n+1} \mid A_{1,n}) = \frac{1}{P(A_{1,n})}\left\{ \sum_{i=1}^{N} \frac{i^2}{N^2}[\pi + (1-\theta)(1-\pi)]P(C_{i,n}) + \sum_{i=1}^{N} \frac{i(i-1)}{N^2}\theta(1-\pi)P(C_{i,n})\right\};$$

and using (2.3.22) as $n \to \infty$, we have

$$\lim P(A_{1,n+1} \mid A_{1,n}) = \frac{\pi + (1-\theta)(1-\pi)}{\pi N^2}\sum_i i^2\binom{N}{i}\pi^i(1-\pi)^{N-i} + \frac{\theta(1-\pi)}{\pi N^2}\sum_i i(i-1)\binom{N}{i}\pi^i(1-\pi)^{N-i}.$$

The first summation is just the second raw moment of the binomial distribution with parameter π, and the second summation is the second raw moment minus the mean $N\pi$. Hence

$$\begin{aligned}\lim P(A_{1,n+1} \mid A_{1,n}) &= \frac{\pi + (1-\theta)(1-\pi)}{\pi N^2}[N^2\pi^2 + N\pi(1-\pi)] \\ &\quad + \frac{\theta(1-\pi)}{\pi N^2}[N^2\pi^2 + N\pi(1-\pi) - N\pi] \\ &= \frac{1}{N}\{\pi(N\pi + 1 - \pi) \\ &\quad + (1-\pi)[(1-\theta)(N\pi + 1 - \pi) + \theta\pi(N-1)]\} \\ &= \frac{1}{N}(N\pi + 1 - \theta - \pi + \theta\pi) \\ &= \pi + \frac{(1-\theta)(1-\pi)}{N},\end{aligned}$$

which establishes (2.3.17). The argument for (2.3.18) is completely symmetrical with π and $1-\pi$ interchanged. This completes the proof.

Applying the methods just used, we may establish for the N-element model that

(2.3.24)
$$\lim P(A_{1,n+1} \mid E_{1,n}) = \pi + \frac{\theta(1-\pi)}{N}$$

and that

(2.3.25)
$$\lim P(A_{2,n+1} \mid E_{2,n}) = 1 - \pi + \frac{\theta\pi}{N}.$$

Let

(2.3.26)
$$x = \frac{1-\theta}{N}, \qquad y = \frac{\theta}{N}.$$

Then $N = 1/(x+y)$ and $\theta = y/(x+y)$. We may use (2.3.17) and (2.3.18) to make a pseudomaximum-likelihood estimate of x, and (2.3.24) and (2.3.25) to estimate y. For x we have

(2.3.27)

	$A_{1,n+1}$	$A_{2,n+1}$
$A_{1,n}$	$\pi + x(1-\pi)$	$(1-x)(1-\pi)$
$A_{2,n}$	$(1-x)\pi$	$1 - \pi + x\pi$

,

which is just the transition matrix (1.3.4) for the one-element model with θ replaced by $1-x$, whence the quadratic equation (2.2.8) yields the pseudo-maximum-likelihood estimate of $1-x$, and thus of x. For y we have

$$\text{(2.3.28)}\qquad \begin{array}{c|cc} & A_{1,n+1} & A_{2,n+1} \\ \hline E_{1,n} & \pi + y(1-\pi) & (1-y)(1-\pi) \\ E_{2,n} & (1-y)\pi & 1-\pi+y\pi \end{array} .$$

Because the matrices for x and y are precisely alike, the same formal remarks hold: i.e., the quadratic equation (2.2.8) may yield the estimate of $1-y$, and thus of y. Of course in this case the transition numbers n_{ij} are different from those used in making the maximum-likelihood estimate of θ for the one-element model. In fact, it should be noticed that (2.3.28) does not even approximate a chain of infinite order, for the "transitions" are from reinforcements to responses. Applications of this method for jointly estimating θ and N are discussed in Sec. 10.6.

2.4. Some Statistical Tests for Markov Chains

In analyzing experimental data for Markov characteristics we shall use four standard tests. The first tests the hypothesis that the Markov process is stationary, i.e., that the transition probabilities are independent of the trial number n. The second concerns the order of the Markov chain, for example, first order versus second order. The third tests goodness-of-fit. The fourth tests the hypothesis that response protocols for different subjects can be viewed as a collection of samples from the same rth-order Markov chain. The derivation of these tests and an analysis of their statistical properties are given in Anderson and Goodman [1957]; also relevant are Bartlett [1951] and Hoel [1954]. For our purposes it will be sufficient to present a brief description of the tests.

Stationarity. Let $p_{ij}(t)$ be the probability of transition from state i on trial $t-1$ to state j on trial t. The null hypothesis is that $p_{ij}(t) = p_{ij}$, i.e., that the transition probabilities of the Markov process are independent of t. The χ^2 test of homogeneity is appropriate, and for each row i of the transition matrix we calculate

$$\chi_i^2 = \sum_{t,j} n_i(t-1)\frac{\left[\dfrac{n_{ij}(t)}{n_i(t-1)} - \dfrac{n_{ij}}{n_i}\right]^2}{n_{ij}/n_i}, \tag{2.4.1}$$

where $n_{ij}(t)$ denotes the observed number of cases in state i at $t-1$ and state j at t. Further, let

$$n_i(t-1) = \sum_j n_{ij}(t), \qquad n_{ij} = \sum_t n_{ij}(t), \qquad n_i = \sum_j n_{ij}.$$

If the null hypothesis is true, χ_i^2 has the usual limiting distribution with $(m-1)(T-1)$ degrees of freedom, where m is the number of states and T is the number of trials. In the experiments reported in subsequent chapters the number of pairs of subjects, or triples, as the case may be, ranges between 20 and 40, which is not sufficient to yield stable estimates of $\hat{p}_{ij}(t) = n_{ij}(t)/n_i(t-1)$ for individual trials t. Consequently, our procedure is to sum over blocks of trials, in which case t now represents a particular block of trials.

Finally, it may be shown that the several χ_i^2 are asymptotically independent, whence the sum $\chi^2 = \sum_i \chi_i^2$ has the usual limiting distribution with $m(m-1)(T-1)$ degrees of freedom.

Order. The Markov character of the sequence of response random variables, or of other sequences of random variables, may be tested directly without recourse to details of stimulus-sampling theory. Again a χ^2 test of homogeneity is appropriate. To begin with, we can test the null hypothesis that the outcomes of trials are statistically independent (zero-order process) against the alternative hypothesis that the process is a first-order Markov chain by computing the sum

$$\chi^2 = \sum_{i,j} n_i \frac{\left(\dfrac{n_{ij}}{n_i} - \dfrac{n_j}{N}\right)^2}{n_j/N}, \tag{2.4.2}$$

where $n_j = \sum_i n_{ij}$, $N = \sum_{i,j} n_{ij}$, and n_{ij} and n_i are as defined above. Again, χ^2 has the usual limiting distribution with $(m-1)^2$ degrees of freedom.

A second null hypothesis is that the process is a first-order Markov chain against the hypothesis that it is a second-order chain. Rejection of the null hypothesis in this case would mean that the response probabilities can be predicted better by observing the two immediately preceding responses rather than simply the single immediately preceding response. The results of this test are of particular psychological interest because they suggest the two main ways of developing models that are more adequate than the one-element ones. The possibility most discussed by us is the development of multi-stimulus-element models that are first-order Markov chains in the unobservable states of conditioning but not in the subject's responses. The other possibility is to develop higher-order Markov chains that still have observable states.

The first-order versus second-order hypothesis can be tested by computing the sum

$$\chi^2 = \sum_{i,j,k} n_{ij} \frac{\left(\dfrac{n_{ijk}}{n_{ij}} - \dfrac{n_{jk}}{n_j}\right)^2}{n_{jk}/n_j}, \tag{2.4.3}$$

where n_{ijk} is defined similarly to n_{ij}. If the null hypothesis is true, χ^2 has a limiting distribution with $m(m-1)^2$ degrees of freedom. It is straightforward to generalize (2.4.3) to a test of $r-1$ order versus r order (see Anderson and Goodman [1957]). It may be noted that the various ratios which occur in these tests are actually the maximum-likelihood estimates of the transition probabilities. Thus $\hat{p}_{ij} = n_{ij}/n_i$, $\hat{p}_{ijk} = n_{ijk}/n_{ij}$, and so on. Also, these order tests are predicated on the assumption of stationarity, although this is not necessary.

Goodness of fit. For all the Markov chains considered in this book the transition matrix depends on at least one conditioning parameter. However, after the parameters have been estimated by the methods discussed in preceding sections, it is possible to make a χ^2 test of the goodness-of-fit of the predicted transition matrix to the observed transition matrix. This test would seem to be restricted to one-element models, which are the only ones having observable states. However, in the next section it is shown that this is not the case, and that the transition numbers for multi-element models may be inferred from observed data even though they are not directly observable.

For simplicity, let $\hat{\theta}$ be the estimated parameter, and $p_{ij}(\hat{\theta})$ be the theoretical transition probabilities based on $\hat{\theta}$. As before, let $\hat{p}_{ij}$ represent the observed transition probabilities as estimated from the transition numbers n_{ij}. Then under the null hypothesis

$$\chi^2 = \sum_{i,j} n_i \frac{[\hat{p}_{ij} - p_{ij}(\hat{\theta})]^2}{p_{ij}(\hat{\theta})} \tag{2.4.4}$$

has the usual limiting distribution with $m(m-1)-1$ degrees of freedom. (If q parameters are estimated, then there are $m(m-1)-q$ degrees of freedom.) When it is convenient, $p_{ij}(\hat{\theta})$ in the denominator may be replaced by $\hat{p}_{ij}$ without seriously affecting the test. For example, if $\hat{p}_{ij} \neq 0$ and $p_{ij}(\hat{\theta}) = 0$, such a substitution is convenient. This situation occurs in the transition matrix (1.4.3) for zero-sum, two-person games.

Although this goodness-of-fit test has the virtue of providing an over-all measure of the adequacy of a particular Markov chain to a given experiment, it would be a mistake to construe it as providing a test of the goodness-of-fit of stimulus-sampling theory to the data of the experiment. Consider, for example, the sequence of response random variables in the one-element model with noncontingent reinforcement. As we shall see in Chapter 10, the fit of this Markov chain to data may be exceptionally good, and yet the theory's additional prediction that $P(A_{1,n+1} \mid E_{1,n}A_{1,n}) = 1$ will be clearly contradicted. The implication of this prediction is that the Markov chain whose states are the possible pairs of responses and reinforcements will not fit the data as well as the Markov chain consisting only of the sequence of response random variables. On the other hand, the poor fit of the sequence of response random variables in the one-element model does not entail rejection of the theory, for a multi-element model may fit considerably better. In short, one should avoid making too simple an interpretation of the relation between the χ^2 goodness-of-fit test and the fundamental theory of stimulus sampling.

Identical Processes. Often a question of major interest in analyzing learning studies is whether or not the subjects in a particular experimental group can be considered equivalent—more specifically, whether or not the response protocols for different subjects can be viewed simply as a collection of samples from the same rth-order Markov chain. We will consider the test only for first-order chains, but the generalization to rth-order chains is obvious.

Let $\hat{p}_{ij}^{(h)} = n_{ij}^{(h)}/n_i^{(h)}$ denote the maximum-likelihood estimate of the first-order transition probability $p_{ij}^{(h)}$ for the process from which sample h ($h = 1, 2, \cdots, s$) was obtained. We wish to test the null hypothesis that $p_{ij}^{(h)} = p_{ij}$ for $h = 1, 2, \cdots, s$; that is, that the s processes are identical.

Again, the χ^2 test of homogeneity is appropriate; that is, to test this hypothesis, we calculate

$$\chi_i^2 = \sum_{h,j} \frac{n_i^{(h)}[\hat{p}_{ij}^{(h)} - \hat{p}_{ij}^{(\cdot)}]^2}{\hat{p}_{ij}^{(\cdot)}}, \tag{2.4.5}$$

where $n_{ij}^{(\cdot)} = \sum_h n_{ij}^{(h)}$ and $\hat{p}_{ij}^{(\cdot)} = n_{ij}^{(\cdot)}/\sum_j n_{ij}^{(\cdot)}$. Here χ_i^2 has the usual limiting distribution with $(s-1)(m-1)$ degrees of freedom. Finally,

$$\chi^2 = \sum_i \chi_i^2 \tag{2.4.6}$$

has a limiting χ^2 distribution with $m(m-1)(s-1)$ degrees of freedom.

Prior to the application of the various χ^2 tests discussed in this section, certain explicit reservations and restrictions must be made. The central difficulty may be illustrated with respect to the goodness-of-fit test. The validity of this test depends upon the assumption that the observations are statistically independent. Here the observations are the transitions from one state to another. If the sequence of observed states is not a first-order Markov chain, then the observations in general are not statistically independent and only an approximate validity can be claimed for the test. This lack of independence, which may be expressed by saying that equation (2.2.2) no longer holds, is also the reason why the pseudomaximum-likelihood estimates of Sec. 2.3 are not genuine maximum-likelihood estimates.

Moreover, on several occasions we apply the goodness-of-fit test to sequences of random variables which, even in theory, are not first- or second-order Markov chains, but chains of infinite order. To mitigate this fault, we emphasize that all these chains of infinite order are ergodic in character and that the conditional probabilities, given one or two preceding outcomes, are approximately independent. In nearly every case, comparisons of predicted and observed quantities are tabled; readers who are made uneasy by the approximations may inspect the tables and ignore the χ^2 tests. However, we have found the χ^2 computations useful for indicating qualitatively the directions in which new theoretical developments are most needed.

Finally, another gap in statistical methodology concerns the reduction of the number of degrees of freedom in the χ^2 goodness-of-fit test when the estimated parameter or parameters have not been estimated in an optimal fashion, by a maximum-likelihood method, for instance. To our knowledge the problems of mathematical statistics at issue here are as yet unsolved. It is fairly certain, however, that in always reducing the degrees of freedom by 1 for each estimated parameter we have followed a conservative course which is likely to err only on the side of caution.

*2.5. Estimation of Transition Numbers in Multi-Element Models

It has already been remarked several times that the states of conditioning of stimuli are not observable when more than one stimulus is available for sampling. The literature of stimulus-sampling theory would seem to suggest that the transition numbers associated with these unobservable states of conditioning also are not identifiable: that is, that they are not uniquely determined by the observed response data. Fortunately this is not so for the models we consider in this book, and in fact the transition numbers for the conditioning states can be estimated independent of θ. This is a particularly desirable state of affairs, for the estimated numbers may then themselves be used to estimate θ by the maximum-likelihood methods of Sec. 2.2. It is not difficult to show (see Anderson and Goodman [1957]) that these transition numbers form a set of sufficient statistics.

We now carry through this analysis in detail for the two-element models in the noncontingent situation and also in the zero-sum, two-person situation. The noncontingent case, the simpler of the two, will be considered first. We shall continually refer to the transition matrix (2.3.2) for the two-element model. For uniformity of notation, let n_{ij} be the transition numbers for the observed responses A_1 and A_2, and let m_{ij} be the transition numbers for the unobserved states 0, 1, and 2 of the Markov chain.

First we must write an equation for each n_{ij} in terms of the m_{ij}. In order to do this, it is necessary to compute the conditional probabilities $P(A_{i,n+1}A_{j,n}\mid C_{k,n+1}C_{l,n})$, where $i, j = 1, 2$ and $k, l = 0, 1, 2$. For simplicity, we replace $C_{1,n}$ by 1_n to designate the state with exactly one element conditioned to A_1, and similarly for $C_{0,n}$ and $C_{2,n}$. Beginning with the transition number n_{11} for $A_{1,n+1}A_{1,n}$, we note that the state of conditioning cannot be 0 on either trial, whence

$$n_{11} = \sum_{i,j=1}^{2} m_{ij}P(A_{1,n+1}A_{1,n}\mid j_{n+1}i_n) \,. \tag{2.5.1}$$

Clearly

$$P(A_{1,n+1}A_{1,n}\mid 2_{n+1}2_n) = 1 \,, \tag{2.5.2}$$

and by elementary probability theory, together with the independence-of-path axiom,

$$P(A_{1,n+1}A_{1,n}\mid 1_{n+1}2_n) = \frac{P(A_{1,n+1}\mid 1_{n+1})P(1_{n+1}\mid A_{1,n}2_n)P(A_{1,n}\mid 2_n)}{P(1_{n+1}\mid 2_n)} \,. \tag{2.5.3}$$

Now $P(1_{n+1}\mid 2_n)$ is given by the matrix (2.3.2); and by (2.3.8), $P(1_{n+1}\mid A_{1,n}2_n) = \theta(1-\pi)$, whence

$$P(A_{1,n+1}A_{1,n}\mid 1_{n+1}2_n) = \frac{1}{2}\theta(1-\pi)\frac{1}{\theta(1-\pi)} = \frac{1}{2} \,. \tag{2.5.4}$$

On the other hand, in order to make an A_1 response on trial n, the one element conditioned to this response must be sampled and thus the other element (which is neither sampled nor conditioned to A_1) cannot change its conditioning. Therefore, $P(2_{n+1} \mid A_{1,n}1_n) = 0$, and

$$P(A_{1,n+1}A_{1,n} \mid 2_{n+1}1_n) = 0\,. \tag{2.5.5}$$

For the fourth term, we have

$$P(A_{1,n+1}A_{1,n} \mid 1_{n+1}1_n) = \frac{P(A_{1,n+1} \mid 1_{n+1})P(1_{n+1} \mid A_{1,n}1_n)P(A_{1,n} \mid 1_n)}{P(1_{n+1} \mid 1_n)}\,.$$

Now by virtue of (2.3.7) we have $P(1_{n+1} \mid A_{1,n}1_n) = \pi + (1-\theta)(1-\pi)$, and we conclude that

$$P(A_{1,n+1}A_{1,n} \mid 1_{n+1}1_n) = \tfrac{1}{2}\cdot[\pi + (1-\theta)(1-\pi)]\cdot\tfrac{1}{2}/(1-\tfrac{1}{2}\theta)\,. \tag{2.5.6}$$

Combining equations (2.5.1–6), we obtain the equation for n_{11}, and by similar arguments the equations for n_{12}, n_{21}, and n_{22}:

$$\begin{aligned}
n_{11} &= m_{22} + \frac{1}{2}m_{21} + \frac{1}{4}\left[\frac{\pi + (1-\theta)(1-\pi)}{1-\frac{1}{2}\theta}\right]m_{11}\,,\\
n_{12} &= \frac{1}{2}m_{21} + m_{10} + \frac{1}{4}\left[\frac{\pi + (1-\theta)(1-\pi)}{1-\frac{1}{2}\theta}\right]m_{11}\,,\\
n_{21} &= m_{12} + \frac{1}{2}m_{01} + \frac{1}{4}\left[\frac{1-\theta\pi}{1-\frac{1}{2}\theta}\right]m_{11}\,,\\
n_{22} &= m_{00} + \frac{1}{2}m_{01} + \frac{1}{4}\left[\frac{1-\theta\pi}{1-\frac{1}{2}\theta}\right]m_{11}\,.
\end{aligned} \tag{2.5.7}$$

First we eliminate θ by combining the four equations of (2.5.7) as follows:

$$\begin{aligned}
n_{11} + n_{21} &= m_{22} + \tfrac{1}{2}m_{21} + m_{12} + \tfrac{1}{2}m_{01} + \tfrac{1}{2}m_{11}\,,\\
n_{11} + n_{22} &= m_{22} + \tfrac{1}{2}m_{21} + m_{00} + \tfrac{1}{2}m_{01} + \tfrac{1}{2}m_{11}\,,\\
n_{12} + n_{21} &= \tfrac{1}{2}m_{21} + m_{10} + m_{12} + \tfrac{1}{2}m_{01} + \tfrac{1}{2}m_{11}\,,\\
n_{12} + n_{22} &= \tfrac{1}{2}m_{21} + m_{10} + m_{00} + \tfrac{1}{2}m_{01} + \tfrac{1}{2}m_{11}\,.
\end{aligned} \tag{2.5.8}$$

Second, note that for a large number of observations the number of transitions into a given row of (2.3.2) must approximate very closely the number of transitions out of the row, independent of the value of θ. Whence we have the two equations

$$m_{01} = m_{10}\,, \qquad m_{12} = m_{21}\,. \tag{2.5.9}$$

Third, since the ratio of p_{10} to p_{12} is independent of θ, we have

$$m_{10} = \frac{1-\pi}{\pi}m_{12}\,. \tag{2.5.10}$$

Finally, two additional linear relations can be obtained by comparing transition numbers of different rows. From (2.3.2) we see that $p_{01}/p_{12} = 2$, whence

$$\frac{m_{01}}{m_{00} + m_{01}} = \frac{2m_{12}}{m_{10} + m_{11} + m_{12}}\,, \tag{2.5.11}$$

and $p_{21}/p_{10} = 2$, whence

$$\frac{m_{21}}{m_{21} + m_{22}} = \frac{2m_{10}}{m_{10} + m_{11} + m_{12}}\,. \tag{2.5.12}$$

From (2.5.9–12) we infer

$$(1-\pi)(m_{10}+m_{11}+m_{12}) = 2\pi(m_{00}+m_{01}) \, ,$$
$$(1-\pi)(m_{10}+m_{11}+m_{12}) = 2\pi(m_{21}+m_{22}) \, ,$$

which pair of equations is equivalent to

$$\begin{gathered} m_{00}+m_{01} = m_{21}+m_{22} \\ (1-\pi)(m_{10}+m_{11}+m_{12}) = 2\pi(m_{00}+m_{01}) \, . \end{gathered} \tag{2.5.13}$$

The nine equations (2.5.8–10) and (2.5.13) in seven unknowns may be expressed by the matrix equation

$$\begin{bmatrix} & \frac{1}{2} & & \frac{1}{2} & 1 & \frac{1}{2} & 1 \\ 1 & \frac{1}{2} & & \frac{1}{2} & & \frac{1}{2} & 1 \\ & \frac{1}{2} & 1 & \frac{1}{2} & 1 & \frac{1}{2} & \\ 1 & \frac{1}{2} & 1 & \frac{1}{2} & & \frac{1}{2} & \\ & 1 & -1 & & & & \\ & & & & 1 & -1 & \\ & & \pi & & \pi-1 & & \\ 1 & 1 & & & & -1 & -1 \\ 2\pi & 2\pi & \pi-1 & \pi-1 & \pi-1 & & \end{bmatrix} \begin{bmatrix} m_{00} \\ m_{01} \\ m_{10} \\ m_{11} \\ m_{12} \\ m_{21} \\ m_{22} \end{bmatrix} = \begin{bmatrix} n_{11}+n_{21} \\ n_{11}+n_{22} \\ n_{12}+n_{21} \\ n_{12}+n_{22} \\ 0 \\ 0 \\ 0 \\ 0 \\ 0 \end{bmatrix} . \tag{2.5.14}$$

We want to use the method of least squares to solve this system of equations for the overdetermined variables m_{ij}. However, the total number $N = \sum n_{ij} = \sum m_{ij}$ of transitions is known with probability 1, and thus is not to be estimated in the least-squares procedure. In order to avoid having $\sum m_{ij} \neq N$, we replace the last equation of (2.5.8) by

$$N = \sum_{i,j} m_{ij} \, . \tag{2.5.15}$$

Each equation is a consequence of the other, together with the remaining three equations of (2.5.8). We then use (2.5.15) to eliminate one of the seven variables, say m_{00}, thereby guaranteeing exact satisfaction of (2.5.15) by the least-squares estimate of the m_{ij}. However, when m_{00} is eliminated we find that the second and third equations of (2.5.14) are identical for $N - n_{11} - n_{22} = n_{12} + n_{21}$. Hence, deleting the redundant expression, we obtain the matrix equation

$$\begin{bmatrix} \frac{1}{2} & & \frac{1}{2} & 1 & \frac{1}{2} & 1 \\ \frac{1}{2} & 1 & \frac{1}{2} & 1 & \frac{1}{2} & \\ 1 & -1 & & & & \\ & & & 1 & -1 & \\ & \pi & & \pi-1 & & \\ & 1 & 1 & 1 & 2 & 2 \\ & \frac{1+\pi}{2\pi} & \frac{1+\pi}{2\pi} & \frac{1+\pi}{2\pi} & 1 & 1 \end{bmatrix} \begin{bmatrix} m_{01} \\ m_{10} \\ m_{11} \\ m_{12} \\ m_{21} \\ m_{22} \end{bmatrix} = \begin{bmatrix} n_{11}+n_{21} \\ n_{12}+n_{21} \\ 0 \\ 0 \\ 0 \\ N \\ N \end{bmatrix} , \tag{2.5.16}$$

which can be written as $QM = K$, where Q is the 7×6 matrix, M the column vector of m_{ij}, and K the column vector on the right of the equality sign. For the moment it is convenient to renumber the variables m_{ij} as $m_1, \cdots, m_6$. Applying now the method of least squares, we seek to minimize

$$S(m_1, \cdots, m_6) = \sum_{i=1}^{7} \left(k_i - \sum_{j'=1}^{6} q_{ij'} m_{j'}\right)^2 .$$

Now for $j = 1, \cdots, 6$ we have

$$\frac{\partial S}{\partial m_j} = -2 \sum_i \left(k_i - \sum_{j'} q_{ij'} m_{j'}\right) q_{ij} .$$

Setting these partial derivatives equal to zero, to minimize $S(m_1, \cdots, m_6)$ we solve the system of six equations, which may be written

$$BM = A , \tag{2.5.17}$$

where $B = Q'Q$, $A = Q'K$, and Q' is the transpose of the matrix Q.

The above solution for m_{ij} is based on the nine relations given by (2.5.8–10) and (2.5.13). However, the two equations of (2.5.9) are different from the other equations in that the identities hold with probability 1 independent of the subject's particular sequence of reinforcement.[3] For this reason it may be desirable in some empirical applications of this method to ensure that the in-out relations of (2.5.9) hold exactly in the least-squares solution. When this is the case we simply replace m_{10} by m_{01} and m_{21} by m_{12} in (2.5.16) and obtain

$$\begin{bmatrix} \frac{1}{2} & \frac{1}{2} & 1\frac{1}{2} & 1 \\ 1\frac{1}{2} & \frac{1}{2} & 1\frac{1}{2} & 0 \\ \pi & 0 & \pi - 1 & 0 \\ 1 & 1 & 3 & 2 \\ \dfrac{1+\pi}{2\pi} & \dfrac{1+\pi}{2\pi} & \dfrac{1+3\pi}{2\pi} & 1 \end{bmatrix} \begin{bmatrix} m_{01} \\ m_{11} \\ m_{12} \\ m_{22} \end{bmatrix} = \begin{bmatrix} n_{11} + n_{21} \\ n_{12} + n_{21} \\ 0 \\ N \\ N \end{bmatrix} . \tag{2.5.18}$$

The least-squares solution for m_{ij} is exactly like (2.5.17).

Given these estimated transition numbers m_{ij} [in terms of either (2.5.16) or (2.5.18)], we may now make a maximum-likelihood estimate of θ. From the transition matrix (2.3.2), we see at once that

$$\frac{dL(\theta)}{d\theta} = -\frac{m_{00}\pi}{1 - \theta\pi} + \frac{m_{01} + m_{10} + m_{12} + m_{21}}{\theta} - \frac{m_{11}}{2 - \theta} - \frac{m_{22}(1 - \pi)}{1 - \theta(1 - \pi)} .$$

Setting this derivative equal to 0, we obtain $\hat{\theta}$ as a root of the following cubic equation:

[3] More exactly, for each subject $|n_{12} - n_{21}| \leqq 1$ with probability 1.

$$
\begin{aligned}
(-m_{00} - m_{11} - m_{22} - m_{01} - m_{10} - m_{12} - m_{21})\pi(1-\pi)\theta^3 \\
+ [(3m_{00} + m_{22})\pi + (m_{01} + m_{10} + m_{12} + m_{21})(1 + 2\pi - 2\pi^2) \\
+ m_{11} + m_{22} - 2(m_{00} + m_{22})\pi^2]\theta^2 \\
+ [2(m_{22} - m_{00})\pi - 3(m_{01} + m_{10} + m_{12} + m_{21}) - m_{11} - 2m_{22}]\theta \\
+ 2(m_{01} + m_{10} + m_{12} + m_{21}) = 0 \,.
\end{aligned} \tag{2.5.19}
$$

This estimate $\hat{\theta}$ may be compared with the first-order pseudomaximum-likelihood estimate θ^* of Sec. 2.3. Moreover, the χ^2 goodness-of-fit test (2.4.4) may be applied to the 2×2 table of transition probabilities $P(A_{i,n+1} \mid A_{j,n})$ for $i, j = 1, 2$ at asymptote. The theoretical values $p_{ij}(\hat{\theta})$ are given by (2.3.12) and (2.3.13). More interesting still is the application of the χ^2 test to the fit at asymptote of the theoretical probabilities $P(A_{i,n+1} \mid E_{k,n}A_{j,n})$, for which the one-element model is particularly bad because the occurrence of a reinforced response is predicted with probability 1. Given $\hat{\theta}$, we may apply the χ^2 test to the following 4×2 table, whose entries are computed at asymptote by the methods of Sec. 2.3:

(2.5.20)

	A_1	A_2
A_1E_1	$\frac{1}{2} + \frac{\pi}{2}$	$\frac{1}{2} - \frac{\pi}{2}$
A_1E_2	$\frac{\pi}{2} + \frac{1-\theta}{2}$	$\frac{1}{2} + \frac{\theta}{2} - \frac{\pi}{2}$
A_2E_1	$\frac{\pi}{2} + \frac{\theta}{2}$	$1 - \frac{\pi}{2} - \frac{\theta}{2}$
A_2E_2	$\frac{\pi}{2}$	$1 - \frac{\pi}{2}$

The test is stringent, for neither the first nor the fourth row has any probabilities that depend on θ. The results of this test are compared in Chapter 10 with the results of a similar test for the generalized conditioning model discussed in the next section. In the two-element model the sequence of random variables $\langle \boldsymbol{A}_1\boldsymbol{E}_1, \boldsymbol{A}_2\boldsymbol{E}_2, \cdots, \boldsymbol{A}_n\boldsymbol{E}_n, \cdots \rangle$ is not a Markov chain, but a chain of infinite order. Consequently the χ^2 goodness-of-fit test (2.4.4), when applied to the table (2.5.20), has the defect mentioned at the end of Sec. 2.4.

The above method for estimating the transition numbers m_{ij} in the noncontingent case may also be applied to the zero-sum, two-person situation. Here the method is more practical because of the difficulty of obtaining expressions for the asymptotic probabilities of the conditioning states in the two-element model.

Corresponding to the transition matrix (1.4.3) for the one-element model, the transition matrix for the two-element model of the zero-sum, two-person case is given in (2.5.21), where the state ij means that player A has i stimuli conditioned to response A_1 and player B has j stimuli conditioned to response B_1:

(2.5.21)

	22	21	20	12	11	10	02	01	00
22	$(1-\theta)$	θa_1		$\theta(1-a_1)$					
21	$\frac{\theta a_2}{2}$	$(1-\theta)$	$\frac{\theta a_1}{2}$		$\frac{\theta(2-a_1-a_2)}{2}$				
20		θa_2	$(1-\theta)$			$\theta(1-a_2)$			
12	$\frac{\theta(1-a_3)}{2}$			$(1-\theta)$	$\frac{\theta(a_1+a_3)}{2}$		$\frac{\theta(1-a_1)}{2}$		
11		$\frac{\theta(2-a_3-a_4)}{4}$		$\frac{\theta(a_2+a_4)}{4}$	$(1-\theta)$	$\frac{\theta(a_1+a_3)}{4}$		$\frac{\theta(2-a_1-a_2)}{4}$	
10			$\frac{\theta(1-a_4)}{2}$		$\frac{\theta(a_2+a_4)}{2}$	$(1-\theta)$			$\frac{\theta(1-a_2)}{2}$
02				$\theta(1-a_3)$			$(1-\theta)$	θa_3	
01					$\frac{\theta(2-a_3-a_4)}{2}$		$\frac{\theta a_4}{2}$	$(1-\theta)$	$\frac{\theta a_3}{2}$
00						$\theta(1-a_4)$		θa_4	$(1-\theta)$

On the basis of this matrix we proceed exactly as for the noncontingent case. By analogy with (1.4.13–15), we introduce the events X_n, Y_n, Z_n, W_n:

$$X_n = A_{1,n}B_{1,n}, \qquad Z_n = A_{2,n}B_{1,n},$$
$$Y_n = A_{1,n}B_{2,n}, \qquad W_n = A_{2,n}B_{1,n}.$$

We then need to compute the conditional probabilities $P(X_{n+1}X_n \mid 22_{n+1}22_n)$, etc., of which there are $4 \times 4 \times 9 \times 9 = 1296$, most of which are 0. To indicate the method, we shall consider in detail the probabilities for $X_{n+1}X_n$ to obtain an expression for the observed transition number n_{XX} in terms of the unobserved $m_{ij,i'j'}$. By analogy with (2.5.1)

$$n_{XX} = \sum_{i,j=1}^{2} \sum_{i',j'=1}^{2} m_{ij,i'j'} P(X_{n+1}X_n \mid i'j'_{n+1}ij_n), \tag{2.5.22}$$

where summation over $i, j, i', j' = 0$ is omitted because these values prohibit an X response by the pair of players. Of the sixteen conditional probabilities occurring in the summation on the right, four are 0 because of zeros in the transition matrix (2.5.21), namely, the transitions 22_n11_{n+1}, 21_n12_{n+1}, 12_n21_{n+1}, and 11_n22_{n+1}. Moreover, the same argument that established (2.5.5) may be used to show that

$$\begin{aligned} P(X_{n+1}X_n \mid 22_{n+1}21_n) &= P(X_{n+1}X_n \mid 22_{n+1}12_n) \\ &= P(X_{n+1}X_n \mid 21_{n+1}11_n) = P(X_{n+1}X_n \mid 12_{n+1}11_n) = 0\,. \end{aligned}$$

The remaining eight transitions all contribute to n_{XX}, and we now proceed to compute them. The important and interesting observation is that all eight conditional probabilities are independent of θ. First, it is immediately clear that

$$P(X_{n+1}X_n \mid 22_{n+1}22_n) = 1 ,$$

and that

$$P(X_{n+1}X_n \mid 12_{n+1}22_n) = P(X_{n+1}X_n \mid 21_{n+1}22_n) = \tfrac{1}{2} .$$

Now by elementary probability theory and the independence-of-path axiom

$$P(X_{n+1}X_n \mid 21_{n+1}21_n) = \frac{P(X_{n+1} \mid 21_{n+1})P(21_{n+1} \mid X_n21_n)P(X_n \mid 21_n)}{P(21_{n+1} \mid 21_n)}$$
$$= \frac{\frac{1}{2}\cdot(1-\theta)\cdot\frac{1}{2}}{1-\theta} = \frac{1}{4} .$$

The setting of $P(21_{n+1} \mid X_n21_n)$ equal to $1-\theta$ is explained by Fig. 2.1, where it is clear that $P(21_{n+1} \mid X_n21_n) = a_1(1-\theta_B) + (1-a_1)(1-\theta_A)$; the right-hand term reduces to $1-\theta$ on the assumption that $\theta_A = \theta_B$. Also, for reasons of symmetry, $P(X_{n+1}X_n \mid 12_{n+1}12_n) = \frac{1}{4}$.

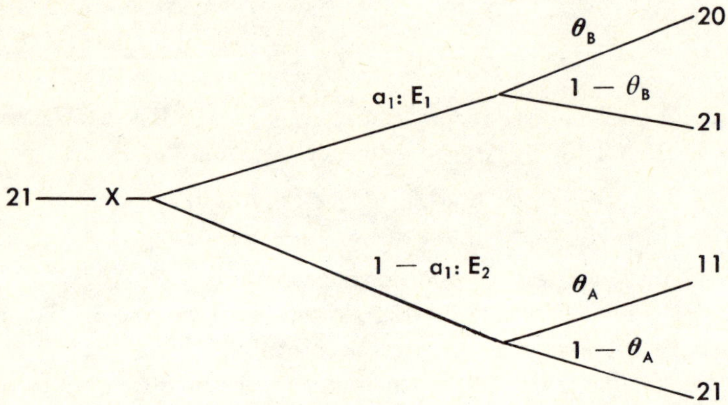

FIGURE 2.1.

We now consider the sixth term:

$$P(X_{n+1}X_n \mid 11_{n+1}21_n) = \frac{P(X_{n+1} \mid 11_{n+1})P(11_{n+1} \mid X_n21_n)P(X_n \mid 21_n)}{P(11_{n+1} \mid 21_n)}$$
$$= \frac{\frac{1}{4}\cdot\theta(1-a_1)\cdot\frac{1}{2}}{\frac{1}{2}\theta(2-a_1-a_2)} = \frac{\frac{1}{4}(1-a_1)}{2-a_1-a_2} .$$

The fact that $P(11_{n+1} \mid X_n21_n) = \theta(1-a_1)$ may be seen from Fig. 2.1, and $P(11_{n+1} \mid 21_n)$ may be read off directly from the transition matrix (2.5.21). Moreover, in exactly similar fashion it can be shown that

$$P(X_{n+1}X_n \mid 11_{n+1}12_n) = \frac{1}{4}\left(\frac{a_1}{a_1+a_3}\right) .$$

Finally for the last and eighth term we have

$$P(X_{n+1}X_n \mid 11_{n+1}11_n) = \frac{P(X_{n+1} \mid 11_{n+1})P(11_{n+1} \mid X_n11_n)P(X_n \mid 11_n)}{P(11_{n+1} \mid 11_n)}$$
$$= \frac{\frac{1}{4}\cdot(1-\theta)\cdot\frac{1}{4}}{1-\theta} = \frac{1}{16},$$

where the tree for $P(11_{n+1} \mid X_n11_n)$ is given in Fig. 2.2. (The subscripts A and

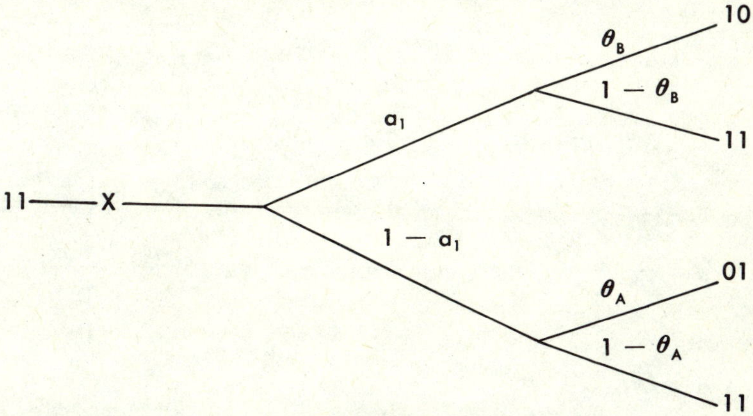

FIGURE 2.2.

B have been placed on the θ's to indicate which player is affected on a given branch by not being reinforced in his response.) Combining these eight results we then have for n_{XX} the following linear equation, which is independent of θ and holds not only at asymptote but for all trials:

(2.5.23)
$$n_{XX} = m_{22.22} + \tfrac{1}{2}(m_{22.21} + m_{22.12}) + \tfrac{1}{4}m_{21.21} + \frac{1-a_1}{4(2-a_1-a_2)}m_{21.11} + \tfrac{1}{4}m_{12.12} + \frac{a_1}{4(a_1+a_3)}m_{12.11} + \tfrac{1}{16}m_{11.11}.$$

By similar methods we obtain for the other fifteen observed transitions in X, Y, Z, and W the linear equations:

(2.5.24)
$$n_{XY} = \tfrac{1}{2}m_{22.21} + \tfrac{1}{4}m_{21.21} + m_{21.20} + \frac{1-a_1}{4(2-a_1-a_2)}m_{21.11} + \frac{a_1}{4(a_1+a_3)}m_{12.11} + \tfrac{1}{16}m_{11.11} + \frac{a_1}{2(a_1+a_3)}m_{11.10},$$

$$n_{XZ} = \tfrac{1}{2}m_{22.12} + \frac{1-a_1}{4(2-a_1-a_2)}m_{21.11} + \tfrac{1}{4}m_{12.12} + \frac{a_1}{4(a_1+a_3)}m_{12.11} + m_{12.02} + \tfrac{1}{16}m_{11.11} + \frac{1-a_1}{2(2-a_1-a_2)}m_{11.01},$$

$$n_{XW} = \frac{1-a_1}{4(2-a_1-a_2)}m_{21.11} + \frac{a_1}{4(a_1+a_3)}m_{12.11} + \tfrac{1}{16}m_{11.11} + \frac{a_1}{2(a_1+a_3)}m_{11.10} + \frac{1-a_1}{2(2-a_1-a_2)}m_{11.01};$$

$$n_{YX} = m_{21.22} + \tfrac{1}{4}m_{21.21} + \frac{1-a_2}{4(2-a_1-a_2)}m_{21.11} + \tfrac{1}{2}m_{20.21} + \frac{a_2}{2(a_2+a_4)}m_{11.12} + \tfrac{1}{16}m_{11.11} + \frac{a_2}{4(a_2+a_4)}m_{10.11}\,,$$

$$n_{YY} = \tfrac{1}{4}m_{21.21} + \frac{1-a_2}{4(2-a_1-a_2)}m_{21.11} + \tfrac{1}{2}m_{20.21} + m_{20.20} + \tfrac{1}{2}m_{20.10} + \tfrac{1}{16}m_{11.11} + \frac{a_2}{4(a_2+a_4)}m_{10.11} + \tfrac{1}{4}m_{10.10}\,,$$

$$n_{YZ} = \frac{1-a_2}{4(2-a_1-a_2)}m_{21.11} + \frac{a_2}{2(a_2+a_4)}m_{11.12} + \tfrac{1}{16}m_{11.11} + \frac{1-a_2}{2(2-a_1-a_2)}m_{11.01} + \frac{a_2}{4(a_2+a_4)}m_{10.11}\,,$$

$$n_{YW} = \frac{1-a_2}{4(2-a_1-a_2)}m_{21.11} + \tfrac{1}{2}m_{20.10} + \tfrac{1}{16}m_{11.11} + \frac{1-a_2}{2(2-a_1-a_2)}m_{11.01} + \frac{a_2}{4(a_2+a_4)}m_{10.11} + \tfrac{1}{4}m_{10.10} + m_{10.00}\,;$$

$$n_{ZX} = m_{12.22} + \tfrac{1}{4}m_{12.12} + \frac{a_3}{4(a_1+a_3)}m_{12.11} + \frac{1-a_3}{2(2-a_3-a_4)}m_{11.21} + \tfrac{1}{16}m_{11.11} + \tfrac{1}{2}m_{02.12} + \frac{1-a_3}{4(2-a_3-a_4)}m_{01.11}\,,$$

$$n_{ZY} = \frac{a_3}{4(a_1+a_3)}m_{12.11} + \frac{1-a_3}{2(2-a_3-a_4)}m_{11.21} + \tfrac{1}{16}m_{11.11} + \frac{a_3}{2(a_1+a_3)}m_{11.10} + \frac{1-a_3}{4(2-a_3-a_4)}m_{01.11}\,,$$

$$n_{ZZ} = \tfrac{1}{4}m_{12.12} + \frac{a_3}{4(a_1+a_3)}m_{12.11} + \tfrac{1}{16}m_{11.11} + \tfrac{1}{2}m_{02.12} + m_{02.02} + \tfrac{1}{2}m_{02.01} + \frac{1-a_3}{4(2-a_3-a_4)}m_{01.11} + \tfrac{1}{4}m_{01.01}\,,$$

$$n_{ZW} = \frac{a_3}{4(a_1+a_3)}m_{12.11} + \tfrac{1}{16}m_{11.11} + \frac{a_3}{2(a_1+a_3)}m_{11.10} + \tfrac{1}{2}m_{02.01} + \frac{1-a_3}{4(2-a_3-a_4)}m_{01.11} + \tfrac{1}{4}m_{01.01} + m_{01.00}\,;$$

$$n_{WX} = \frac{1-a_4}{2(2-a_3-a_4)}m_{11.21} + \frac{a_4}{2(a_2+a_4)}m_{11.12} + \tfrac{1}{16}m_{11.11} + \frac{a_4}{4(a_2+a_4)}m_{10.11} + \frac{1-a_4}{4(2-a_3-a_4)}m_{01.11}\,,$$

$$n_{WY} = \frac{1-a_4}{2(2-a_3-a_4)}m_{11.21} + \tfrac{1}{16}m_{11.11} + m_{10.20} + \frac{a_4}{4(a_2+a_4)}m_{10.11} + \tfrac{1}{4}m_{10.10} + \frac{1-a_4}{4(2-a_3-a_4)}m_{01.11} + \tfrac{1}{2}m_{00.10}\,,$$

$$n_{WZ} = \frac{a_4}{2(a_2 + a_4)} m_{11.12} + \tfrac{1}{16} m_{11.11} + \frac{a_4}{4(a_2 + a_4)} m_{10.11} + \frac{1 - a_4}{4(2 - a_3 - a_4)} m_{01.11} + m_{01.02} + \tfrac{1}{4} m_{01.01} + \tfrac{1}{2} m_{00.01} \,,$$

$$n_{WW} = \tfrac{1}{16} m_{11.11} + \frac{a_4}{4(a_2 + a_4)} m_{10.11} + \tfrac{1}{4} m_{10.10} + \frac{1 - a_4}{4(2 - a_3 - a_4)} m_{01.11} + \tfrac{1}{4} m_{01.01} + \tfrac{1}{2} m_{00.10} + \tfrac{1}{2} m_{00.01} + m_{00.00} \,.$$

Next, from the necessary approximate equality between the number of transitions in and out of any row, we have immediately from the transition matrix (2.5.19) the following nine linear equations:

(2.5.25)
$$\begin{aligned}
&m_{12.22} + m_{21.22} = m_{22.21} + m_{22.12} \,, \\
&m_{11.21} + m_{20.21} + m_{22.21} = m_{21.22} + m_{21.20} + m_{21.11} \,, \\
&m_{10.20} + m_{21.20} = m_{20.21} + m_{20.10} \,, \\
&m_{02.12} + m_{11.12} + m_{22.12} = m_{12.22} + m_{12.11} + m_{12.02} \,, \\
&m_{01.11} + m_{10.11} + m_{12.11} + m_{21.11} = m_{11.21} + m_{11.12} + m_{11.10} + m_{11.01} \,, \\
&m_{00.10} + m_{11.10} + m_{20.10} = m_{10.20} + m_{10.11} + m_{10.00} \,, \\
&m_{01.02} + m_{12.02} = m_{02.12} + m_{02.01} \,, \\
&m_{00.01} + m_{02.01} + m_{11.01} = m_{01.11} + m_{01.02} + m_{01.00} \,, \\
&m_{01.00} + m_{10.00} = m_{00.10} + m_{00.01} \,.
\end{aligned}$$

Finally, we have fifteen linear equations like (2.5.10) that arise from linear relationships within a given row and are independent of θ:

(2.5.26)
$$\begin{aligned}
&a_1 m_{22.12} = (1 - a_1) m_{22.21} \,, && (2 - a_3 - a_4) m_{11.01} = (2 - a_1 - a_2) m_{11.21} \,, \\
&a_2 m_{21.20} = a_1 m_{21.22} \,, && (1 - a_4) m_{10.11} = (a_2 + a_4) m_{10.20} \,, \\
&a_2 m_{21.11} = (2 - a_1 - a_2) m_{21.22} \,, && (1 - a_4) m_{10.00} = (1 - a_2) m_{10.20} \,, \\
&a_2 m_{20.10} = (1 - a_2) m_{20.21} \,, && (1 - a_3) m_{02.01} = a_3 m_{02.12} \,, \\
&(1 - a_3) m_{12.11} = (a_1 + a_3) m_{12.22} \,, && (2 - a_3 - a_4) m_{01.02} = a_4 m_{01.11} \,, \\
&(1 - a_3) m_{12.02} = (1 - a_1) m_{12.22} \,, && (2 - a_3 - a_4) m_{01.00} = a_3 m_{01.11} \,, \\
&(2 - a_3 - a_4) m_{11.12} = (a_2 + a_4) m_{11.21} \,, && (1 - a_4) m_{00.01} = a_4 m_{00.10} \,. \\
&(2 - a_3 - a_4) m_{11.10} = (a_1 + a_3) m_{11.21} \,,
\end{aligned}$$

In addition to these 40 equations there are a large number of nonlinear ones which arise from ratio comparisons between rows (cf. 2.5.13). However, the 40 are adequate to give a reasonably good least-squares fit, and further consideration is restricted to them. In making the least-squares fit, one variable is eliminated, as in the noncontingent case, to guarantee that $\sum m_{ij} = \sum n_{ij}$.

Given the estimated m_{ij}, we may then make a maximum-likelihood estimate of θ. This estimate, which surprisingly is simpler than the estimate for the noncontingent case, may be expressed as

(2.5.27)
$$\hat{\theta} = \frac{\delta}{\delta + \delta'} \,,$$

where

$$\delta' = m_{22.22} + m_{21.21} + m_{20.20} + m_{12.12} + m_{11.11} + m_{10.10} + m_{02.02} + m_{01.01} + m_{00.00} ,$$

$$\begin{aligned}\delta = m_{22.21} &+ m_{22.12} + m_{21.22} + m_{21.20} + m_{21.11} + m_{20.21} + m_{20.10} + m_{12.22} + m_{12.11} \\ &+ m_{12.02} + m_{11.21} + m_{11.12} + m_{11.10} + m_{11.01} + m_{10.20} + m_{10.11} + m_{10.00} \\ &+ m_{02.12} + m_{02.01} + m_{01.11} + m_{01.02} + m_{01.00} + m_{00.10} + m_{00.01} .\end{aligned}$$

Given $\hat{\theta}$, we test the goodness-of-fit of asymptotic transition probabilities in the two-element model to the observed values. The particular transition probabilities of interest are $P_\infty(A_{i,n+1}B_{j,n+1} \mid E_{k,n}A_{i',n}B_{j',n})$, where $E_{k,n}$ denotes the reinforcing event for Player A (in the zero-sum case the reinforcing event given to one player uniquely determines the reinforcing event given to the other player). As noted in Chapter 1, in the zero-sum case the one-element models make far too many response predictions with probability 1, and so this extension is of particular interest. These theoretical probabilities are functions of the asymptotic probabilities u_{ij} of the states ij in the Markov chain whose transition matrix is (2.5.21). Consequently, we solve (2.5.21) numerically for u_{ij}, compute the theoretical probabilities $P_\infty(A_{i,n+1}B_{j,n+1} \mid E_{k,n}A_{i',n}B_{j',n})$, and proceed to a χ^2 goodness-of-fit test. The remarks made after (2.5.20) about a similar test in the noncontingent case also apply here. These 32 theoretical probabilities may be arranged in an 8×4 matrix, but owing to their length it is convenient to write them separately:

(2.5.28)
$$\begin{aligned}
P_\infty(X \mid E_1X) &= (u_{22}x + \tfrac{1}{2}u_{21}z + \tfrac{1}{4}u_{12}x + \tfrac{1}{8}u_{11}z)/P_\infty(X) , \\
P_\infty(Y \mid E_1X) &= (u_{22}w + \tfrac{1}{2}u_{21}y + \tfrac{1}{4}u_{12}w + \tfrac{1}{8}u_{11}y)/P_\infty(X) , \\
P_\infty(Z \mid E_1X) &= (\tfrac{1}{4}u_{12}x + \tfrac{1}{8}u_{11}z)/P_\infty(X) , \\
P_\infty(W \mid E_1X) &= (\tfrac{1}{4}u_{12}w + \tfrac{1}{8}u_{11}y)/P_\infty(X) , \\
P_\infty(X \mid E_2X) &= (u_{22}x + \tfrac{1}{4}u_{21}x + \tfrac{1}{2}u_{12}z + \tfrac{1}{8}u_{11}z)/P_\infty(X) , \\
P_\infty(Y \mid E_2X) &= (\tfrac{1}{4}u_{21}x + \tfrac{1}{8}u_{11}z)/P_\infty(X) , \\
P_\infty(Z \mid E_2X) &= (u_{22}w + \tfrac{1}{4}u_{21}w + \tfrac{1}{2}u_{12}y + \tfrac{1}{8}u_{11}y)/P_\infty(X) , \\
P_\infty(W \mid E_2X) &= (\tfrac{1}{4}u_{21}w + \tfrac{1}{8}u_{11}y)/P_\infty(X) ;
\end{aligned}$$

$$\begin{aligned}
P_\infty(X \mid E_1Y) &= (\tfrac{1}{2}u_{21}y + u_{20}w + \tfrac{1}{8}u_{11}y + \tfrac{1}{4}u_{10}w)/P_\infty(Y) , \\
P_\infty(Y \mid E_1Y) &= (\tfrac{1}{2}u_{21}z + u_{20}x + \tfrac{1}{8}u_{11}z + \tfrac{1}{4}u_{10}x)/P_\infty(Y) , \\
P_\infty(Z \mid E_1Y) &= (\tfrac{1}{8}u_{11}y + \tfrac{1}{4}u_{10}w)/P_\infty(Y) , \\
P_\infty(W \mid E_1Y) &= (\tfrac{1}{8}u_{11}z + \tfrac{1}{4}u_{10}x)/P_\infty(Y) , \\
P_\infty(X \mid E_2Y) &= (\tfrac{1}{4}u_{21}x + \tfrac{1}{8}u_{11}z)/P_\infty(Y) , \\
P_\infty(Y \mid E_2Y) &= (\tfrac{1}{4}u_{21}x + u_{20}x + \tfrac{1}{8}u_{11}z + \tfrac{1}{2}u_{10}z)/P_\infty(Y) , \\
P_\infty(Z \mid E_2Y) &= (\tfrac{1}{4}u_{21}w + \tfrac{1}{8}u_{11}y)/P_\infty(Y) , \\
P_\infty(W \mid E_2Y) &= \tfrac{1}{4}u_{21}w + u_{20}w + \tfrac{1}{8}u_{11}y + \tfrac{1}{2}u_{10}y)/P_\infty(Y) ;
\end{aligned}$$

$$\begin{aligned}
P_\infty(X \mid E_2Z) &= (\tfrac{1}{4}u_{12}x + \tfrac{1}{8}u_{11}z)/P_\infty(Z) , \\
P_\infty(Y \mid E_2Z) &= (\tfrac{1}{4}u_{12}w + \tfrac{1}{8}u_{11}y)/P_\infty(Z) , \\
P_\infty(Z \mid E_2Z) &= (\tfrac{1}{4}u_{12}x + \tfrac{1}{8}u_{11}z + u_{02}x + \tfrac{1}{2}u_{01}z)/P_\infty(Z) , \\
P_\infty(W \mid E_2Z) &= (\tfrac{1}{4}u_{12}w + \tfrac{1}{8}u_{11}y + u_{02}w + \tfrac{1}{2}u_{01}y)/P_\infty(Z) ,
\end{aligned}$$

$$P_\infty(X \mid E_1Z) = (\tfrac{1}{2}u_{12}y + \tfrac{1}{8}u_{11}y + u_{02}w + \tfrac{1}{4}u_{01}w)/P_\infty(Z)\,,$$
$$P_\infty(Y \mid E_1Z) = (\tfrac{1}{8}u_{11}y + \tfrac{1}{4}u_{01}w)/P_\infty(Z)\,,$$
$$P_\infty(Z \mid E_1Z) = (\tfrac{1}{2}u_{12}z + \tfrac{1}{8}u_{11}z + u_{02}x + \tfrac{1}{4}u_{01}x)/P_\infty(Z)\,,$$
$$P_\infty(W \mid E_1Z) = (\tfrac{1}{8}u_{11}z + \tfrac{1}{4}u_{01}x)/P_\infty(Z)\,;$$

$$P_\infty(X \mid E_2W) = (\tfrac{1}{8}u_{11}y + \tfrac{1}{4}u_{10}w)/P_\infty(W)\,,$$
$$P_\infty(Y \mid E_2W) = (\tfrac{1}{8}u_{11}z + \tfrac{1}{4}u_{10}x)/P_\infty(W)\,,$$
$$P_\infty(Z \mid E_2W) = (\tfrac{1}{8}u_{11}y + \tfrac{1}{4}u_{10}w + \tfrac{1}{2}u_{01}y + u_{00}w)/P_\infty(W)\,,$$
$$P_\infty(W \mid E_2W) = (\tfrac{1}{8}u_{11}z + \tfrac{1}{4}u_{10}x + \tfrac{1}{2}u_{01}z + u_{00}x)/P_\infty(W)\,,$$
$$P_\infty(X \mid E_1W) = (\tfrac{1}{8}u_{11}y + \tfrac{1}{4}u_{01}w)/P_\infty(W)\,,$$
$$P_\infty(Y \mid E_1W) = (\tfrac{1}{8}u_{11}y + \tfrac{1}{2}u_{10}y + \tfrac{1}{4}u_{01}w + u_{00}w)/P_\infty(W)\,,$$
$$P_\infty(Z \mid E_1W) = (\tfrac{1}{8}u_{11}z + \tfrac{1}{4}u_{01}x)/P_\infty(W)\,,$$
$$P_\infty(W \mid E_1W) = (\tfrac{1}{8}u_{11}z + \tfrac{1}{2}u_{10}z + \tfrac{1}{4}u_{01}x + u_{00}x)/P_\infty(W)\,;$$

where $x = (1-\theta) + \frac{1}{2}\theta$, $y = \theta + \frac{1}{2}(1-\theta)$, $z = \frac{1}{2}(1-\theta)$, $w = \frac{1}{2}\theta$, and

$$P_\infty(X) = u_{22} + \tfrac{1}{2}(u_{21} + u_{12}) + \tfrac{1}{4}u_{11}\,,$$
$$P_\infty(Y) = u_{20} + \tfrac{1}{2}(u_{21} + u_{10}) + \tfrac{1}{4}u_{11}\,,$$
$$P_\infty(Z) = u_{02} + \tfrac{1}{2}(u_{12} + u_{01}) + \tfrac{1}{4}u_{11}\,,$$
$$P_\infty(W) = u_{00} + \tfrac{1}{2}(u_{10} + u_{01}) + \tfrac{1}{4}u_{11}\,.$$

*2.6. Generalized Conditioning Model

On several occasions we have noted that the mathematical advantages of one-element models are partially offset by some of the unrealistic predictions they generate. An example is the set of four zeros in the transition matrix (1.4.3) for the zero-sum, two-person game. It is virtually certain that transitions will be observed for these cells, and this expectation is supported by data in Chapters 3 and 4. The source of the difficulty is the one mentioned in the preceding section, i.e., the fact that in the one-element model

$$P(A_{i,n+1} \mid E_{i,n}A_{i,n}) = 1\,. \tag{2.6.1}$$

That is, conditioning to the A_i response cannot change when that response is reinforced. In the two-person, zero-sum situation only one subject of the pair has his actual response reinforced, and thus by virtue of (2.6.1) both subjects cannot change conditioning states. Consequently, the anti-diagonal of the transition matrix must be uniformly 0. As noted earlier, this difficulty may be avoided by assuming a multi-element model, but serious mathematical and statistical difficulties result from this shift.

There is fortunately a second alternative, which we pursue in this section. The essence of this alternative is to generalize the assumptions about conditioning that are embodied in Axiom C2. For simplicity, this more general axiom, designated C2′, will be stated for the two-response case only.

AXIOM C2′. *If a stimulus element is sampled on a trial, and if response A_i is made and then followed by reinforcement E_j, there is a probability c_{ij} that the stimulus is conditioned to A_1.*

In the one-element model the state of conditioning may be identified with the response to which the single stimulus is conditioned. Consequently, for every trial n

$$c_{ij} = P(A_{1,n+1} \mid E_{j,n}A_{i,n}) , \qquad 1 - c_{ij} = P(A_{2,n+1} \mid E_{j,n}A_{i,n}) .$$

The θ formulation of Axiom C2 may be expressed as the following special case of C2′:

$$c_{11} = 1 , \qquad c_{12} = 1 - \theta , \qquad c_{21} = \theta , \qquad c_{22} = 0 . \tag{2.6.2}$$

Before examining the formal consequences of this generalized axiom, we consider what psychological arguments can be proposed to support its introduction. In the first place, the experimenter-defined events E_k are not necessarily events that reinforce the possible responses for the subject in the manner intended by the experimenter. The notation introduced in Estes and Suppes [1959*a*], [1959*b*] made this point explicit by distinguishing between the observable experimenter-defined outcomes O_j and the unobservable subjective reinforcing events E_k. (Bush and Mosteller [1955] made a similar differentiation but did not actually use it much in their formal developments.) Since the detailed nature of conditioning is not yet well understood, there are advantages to a model that permits direct estimation of the coefficients c_{ij} without the major constraints imposed by (2.6.2). There is, for instance, the possibility that conditioning of the stimulus is more affected by the occurrence on the preceding trial of the response or reinforcement that has on the average occurred more often. In succeeding chapters we scrutinize the data for precisely this effect. The ability of the generalized conditioning model to analyze such an effect has the virtue of incorporating into a one-stage Markov chain what is very possibly an important time-dependent "historical" phenomenon that accumulates over trials. The intuitive idea that gave rise to Axiom C2′ was that the conditioning parameters c_{ij} may be changing over trials as information accrues to the subject and may approach a stable value only in the long run. If one were willing to make this assumption, then it would also seem reasonable to postulate that the rate of change and final asymptotic value of c_{ij} are determined in large part by the particular reinforcement schedule; that is, c_{ij} would be a function not only of organismic factors but also of the reinforcement sequence. In Sec. 10.5 we actually examine the change in c_{ij} over trials as a function of both magnitude of monetary reward and reinforcement schedules.

If Axiom C2′ is to be applied to data over all trials of an experiment, one would require, in addition to the present set of axioms, a theory describing changes in the conditioning parameters. We have examined several approaches to the construction of such an extended version of the model, but they go beyond the scope of this book. In view of these comments regarding a change in c_{ij} on early trials, it should be clear that the present model as embodied in

Axiom C2′ would be expected to describe only asymptotic behavior; for only under this condition could we assume that the c_{ij}'s were constant. This is why we have restricted our applications of the generalized conditioning model almost always to data for which certain types of response stability obtain. However, as noted before, in Sec. 10.5 some surprising and paradoxical results are obtained when the c_{ij}'s are estimated at several points during the pre-asymptotic stage of learning.

We turn now to the formal development of the generalized conditioning model. Once again, we assume for simplicity that each subject has only one stimulus element available for sampling. We begin, as in Chapter 1, with the noncontingent case.

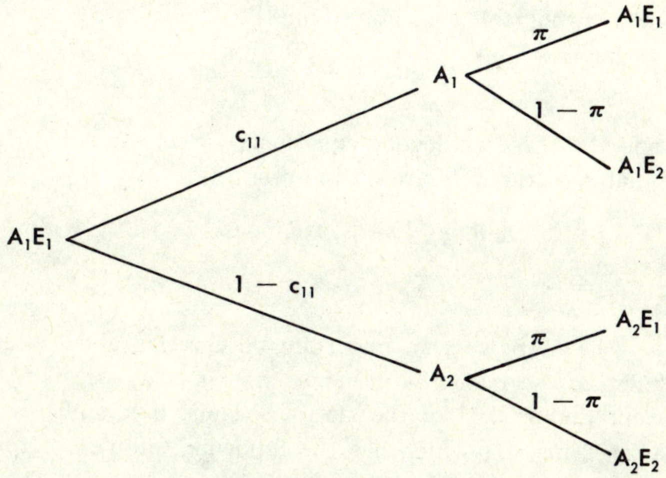

FIGURE 2.3

It is obvious that the 2×2 transition matrix (1.3.4) for the noncontingent case may be rewritten in terms of the coefficients c_{ij}. However, it is more to the point to use the generalized conditioning axiom to analyze the modifications in (2.6.1), i.e., the probability that an A_1 response will occur on trial $n+1$ given an E_1 reinforcement and an A_1 response on trial n. It has already been remarked that the prediction of this probability does not follow from the Markov chain consisting of the sequence of response random variables, and this fact suggests that the reinforcement random variables be included in the chain. With this inclusion the states of the chain are the ordered pairs A_iE_j representing possible response and reinforcement combinations on a given trial. For all the one-element models considered in this book, it may be shown (see Estes and Suppes [1959*b*]) that the inclusion of the reinforcing events in the states of the chain does not disturb its Markovian character. A typical tree for the noncontingent case is given in Fig. 2.3. The three other trees are similar in form; together they yield the following transition matrix:

(2.6.3)

	A_1E_1	A_1E_2	A_2E_1	A_2E_2
A_1E_1	$c_{11}\pi$	$c_{11}(1-\pi)$	$(1-c_{11})\pi$	$(1-c_{11})(1-\pi)$
A_1E_2	$c_{12}\pi$	$c_{12}(1-\pi)$	$(1-c_{12})\pi$	$(1-c_{12})(1-\pi)$
A_2E_1	$c_{21}\pi$	$c_{21}(1-\pi)$	$(1-c_{21})\pi$	$(1-c_{21})(1-\pi)$
A_2E_2	$c_{22}\pi$	$c_{22}(1-\pi)$	$(1-c_{22})\pi$	$(1-c_{22})(1-\pi)$

The rows indicate the response and reinforcing event on trial n, and the columns the response and reinforcing event on trial $n+1$.

The maximum-likelihood estimates of the coefficients of conditioning c_{ij} assume a particularly simple form, for the partial derivative with respect to any c_{ij} of the likelihood function $L(c_{11}, c_{12}, c_{21}, c_{22})$, corresponding to $L(\theta)$ of (2.2.5), is a function only of c_{ij}. For example,

$$\frac{\partial L}{\partial c_{11}} = \frac{n_{11}+n_{12}}{c_{11}} - \frac{n_{13}+n_{14}}{1-c_{11}}. \tag{2.6.4}$$

From (2.6.4) and the three analogous equations, we conclude that the maximum-likelihood estimates of the c_{ij}'s are as follows:

$$\begin{aligned} \hat{c}_{11} &= (n_{11}+n_{12})/\sum_j n_{1j}, & \hat{c}_{21} &= (n_{31}+n_{32})/\sum_j n_{3j}, \\ \hat{c}_{12} &= (n_{21}+n_{22})/\sum_j n_{2j}, & \hat{c}_{22} &= (n_{41}+n_{42})/\sum_j n_{4j}, \end{aligned} \tag{2.6.5}$$

where $j = 1, \cdots, 4$. Moreover, an important observation about these estimates is that in tabulating the transition numbers n_{ij}, it is unnecessary to record the reinforcing event that is part of the state j. Thus, it is sufficient to tabulate the data in a 4×2 matrix, which has the following simple theoretical form:

(2.6.6)

	A_1	A_2
A_1E_1	c_{11}	$1-c_{11}$
A_1E_2	c_{12}	$1-c_{12}$
A_2E_1	c_{21}	$1-c_{21}$
A_2E_2	c_{22}	$1-c_{22}$

The maximum-likelihood estimates $\hat{c}_{ij}$ are then just the estimated conditional probabilities $P(A_1 \mid E_jA_i)$, which are constant over trials.

Moreover, the tabulations indicated for (2.6.6) are sufficient for the χ^2 goodness-of-fit test, provided we assume that

$$\pi(n_{i1}+n_{i2}) = n_{i1}, \qquad \pi(n_{i3}+n_{i4}) = n_{i3}, \qquad (i = 1, \cdots, 4). \tag{2.6.7}$$

The assumption of these equalities implies that the maximum-likelihood estimate of the probability of reinforcement is π. Obviously, for properly selected reinforcement sequences this assumption can be wrong only by a small experimental error. To clarify this point concerning the goodness-of-fit test, we need consider only the computation for the first two cells of the 4×4 matrix, i.e., for the n_{11} and n_{12} transition numbers. Following (2.4.4), the contribution η of these two cells to the value of χ^2 is

$$\eta = \frac{n_1\left[\dfrac{n_{11}}{n_1} - c_{11}\pi\right]^2}{c_{11}\pi} + \frac{n_1\left[\dfrac{n_{12}}{n_1} - c_{11}(1-\pi)\right]^2}{c_{11}(1-\pi)}, \tag{2.6.8}$$

and using (2.6.7), we have

$$\begin{aligned}\eta &= \frac{n_1\left[\dfrac{\pi(n_{11}+n_{12})}{n_1} - c_{11}\pi\right]^2}{c_{11}\pi} + \frac{n_1\left[\dfrac{(1-\pi)(n_{11}+n_{12})}{n_1} - c_{11}(1-\pi)\right]^2}{c_{11}(1-\pi)} \\ &= \frac{n_1}{c_{11}}\left[\pi\left(\frac{n_{11}+n_{12}}{n_1} - c_{11}\right)^2 + (1-\pi)\left(\frac{n_{11}+n_{12}}{n_1} - c_{11}\right)^2\right] \\ &= \frac{n_1\left[\dfrac{n_{11}+n_{12}}{n_1} - c_{11}\right]^2}{c_{11}},\end{aligned}$$

which is just the expression for computing the contribution to a χ^2 test of the first cell of (2.6.6). This line of argument directly establishes that the χ^2 goodness-of-fit (2.4.4) for the 4×4 transition matrix (2.6.3) may be replaced by a similar test for the 4×2 matrix (2.6.6). However by now the reader may have realized that this particular goodness-of-fit test is vacuous, for the four estimated parameters c_{ij} guarantee that the fit is exact, each row being exactly fitted by one estimated c_{ij}. The upshot is that without the imposition of constraints that specify relations between the c_{ij}'s, a test of the goodness-of-fit of the generalized conditioning model for the noncontingent case is not provided by this particular test.

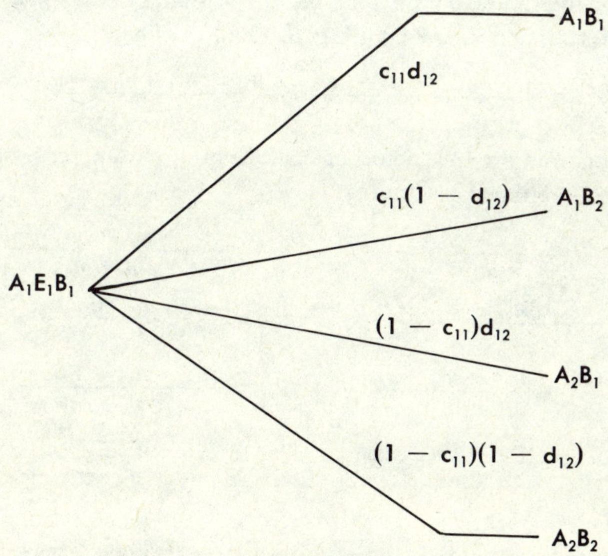

FIGURE 2.4.

This same situation does not obtain in the two-person situations. Again we shall restrict ourselves to the zero-sum case. By analogy with the transition matrix (2.6.3) for the noncontingent case, the chain for the two-person situation, which includes the reinforcing events in its states, has 16 states. However, in the zero-sum case this number may be reduced to 8, for on a given trial the responses of both subjects and the reinforcing event given one subject uniquely determine the reinforcing event given the other subject; that is, as remarked earlier, exactly one of the two subjects has his actual response reinforced. Also, the arguments given above to justify a rectangular 4×2 matrix for the noncontingent case apply here *mutatis mutandis*. So we need consider only an 8×4 matrix. The tree for the first row is given in Fig 2.4, where E_1 designates the reinforcement given subject A, and c_{ij} and d_{ij} are the conditioning parameters for subjects A and B, respectively. The 8×4 matrix has the following form:

(2.6.9)

	A_1B_1	A_1B_2	A_2B_1	A_2B_2
$A_1E_1B_1$	$c_{11}d_{12}$	$c_{11}(1-d_{12})$	$(1-c_{11})d_{12}$	$(1-c_{11})(1-d_{12})$
$A_1E_1B_2$	$c_{11}d_{21}$	$c_{11}(1-d_{21})$	$(1-c_{11})d_{21}$	$(1-c_{11})(1-d_{21})$
$A_1E_2B_1$	$c_{12}d_{11}$	$c_{12}(1-d_{11})$	$(1-c_{12})d_{11}$	$(1-c_{12})(1-d_{11})$
$A_1E_2B_2$	$c_{12}d_{22}$	$c_{12}(1-d_{22})$	$(1-c_{12})d_{22}$	$(1-c_{12})(1-d_{22})$
$A_2E_1B_1$	$c_{21}d_{11}$	$c_{21}(1-d_{11})$	$(1-c_{21})d_{11}$	$(1-c_{21})(1-d_{11})$
$A_2E_1B_2$	$c_{21}d_{22}$	$c_{21}(1-d_{22})$	$(1-c_{21})d_{22}$	$(1-c_{21})(1-d_{22})$
$A_2E_2B_1$	$c_{22}d_{12}$	$c_{22}(1-d_{12})$	$(1-c_{22})d_{12}$	$(1-c_{22})(1-d_{12})$
$A_2E_2B_2$	$c_{22}d_{21}$	$c_{22}(1-d_{21})$	$(1-c_{22})d_{21}$	$(1-c_{22})(1-d_{21})$

Let the transition numbers n_{ij} be for this 8×4 matrix, and let $L(c_{11}, \cdots, d_{22})$ be the likelihood function corresponding to (2.2.5). Then

$$\frac{\partial L}{\partial c_{11}} = \frac{n_{11} + n_{12} + n_{21} + n_{22}}{c_{11}} - \frac{n_{13} + n_{14} + n_{23} + n_{24}}{1 - c_{11}} .$$

Similar expressions obtain for the other conditioning parameters. Thus we have for the maximum-likelihood estimates

$$\hat{c}_{11} = \frac{n_{11} + n_{12} + n_{21} + n_{22}}{n_1 + n_2}, \qquad \hat{d}_{11} = \frac{n_{31} + n_{33} + n_{51} + n_{53}}{n_3 + n_5},$$

$$\hat{c}_{12} = \frac{n_{31} + n_{32} + n_{41} + n_{42}}{n_3 + n_4}, \qquad \hat{d}_{12} = \frac{n_{11} + n_{13} + n_{71} + n_{73}}{n_1 + n_7},$$

(2.6.10)

$$\hat{c}_{21} = \frac{n_{51} + n_{52} + n_{61} + n_{62}}{n_5 + n_6}, \qquad \hat{d}_{21} = \frac{n_{21} + n_{23} + n_{81} + n_{83}}{n_2 + n_8},$$

$$\hat{c}_{22} = \frac{n_{71} + n_{72} + n_{81} + n_{82}}{n_7 + n_8}, \qquad \hat{d}_{22} = \frac{n_{41} + n_{43} + n_{61} + n_{63}}{n_4 + n_6},$$

where as before $n_i = \sum_j n_{ij}$.

Given these eight estimated parameters, the χ^2 goodness-of-fit test (2.4.4) applied to the 8×4 matrix (2.6.9) has $8 \times 3 - 8 = 16$ degrees of freedom and obviously the test for this two-person situation provides a real check on the empirical adequacy of the model. In subsequent chapters we shall compare the fit of the model with these eight conditioning parameters to the fit when it is assumed that $c_{ij} = d_{ij}$. On this latter assumption, the maximum-likelihood estimates of the c_{ij}'s are, of course, different from those given above; these new estimates are presented below:

(2.6.11)
$$\hat{c}_{11} = \frac{n_{11} + n_{12} + n_{21} + n_{22} + n_{31} + n_{33} + n_{51} + n_{53}}{n_1 + n_2 + n_3 + n_5},$$
$$\hat{c}_{12} = \frac{n_{11} + n_{13} + n_{31} + n_{32} + n_{41} + n_{42} + n_{71} + n_{73}}{n_1 + n_3 + n_4 + n_7},$$
$$\hat{c}_{21} = \frac{n_{21} + n_{23} + n_{51} + n_{52} + n_{61} + n_{62} + n_{81} + n_{83}}{n_2 + n_5 + n_6 + n_8},$$
$$\hat{c}_{22} = \frac{n_{41} + n_{43} + n_{61} + n_{63} + n_{71} + n_{72} + n_{81} + n_{82}}{n_4 + n_6 + n_7 + n_8}.$$

2.7. Summary

This chapter describes various methods of data analysis that are specific to the Markov learning models introduced in Chapter 1. We begin by considering maximum-likelihood estimates of the conditioning parameter θ. Unfortunately, it turns out that these methods are limited to one-element models, and therefore it is necessary to introduce other estimation procedures that are appropriate for multi-element models. The methods developed formally resemble the maximum-likelihood method but in fact do not have the maximum-likelihood property. In this context a method for estimating jointly the number of stimulus elements and the conditioning parameter θ is introduced.

In Sec. 2.4 we discuss various χ^2 tests for testing hypotheses concerning Markov chains. In Sec. 2.5 we present some methods of estimation peculiar to our particular models. These methods are of more theoretical than practical interest; they depend upon first making a least-squares estimate of unobservable transition frequencies. Finally, in Sec. 2.6 a generalized conditioning model is developed in which the single parameter θ is replaced by several conditioning parameters. The model is developed in detail for the simple noncontingent situation and for the two-person, zero-sum game situation; maximum-likelihood estimates of the conditioning parameters are derived.

3

Some Zero Sum, Two-Person Situations

3.1. Introduction

In the preceding two chapters we have presented our basic theory and method of analysis. In the next eight chapters we shall apply this theory and method to the data of a number of multiperson experiments. Each succeeding experiment in this series is designed (by the inclusion of a new independent variable) to approximate a genuine game more closely than does its predecessor. We shall therefore begin each chapter by discussing the specific experimental situation that we are concerned with, and the implications of the new independent variable introduced. We then present the appropriate one-element model and details of the experimental method used to test it. The remaining sections are devoted to analyzing and discussing the various experimental findings. In general, the final two sections analyze the results from the standpoint of the multi-element model and the generalized conditioning model. However, these particular analyses do not appear in every chapter because of the rather staggering computational problems that arise when some of our experiments are considered from the standpoint of these additional models.

The conceptual scheme behind the experiments reported in this chapter is described in some detail in Sec. 1.4. It should be added that in the experiments discussed in this chapter, which were the first ones performed, we deliberately took a conservative attitude. That is, we tried to hold the experimental situation as close as possible to the one used in a number of earlier one-person studies testing the adequacy of stimulus-sampling theory to account for behavior in verbal conditioning situations (e.g., Estes and Straughan [1954], Burke, Estes, and Hellyer [1954], Detambel [1955], Atkinson [1956], Neimark [1956], Brand, Woods, and Sakoda [1956] and Anderson and Grant [1957]). The interaction of the two subjects forming a pair was limited in at least three ways: (1) neither subject was shown the payoff matrix (1.4.1); (2) neither subject was directly informed of the responses of the other; (3) subjects were in fact told that they were not interacting at all, but were being run in pairs only as a matter of experimental convenience. Each of these restrictions was removed in later experiments, which are reported in subsequent chapters. With these restrictions operative, the experimental situation certainly was not viewed as a zero-sum, two-person game by the subjects. For this reason the word *game* was not used in the title of this chapter. On the other hand, the remarks made in Sec. 1.6. still apply, and later in the chapter we consider the extent to which the subjects tended asymptotically to approach game-theory strategies.

From the theoretical standpoint, a given experimental case of the zero-sum situation is specified by setting the values of the probabilities a_i in the payoff matrix (1.4.1). To facilitate comparisons with game theory, three sets of a_i values were used corresponding to three standard cases of 2×2 games. The first case is labeled the *Mixed Group* because both players have mixed minimax strategies. The a_i values are given by the payoff matrix

(3.1.1)

	B_1	B_2
A_1	$\frac{1}{3}$	1
A_2	$\frac{1}{2}$	$\frac{1}{6}$

The minimax strategy for Player A is to choose A_1 with probability $\frac{1}{3}$, and the minimax strategy for B is to choose B_1 with probability $\frac{5}{6}$. (The method of finding these strategies is given in Sec. 1.6.) From (1.4.7) and (3.1.1) we obtain the following results for the asymptotic probabilities u_j of the four states of the Markov chain (1.4.3), that is, the asymptotic probabilities of the response pairs (A_1B_1), (A_1B_2), (A_2B_1), and (A_2B_2):

$$u_1 = \frac{15\varepsilon + 18}{50\varepsilon + 40}, \qquad u_3 = \frac{20\varepsilon + 4}{50\varepsilon + 40}, \tag{3.1.2}$$
$$u_2 = \frac{15\varepsilon + 6}{50\varepsilon + 40}, \qquad u_4 = \frac{12}{50\varepsilon + 40}.$$

Here, and henceforth in this section, $\varepsilon = \theta_A/\theta_B$. Recalling (1.2.11), (1.4.10), and (1.4.11), we have immediately from (3.1.2)

$$\alpha = .6\,, \qquad \beta = \frac{35\varepsilon + 22}{50\varepsilon + 40}, \tag{3.1.3}$$

which means that the mean probability at asymptote of an A_1 response by Player A is independent of the estimation of the learning parameters. (This result for α is independent of the number of stimuli assumed and also holds for the linear model formulated in Sec. 1.5.) The linear relation (1.4.12) is just the line $\alpha = .6$ for the Mixed Group.

For the five quantities above that do depend on ε we may obtain bounds that are independent of ε by letting ε vary from 0 to ∞:[1]

[1] Because in the notation of (1.4.8) $du_j/d\varepsilon = (dc_j - d_jc)/(c\varepsilon + d)^2$, the rate of change of u_j with respect to ε is either always positive or always negative for $0 < \varepsilon < \infty$; whence the maximum occurs at $\varepsilon = 0$ or $\varepsilon = \infty$, and similarly for the minimum.

$$
\begin{array}{ll}
.300 < u_1 < .450 \,, & .100 < u_3 < .400 \,, \\
.150 < u_2 < .300 \,, & 0 < u_4 < .300 \,, \\
\multicolumn{2}{c}{.550 < \beta < .700 \,.}
\end{array} \tag{3.1.4}
$$

Also, if it is assumed that $\theta = \theta_A = \theta_B$ (i.e., $\varepsilon = 1$), we obtain the specific predictions

$$
\begin{array}{lll}
u_1 = .367 \,, & u_3 = .267 \,, & \alpha = .600 \,, \\
u_2 = .233 \,, & u_4 = .133 \,, & \beta = .633 \,.
\end{array} \tag{3.1.5}
$$

The second case is labeled the *Pure Group* because both players have pure minimax strategies. The a_i values are given by the payoff matrix

(3.1.6)

	B_1	B_2
A_1	$\frac{1}{2}$	1
A_2	$\frac{1}{2}$	$\frac{1}{4}$

Here $a_1 = \frac{1}{2}$ is a saddle point of the matrix; thus from the standpoint of game theory the optimal strategies are A_1 and B_1, respectively, for the two players. For the asymptotic probabilities u_j, α, and β, we have

$$
\begin{array}{lll}
u_1 = \dfrac{3\varepsilon + 4}{9\varepsilon + 9} \,, & u_3 = \dfrac{3\varepsilon + 1}{9\varepsilon + 9} \,, & \alpha = \dfrac{2}{3} \,, \\[2ex]
u_2 = \dfrac{3\varepsilon + 2}{9\varepsilon + 9} \,, & u_4 = \dfrac{2}{9\varepsilon + 9} \,, & \beta = \dfrac{6\varepsilon + 5}{9\varepsilon + 9} \,.
\end{array} \tag{3.1.7}
$$

Again, the linear "interaction" line (1.4.12) is simply the vertical line satisfying the equation $\alpha = \frac{2}{3}$. As bounds for the quantities depending on ε, we have

$$
\begin{array}{ll}
.333 < u_1 < .444 \,, & .111 < u_3 < .333 \,, \\
.222 < u_2 < .333 \,, & 0 < u_4 < .222 \,, \\
\multicolumn{2}{c}{.555 < \beta < .667 \,.}
\end{array} \tag{3.1.8}
$$

And for $\theta_A = \theta_B$

$$
\begin{array}{lll}
u_1 = .389 \,, & u_3 = .222 \,, & \alpha = .667 \,, \\
u_2 = .278 \,, & u_4 = .111 \,, & \beta = .611 \,.
\end{array} \tag{3.1.9}
$$

The third case is labeled the *Sure Group* because both players have sure-thing strategies, namely A_1 and B_1, respectively. The a_i values in terms of the payoff matrix are

(3.1.10)

	B_1	B_2
A_1	$\frac{1}{2}$	1
A_2	$\frac{1}{4}$	$\frac{3}{4}$

For the asymptotic probabilities, we have

$$
\begin{array}{lll}
u_1 = \dfrac{3\varepsilon + 10}{9\varepsilon + 23} \,, & u_3 = \dfrac{2\varepsilon + 6}{9\varepsilon + 23} \,, & \alpha = \dfrac{7\varepsilon + 15}{9\varepsilon + 23} \,, \\[2ex]
u_2 = \dfrac{4\varepsilon + 5}{9\varepsilon + 23} \,, & u_4 = \dfrac{2}{9\varepsilon + 23} \,, & \beta = \dfrac{5\varepsilon + 16}{9\varepsilon + 23} \,.
\end{array} \tag{3.1.11}
$$

In this case, the linear interaction line is determined by the equation $3\alpha = 10\beta - 5$. The inequalities for the Sure Group are the following:

$$
\begin{aligned}
&.333 < u_1 < .435\,, \qquad && 0 < u_4 < .087\,,\\
&.218 < u_2 < .444\,, && .652 < \alpha < .778\,,\\
&.222 < u_3 < .261\,, && .555 < \beta < .696\,.
\end{aligned}
\tag{3.1.12}
$$

Finally, for $\theta_A = \theta_B$

$$
\begin{aligned}
&u_1 = .406\,, \qquad && u_3 = .250\,, \qquad && \alpha = .688\,,\\
&u_2 = .282\,, && u_4 = .062\,, && \beta = .656\,.
\end{aligned}
\tag{3.1.13}
$$

Numerical computations for the two-element model and the generalized conditioning model are presented in Secs. 3.7 and 3.8.

The experimental data for these three groups are reported in Atkinson and Suppes [1958], but the methods of analysis discussed in Chapter 2 were not used in that article. Consequently, most of the analysis of data in this chapter (and especially in Secs. 3.5 through 3.8) is new.

3.2. Experimental Method

Subjects. The subjects were 120 undergraduates obtained from introductory courses in psychology and philosophy at Stanford University. They were randomly assigned to three groups of equal size.

Apparatus. The subjects, run in pairs, sat at opposite ends of an 8 × 3-foot table. Mounted vertically on the table top facing each subject was a 50-inch-wide by 30-inch-high black panel placed 22 inches from the end of the table. The experimenter sat between the two panels and was not visible to either subject. The apparatus, as viewed from the subject's side, consisted of two silent operating keys mounted 8 inches apart on the table top and 12 inches from the end of the table; upon the panel, three milk-glass panel lights were mounted. One of these lights, which served as the signal for the subject to respond, was centered between the keys at a height of 17 inches from the table top. Each of the two remaining lights, the reinforcing signals, was at a height of 11 inches directly above one of the keys. The presentation and duration of the lights were automatically controlled. The subjects were not visible to one another and could not see one another's responses or panel lights.

Procedure. The subjects were read the following instructions: "We always run subjects in pairs because this is the way the equipment has been designed and also because it is the most economical procedure. Actually, however, you are working on two completely different and independent problems.

"The experiment for each of you consists of a series of trials. The top center lamp on your panel will light for about 2 seconds to indicate the start of each trial. Shortly thereafter one or the other of the two lower lamps will light up. Your job is to predict on each trial which one of the two lower lamps

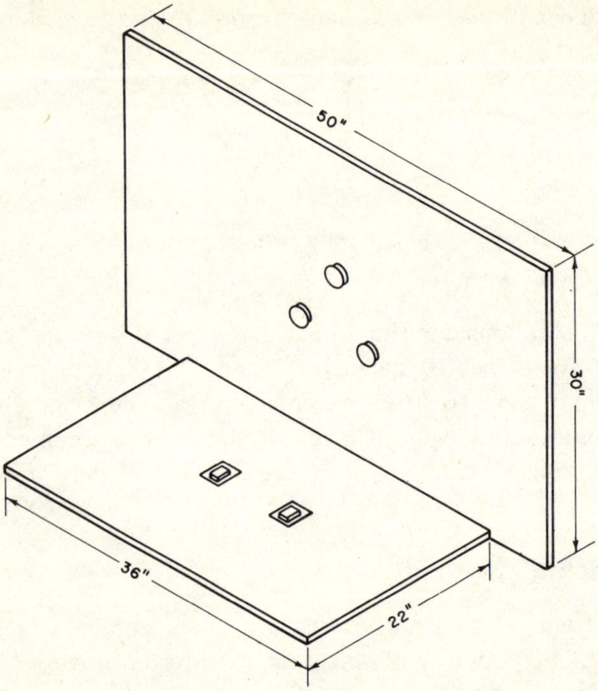

FIGURE 3.1. Diagram of the apparatus as viewed from either subject's side.

will light and to indicate your prediction by pressing the proper key. That is, if you expect the left lamp to light, press the left key; if you expect the right lamp to light, press the right key. On each trial press one or the other of the two keys but never both. If you are not sure which key to press, then guess.

"Be sure to indicate your choice by pressing the proper key immediately after the onset of the signal light. That is, when the signal light goes on press one or the other key down and release it. Then wait until one of the lower lights goes on. If the light above the key you pressed goes on, your prediction was correct; if the light above the key opposite from the one you pressed goes on, you were incorrect and you should have pressed the other key. At times you may feel frustrated or irritated if you cannot understand what the experiment is all about. Nevertheless, continue trying to make the very best prediction you can on each trial."

For each pair, one subject was randomly selected as Player A and the other as Player B. For each subject, one of the two response keys was randomly designated Response 1 and the other Response 2, with the restriction that the following possible combinations occurred equally often in each of the three experimental groups: (*a*) A_1 and B_1 on the right, (*b*) A_1 on the right and B_1 on the left, (*c*) A_1 on the left and B_1 on the right, and (*d*) A_1 and B_1 on the left.

After the instructions were read, 200 trials were run in continuous sequence.

For each pair of subjects, sequences of reinforcing lights were generated in accordance with assigned values of a_i and observed responses.

On all trials the signal light was lighted for 3.5 seconds; the time between successive signal exposures was 10 seconds. The reinforcing light followed the cessation of the signal light by 1.5 seconds and remained on for 2 seconds.

3.3. Mean Learning Curves and Asymptotic Results

Figure 3.2 presents the observed mean proportions of A_1 and B_1 responses in successive blocks of 40 trials for the entire sequence of 200 trials. An inspection of this figure indicates that response proportions are fairly stable over the last 80 trials, except possibly for B_1 responses in the Pure Group. To check the stability of response proportions for individual data, t's for paired measures were computed between response proportions for the first and last halves of the final block of 60 trials. In all cases, the obtained values of t were not significant at the .10 level. In view of these results, it appears reasonable to assume that a fairly stable level of responding has been reached; consequently, the proportions computed over the last 60 trials were used as an estimate of α and β. Table 3.1 presents the observed proportions of A_1 and B_1 responses in the last 60 trials for individual subject-pairs. Table 3.2 presents the average values for these proportions and the standard deviations. The values predicted by the model for $\theta_A = \theta_B$ are also presented [see (3.1.5), (3.1.9), and (3.1.13)].

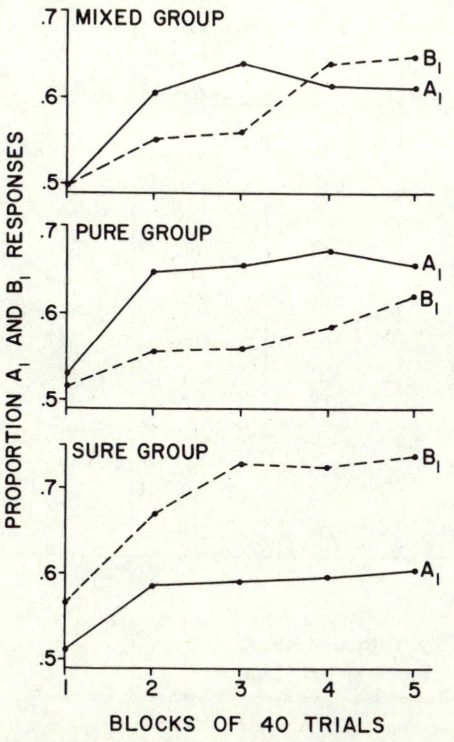

FIGURE 3.2. Observed proportions of A_1 and B_1 responses in 40-trial blocks for the three experimental groups.

Table 3.2 indicates fairly good over-all agreement between observed and predicted data. The Mixed and Sure Groups are extremely close to theoretical values. However, for the Sure Group the observed proportion for A_1's is low, while the proportion for B_1's is somewhat high. We shall defer a statistical test of these results until the proportions for the response pairs (A_1B_1), (A_1B_2), (A_2B_1), and (A_2B_2) are examined.

In Fig. 3.3 theoretical learning curves are presented for the one-element model when $\theta = \theta_A = \theta_B$. The functions give the expected proportion of A_1 and B_1 responses in successive 10-trial blocks for initial values of $u_1^{(1)} = u_2^{(1)} = u_3^{(1)} = u_4^{(1)} = \frac{1}{4}$.

TABLE 3.1

OBSERVED PROPORTIONS OF A_1 AND B_1 RESPONSES OVER THE LAST BLOCK OF 60 TRIALS FOR INDIVIDUAL SUBJECTS

Subject-Pairs	Mixed		Pure		Sure	
	A_1	B_1	A_1	B_1	A_1	B_1
1	.650	.650	.667	.567	.500	.800
2	.733	.650	.700	.550	.683	.767
3	.450	.766	.700	.617	.433	.667
4	.667	.600	.700	.683	.584	.750
5	.583	.667	.667	.533	.733	.683
6	.600	.567	.517	.500	.567	.800
7	.667	.650	.617	.567	.717	.867
8	.483	.600	.433	.550	.483	.767
9	.483	.700	.683	.633	.684	.683
10	.717	.833	.667	.517	.583	.733
11	.600	.600	.750	.583	.733	.767
12	.650	.600	.583	.650	.467	.600
13	.517	.717	.767	.750	.800	.617
14	.517	.683	.633	.700	.684	.767
15	.567	.750	.717	.650	.567	.817
16	.600	.750	.783	.617	.600	.750
17	.717	.666	.767	.583	.667	.700
18	.617	.550	.683	.583	.483	.834
19	.650	.500	.650	.650	.600	.600
20	.633	.483	.717	.550	.550	.650

TABLE 3.2

PREDICTED AND OBSERVED MEAN PROPORTIONS OF A_1 AND B_1 RESPONSES OVER THE LAST BLOCK OF 60 TRIALS

Group	Response A_1			Response B_1		
	Pred.	Obs.	SD	Pred.	Obs.	SD
Mixed	.600	.605	.0794	.633	.649	.0874
Pure	.667	.670	.0832	.611	.602	.0634
Sure	.667	.606	.1005	.700	.731	.0760

For all groups, the curves are virtually at asymptote within the first or second block of trials when θ equals .2 or .6. When appropriately scaled, the curves for $\theta = .05$ compare more favorably with the observed data in Fig. 3.2. However, in all groups for both A_1's and B_1's the theoretical functions for $\theta = .05$ approach the asymptote somewhat more rapidly than do the observed curves. Thus, a value of θ that would provide a good fit to the data would be below .05. We shall return to this point later. (An interesting theoretical prediction about α_n for the Mixed Group is that on early trials it overshoots the asymptote and then approaches .6 from above; the effect may be noted in the observed data of Fig. 3.2. It is too often thought that only exponential learning curves can be derived from stimulus-sampling theory.)

Figure 3.4 presents the observed mean proportions of the response pairs (A_1B_1), (A_1B_2), (A_2B_1), and (A_2B_2) in blocks of 40 trials. The asymptotic predictions in the one-element model for these response pairs are u_1, u_2, u_3, and u_4, respectively; expressions for these quantities are given by (3.1.2), (3.1.7), and (3.1.11). On the far right of Fig. 3.4 the values of u_j are graphed, as given by (3.1.5), (3.1.9), and (3.1.13) for $\theta_A = \theta_B$. The figure indicates good agreement between the u_j's and the observed values for the final 40-trial block.

In order to evaluate statistically the adequacy of the asymptotic predictions we have used a test suggested by Pillai and Ramachandran [1954]. The test involves taking the largest absolute difference between an observed mean value and the predicted value in a collection of samples (in this case three). This difference is then divided by an over-all estimate of the standard error of the mean, i.e., it is assumed that the observations are randomly selected from populations with homogeneous variance. The test was run separately on the (A_1B_1), (A_1B_2), and (A_2B_1) proportions. For the (A_1B_1) measure the largest discrepancy between u_1 and its estimate in the final 40-trial block occurred for the Mixed Group; for the (A_1B_2) and (A_2B_1) measures the largest difference occurred for the Sure Group. The value of the Pillai-

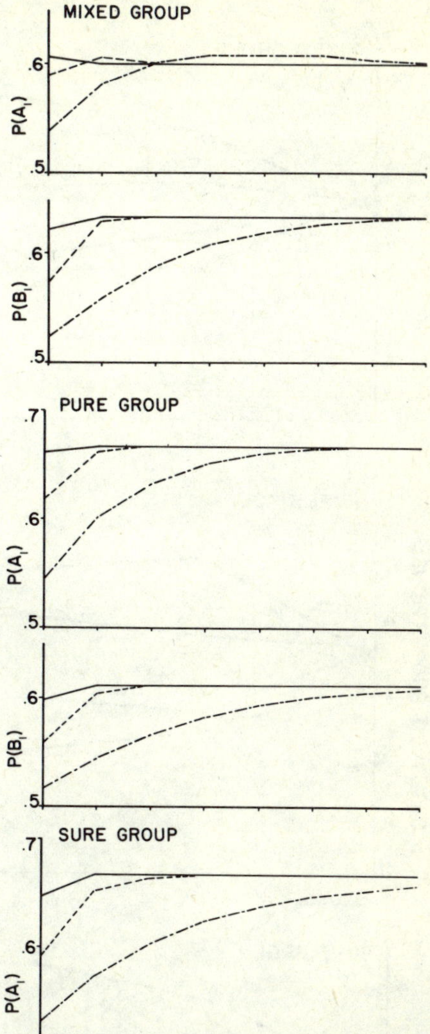

FIGURE 3.3. Theoretical learning curves in successive 10-trial blocks (computed for the one-element model given a uniform distribution on trial 1 and $\theta_A = \theta_B = \theta$).

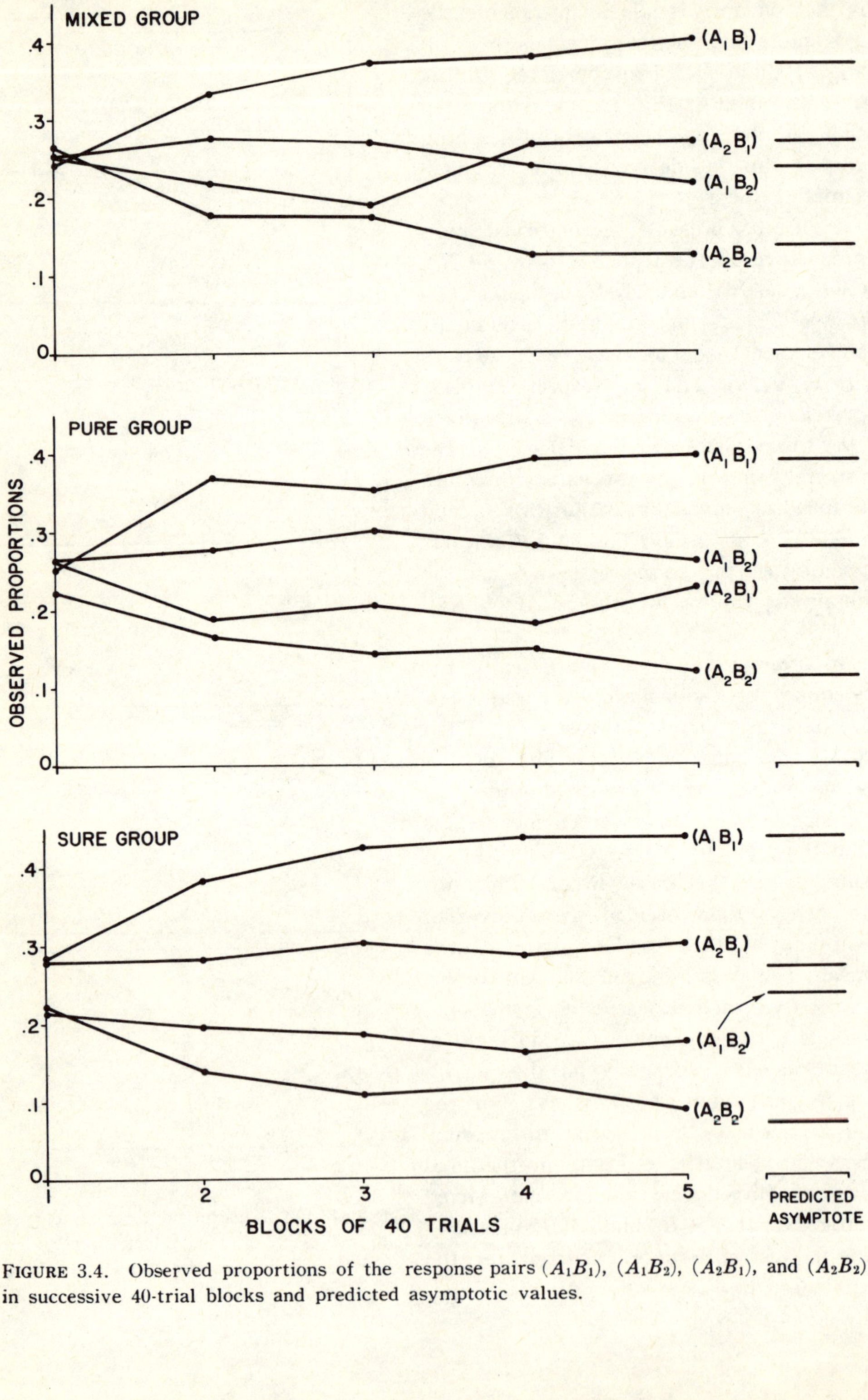

FIGURE 3.4. Observed proportions of the response pairs (A_1B_1), (A_1B_2), (A_2B_1), and (A_2B_2) in successive 40-trial blocks and predicted asymptotic values.

Ramachandran statistic was not significant at the .05 level for any of the three measures. Consequently, in terms of this particular statistical criterion there is no evidence to suggest that we should reject the model.

3.4. Tests for Markov Properties

Table 3.3 presents values for the observed transition matrix $[\hat{p}_{ij}]$ computed over the last 100 trials. The maximum-likelihood estimate $\hat{p}_{ij}$ of the quantity p_{ij} is obtained as follows: (1) Let $n_{ij}(t)$ denote the number of subject-pairs in state i on trial $t-1$ and state j on trial t; and (2) let

$$n_{ij} = \sum_{t=100}^{200} n_{ij}(t)\,;$$

then

$$\hat{p}_{ij} = \frac{n_{ij}}{\sum_j n_{ij}}\,.$$

No statistical test is required to see that the $\hat{p}_{ij}$'s differ significantly from the theoretical values. It suffices to observe that in the theoretical matrix (1.4.3), the anti-diagonal entries and the entry in row 2, column 4 are identically 0 for all three sets of parameter values a_i; whereas in the observed matrices entries in these cells are in some cases markedly different from 0. This result was already anticipated in Chapter 2 and was used to justify the consideration of multi-element models and the generalized conditioning model, to which we turn in later sections.

TABLE 3.3

OBSERVED TRANSITION MATRICES CORRESPONDING TO THE THEORETICAL TRANSITION MATRIX SPECIFIED BY THE MARKOV MODEL

(Computed over the last 100 trials)

	Mixed Group				Pure Group				Sure Group			
n \ $n+1$	(1, 1)	(1, 2)	(2, 1)	(2, 2)	(1, 1)	(1, 2)	(2, 1)	(2, 2)	(1, 1)	(1, 2)	(2, 1)	(2, 2)
(1, 1)	.37	.23	.28	.12	.38	.26	.25	.11	.43	.19	.30	.08
(1, 2)	.54	.25	.15	.06	.50	.31	.11	.08	.52	.19	.22	.07
(2, 1)	.36	.16	.30	.18	.30	.20	.28	.22	.47	.12	.31	.10
(2, 2)	.28	.35	.17	.20	.32	.36	.15	.17	.28	.16	.37	.19

As indicated in Sec. 2.4, a number of relevant questions can be asked about the data without examining a specific model. We begin by testing the hypothesis that the transition probabilities $P(A_{i,n+1}B_{j,n+1} \mid A_{i',n}\,B_{j',n})$ are constant over trials. For each group χ^2 was calculated by (2.4.1) for $t = 1, 2$, and referred to the first and last block of 100 trials. All three values of χ^2 were significant at the .05 level. This finding contradicts the one-element model's predictions. The same test was also run for the last 50-trial block versus the next-to-last

50-trial block. In this case the χ^2 did not approach significance, suggesting that the transition probabilities approached stationarity in the second half of the experiment.

We next test hypotheses concerning the order of the chain. For these tests only the transition data over the final block of 100 trials were used. To begin, consider testing the null hypothesis that the chain is zero-order against the alternative that it is first-order. For each group χ^2 was calculated by (2.4.2), and all three values were highly significant (182.3 for the Mixed Group, 211.5 for the Pure Group, and 136.7 for the Sure Group, with 9 degrees of freedom in all three cases). This result indicates that prediction is significantly improved if one knows the responses of the subject-pair on the preceding trial as against knowing only the over-all level of responding.

The first-order versus second-order hypothesis was tested by computing the χ^2 of (2.4.3). The obtained values, with 36 degrees of freedom, were 81.8 for the Mixed Group, 50.5 for the Pure Group, and 52.9 for the Sure Group. For the Pure and Sure Groups the value of χ^2 is not significant at the .05 level; for the Mixed Group it is. Thus, for the Pure and Sure Groups there is no statistically significant improvement in predictions if one knows the response history of the pair of subjects on the previous two trials as against only the immediately preceding trial. The one-element model, of course, is formulated as a one-stage process, and this finding supports the model. However, this assumption is not necessary for our general theoretical approach; the results for the Mixed Group simply argue for a multi-element analysis of the data.

3.5. Estimate of θ and Predicted Asymptotic Variance.

As indicated in Sec. 2.2, a maximum-likehood estimate of θ can be obtained from the observed transition data by computing (2.2.9). For each pair of subjects n_{ij}'s were obtained separately for the first and last blocks of 100 trials. Two θ's were then estimated for each subject-pair: one based on the first 100 trials and the other on the last 100. The results of these estimations are presented in Table 3.4. There is no evidence to suggest that the value of θ is changing from the first to the second block of trials, or that θ differs from group to group. Paired t tests between θ's estimated from the initial and final 100-trial blocks are in no case significant; the t's (with d.f. = 19) were .65, 1.88, and .42 for the Mixed, Pure, and Sure Groups, respectively. An F test between groups yielded a value of 1.22 (d.f. = 2.57), which also is not significant.

Similar estimates of θ were made in terms of transition frequencies aggregated over all subject-pairs in a given group. The obtained values were .636, .619 (Mixed), .594, .584 (Pure), and .607, .609 (Sure) for the last and first 100-trial blocks, respectively. The mean of the individual estimates and the estimate based on aggregated data are extremely close.

Once a value of θ has been estimated, a number of interesting comparisons can be made between theoretical and observed quantities. We begin by examining the predicted and observed learning curves. As noted above, theoretical graphs of α_n and β_n are given in Fig. 3.3 for $\theta = .05$, .2, and .6. Unfortunately, the functions for $\theta = .6$ would not begin to approximate the observed curves in Fig. 3.2; an adequate fit to these data would require a θ in the vicinity of .05. Thus, the estimated value of θ is far too large to account for the observed learning curves.

TABLE 3.4

MAXIMUM-LIKELIHOOD ESTIMATES OF θ

(Based on the first and last block of 100 trials for individual pairs of subjects)

Subject-Pairs	Mixed		Pure		Sure	
	1–100	101–200	1–100	101–200	1–100	101–200
1	.565	.698	.486	.623	.518	.556
2	.662	.593	.594	.658	.626	.644
3	.654	.644	.590	.550	.694	.605
4	.630	.655	.518	.612	.655	.655
5	.494	.618	.533	.733	.663	.616
6	.724	.630	.634	.549	.581	.514
7	.627	.667	.573	.602	.580	.630
8	.602	.570	.615	.458	.575	.547
9	.531	.667	.632	.626	.531	.617
10	.651	.506	.605	.644	.658	.670
11	.614	.674	.512	.598	.612	.690
12	.430	.493	.602	.626	.670	.696
13	.667	.571	.477	.558	.617	.609
14	.692	.764	.671	.685	.602	.605
15	.568	.704	.543	.631	.663	.573
16	.696	.671	.638	.675	.521	.545
17	.675	.581	.547	.600	.612	.688
18	.686	.725	.590	.612	.570	.544
19	.582	.588	.577	.532	.671	.674
20	.701	.704	.692	.690	.628	.434
Mean	.623	.636	.581	.613	.612	.606

We next consider predicted and observed variance in asymptotic performance. In particular, we are interested in checking the variability predicted by the one-element model against the standard deviations presented in Table 3.1. Theoretical expressions for these quantities can be given in terms of $\text{var}(\bar{A}_N)$ and $\text{var}(\bar{B}_N)$ [see (1.4.20) and (1.4.21)]. Specifically, $\text{var}(\bar{A}_N)$ denotes the variance associated with the number of A_1's in a block of N trials at asymptote, and $\text{var}(\bar{B}_N)$ the variance for B_1's. Hence, the standard deviation for proportions would be $\sqrt{\text{var}(\bar{A}_N)}/N$ and $\sqrt{\text{var}(\bar{B}_N)}/N$. To simplify computation we have assumed that θ is the same for all subjects; given this assumption and $N = 60$, the above quantities correspond to the standard deviations reported in Table 3.1. The predicted values obtained for $\theta = .6$ were as follows for A_1 and B_1 responses, respectively: .264, .239 (Mixed Group); .325, .231 (Pure Group); and .298, .228 (Sure Group). These values are about three times larger than the obtained values in Table 3.1. As one would expect for the one-element model, variability decreases with increasing θ; consequently, to improve the fit between the observed and predicted standard deviations would require a value of θ greater than .6.

Thus, for the maximum-likelihood estimate of θ we have two contrasting results. For the mean learning curves the obtained value of θ was too large, while for asymptotic variability in α and β the estimate of θ was too small. It should be emphasized that this difficulty has also been encountered in many one-person studies. Its existence reflects one of the most serious defects of one-element models.

3.6. Game-Theory Comparisons

Although subjects were not shown the payoff matrix and not even told they were interacting in the experiments reported in this chapter, it may still be maintained that game theory has predictive pertinence (see Sec. 1.6). For it can be argued that after a large number of trials, subjects who are trying to maximize their number of correct predictions will come to approximate an optimal game strategy. Concerning such a conjecture, the results for the Pure and Sure Groups seem decisive. The optimal game strategies are to respond A_1 and B_1, respectively, with probability 1 for both players; these strategies are not even crudely approximated by the observed means shown in Table 3.2. Given this result, it is natural to ask if at least a fair portion of the subjects did not conform to game notions of optimality rather than to stimulus-sampling learning theory. Again, as can be seen from Table 3.1, the results are decisive. The maximum individual value of α in each group of 20 subject-pairs does not approach 1; for the Pure Group max $\alpha = .783$ and max $\beta = .750$, while for the Sure Group max $\alpha = .800$ and max $\beta = .867$.

The results for the Mixed Group also fail to support the hypothesis that subjects will approach an optimal game strategy. The observed α of .605 and β of .649 both differ significantly from the respective minimax strategies of 1/3 and 5/6 beyond the .001 level.

In many zero-sum games, rather wide deviations from minimax strategies do not radically affect the *value* of the game as defined in Sec. 1.6. For the three experimental groups it is of some interest to compare the game-theory value with the learning-theory "value," which we define as the expectation for Player A if both players exactly satisfy the learning-theory postulates with $\theta_A = \theta_B$. There is one difference in the two definitions. It is fundamental to the game theory to assume that the mixed strategies of the two players are statistically independent (naturally this problem does not arise when the optimal strategies are pure strategies). Such statistical independence does not characterize the learning theory; even at asymptote we have for all three groups that $P_\infty(A_iB_j) \neq P_\infty(A_i)P_\infty(B_j)$. Because the joint probabilities u_i are more fundamental to the theory, we define the learning-theory value v_L as $v_L = \sum u_i a_i$. Similarly, we define the observed value v_O of the game as $\sum \hat{u}_i a_i$, where $\hat{u}_i$ is the observed joint probability of response at asymptote.

It was remarked in Sec. 1.4 that (1.4.1) is actually the payoff matrix of a constant-sum game which is strategically equivalent to the zero-sum game obtained by subtracting $\frac{1}{2}$ from each a_i. In Table 3.5 we present the game-theory, learning-theory, and observed values of the game for each of the experimental groups, with this correction made in order to yield a zero-sum game. The observed values are computed on the basis of estimating the u_i's from the last 40 trials.

TABLE 3.5

GAME-THEORY, LEARNING-THEORY, AND OBSERVED VALUES OF THE ZERO-SUM GAMES

Type of Value	Mixed	Pure	Sure
Game Theory	−.056	.000	.000
Learning Theory	.011	.139	.094
Observed	.000	.100	.030

It is worth noting that in all cases the observed value lies between the game-theory value and the learning-theory value, although in two of the three cases it is closer to the latter. This result is particularly striking because it is not true that for all groups the observed α and β lie between the learning-theory asymptotic response probabilities and the optimal game strategies. For example, the observed β in the Pure Group is less than the β predicted by the learning theory, and the game-theory strategy is $\beta = 1$; the observed α in the Sure Group is markedly less than the learning-theory prediction, yet again the game-theory strategy is $\alpha = 1$. Since under the learning theory Player A fares better than Player B, while the reverse is true if optimal game strategies are used, B might be expected to move more than A toward the optimal game strategies.

In order to determine if the facts support such an expectation, the following sign test was run on each group of subjects. Let g_A and g_B be the optimal game strategies for A and B, respectively; l_A and l_B the asymptotic probabilities of response predicted by learning theory; and p_A and p_B the observed asymptotic response probabilities tabulated for individual pairs in Table 3.1. The null hypothesis is that the ratio of the difference between the game strategies and the learning-theory predictions for the two players is equal to the ratio of the difference between the game strategies and the observed response probabilities for the two players; i.e., the null hypothesis is that

$$\frac{g_A - l_A}{g_B - l_B} = \frac{g_A - p_A}{g_B - p_B}.$$

We assign a plus value to those subject-pairs for which the right-hand ratio is greater than the left-hand ratio, for if it is indeed true that p_B approaches g_B more closely than p_A approaches g_A, this fact will be reflected in a large value for the right-hand ratio. Contrariwise, we assign a minus value when the left-hand ratio is greater. The null hypothesis is then that the plus and minus values have a binomial distribution with $p = .5$.

The experimental results for the three groups are as follows: for the Mixed Group, 12 pluses and 8 minuses, which is not significant; for the Pure Group, an exactly even division between pluses and minuses, which obviously is not significant; for the Sure Group, 15 pluses and 5 minuses, which is significant (.02 level). These results may perhaps indicate that the B subjects tended to move toward the optimal game strategy in the Sure Group, but it should be emphasized that even for this group the learning-theory prediction fits the data much better than does the optimal game strategy g_B.

Because a number of experimental studies have found stimulus-sampling theory superior in behavioral validity to simple optimization schemes derived from game theory, various proposals for a *rapprochement* of game theory and learning theory have been made. Simon [1956] proposed one of the better-known examples, which we now consider in some detail. Simon's aim was to explain the asymptotic results for the simple contingent case that was derived and tested experimentally in Estes [1954]. Simple contingent reinforcement is defined by two probabilities π_1 and π_2 such that

$$P(E_{1,n} \mid A_{1,n}) = \pi_1 , \qquad P(E_{2,n} \mid A_{2,n}) = \pi_2 .$$

Corresponding to (1.3.4), the transition matrix in the one-element model for this case is simply

$$\text{(3.6.1)} \qquad \begin{array}{c|cc} & 1 & 2 \\ \hline 1 & 1-\theta(1-\pi_1) & \theta(1-\pi_1) \\ 2 & \theta(1-\pi_2) & 1-\theta(1-\pi_2) \end{array}$$

and by methods already used repeatedly, we have at once

$$u_1 = \frac{1-\pi_2}{(1-\pi_1)+(1-\pi_2)} .$$

It is this last result for the asymptotic mean probability of an A_1 response that Simon wishes to explain. As an alternative to statistical learning theory, Simon proposes the following three postulates.

1. If a subject *persists* in choosing A_1, he assumes the probability of reward to be π_1, and if he persists in choosing A_2, he assumes the probability of reward to be π_2.

2. If the subject *shifts* from the response just chosen, the probability of reward is now unknown and depends on a strategy of "nature."

3. The subject persists or shifts according to his estimates of the chances of reward. Moreover, he measures his success on any trial not in terms of the

actual reward, but in terms of the difference between the reward he has actually obtained and the reward he could have had if he had outguessed "nature." In the terminology of Savage [1957] the subject attempts to minimize his *regret*. Let (I) be the strategy of persisting, (II) the strategy of shifting; and let a and b denote nature's two strategies of reinforcing A_1 or A_2 when a shift occurs. Then the payoff matrix when an A_1 response has just been made is the following:

(3.6.2)

	a	b
(I)	π_1	π_1
(II)	0	1

The regret matrix (r_{ij}) is defined by $r_{ij} = a_{ij} - \max_k a_{kj}$, i.e., the regret of any outcome a_{ij} is defined as the difference between it and the best that could have been done if the strategy j of nature had been known. Thus the regret matrix for (3.6.2) is

(3.6.3)

0	$\pi_1 - 1$
$-\pi_1$	0

We now apply the minimax criterion to the regret matrix, i.e., we seek the mixed strategy that will minimize the maximum regret. Solving by the method of Sec. 1.6 [particularly (1.6.3)], we find that the minimax regret strategy ρ(I) for persisting is π_1. Thus $P(A_1 \mid A_1) = \pi_1$, and by an exactly similar argument, $P(A_2 \mid A_2) = \pi_2$. These conditional probabilities yield the transition matrix (3.6.1) with $\theta = 1$.

We now consider the possibilities of applying Simon's model to the zero-sum situations of this chapter. Suppose Player A makes response A_1. It is not clear how to write the analogue of the matrix (3.6.2). What is needed is an expression for the expectation that A will be reinforced when he makes the A_1 response, i.e., we need to know $P(E_{1,n} \mid A_{1,n})$; but it is not obvious how to compute this quantity in Simon's model. For

$$P(E_{1,n} \mid A_{1,n}) = P(E_{1,n} \mid A_{1,n}B_{1,n})P(B_{1,n} \mid A_{1,n}) + P(E_{1,n} \mid A_{1,n}B_{2,n})P(B_{2,n} \mid A_{1,n})$$
$$= a_1P(B_{1,n} \mid A_{1,n}) + a_2P(B_{2,n} \mid A_{1,n}) ,$$

and his model provides no way to compute the conditional probabilities $P(B_{i,n} \mid A_{1,n})$ for $i = 1, 2$.

It might be objected that Simon's model was not designed to apply to two-person interactions; but since our subjects were not told they were interacting, it is completely reasonable to apply a model designed to analyze behavioral choices in a game against nature. The problem of applying Simon's model here brings out one of its central difficulties. It is not clear how to derive definite results from it for reinforcement schedules of any complexity. For example, for a one-person experiment with reinforcement contingent upon the response given three trials earlier, it is again not clear how to derive the analogue of the matrix (3.6.2).

In connection with the experiment of Estes mentioned above, Simon has this to say ([1957], p. 277): "We need not try to decide whether the subjects who behave in conformity with the predictions of Estes' theory are 'really' minimaxing regret, or whether they are simply behaving in the adaptive fashion implied by the usual learning mechanisms. I would suppose that most economists and statisticians would be tempted to accept the former interpretation, most psychologists the latter. It is not immediately obvious what source, other than introspection, would provide evidence for deciding the issue."

The attempt to apply Simon's model to more complex reinforcement schedules shows the lack of equivalence between his model and stimulus-sampling models. However, it is not even necessary to go beyond the simple contingent case to establish this lack of equivalence and thus to controvert the last sentence just quoted. The Estes model permits a derivation of every conditional probability as a function of the parameters θ, π_1, and π_2; for example, the value of $P(A_{1,n+1} \mid E_{1,n}A_{2,n})$. Simon's three postulates give no method for deriving this quantity or any other sequential statistics of empirical interest. And the reason is obvious. From a psychological standpoint, he provides no theory of the elementary process of interaction between the organism and its environment; this is also the difficulty with game theory as such, as we noted in Sec. 1.6.[2] Finally, the three postulates do not seem to us very plausible. When an organism makes a large number of responses, there seems to be no apparent reason why persisting and shifting response choice should be treated so completely unsymmetrically; this is but one of many game-theory hypotheses which might be plausible to the subject.

That these postulates cannot serve as the basis of a general theory becomes particularly evident when we consider a contingent reinforcement schedule with probabilities of reinforcement dependent on the immediately preceding trial. By appropriate choice of the probabilities it may be advantageous to the subject to shift his response every trial, and this strategy will rapidly become evident to him. Simon's second postulate is particularly unrealistic in this situation. Our own analysis of this situation from the standpoint of stimulus-sampling theory is given in Sec. 1.7 in the discussion of hypothesis models. In summary, it seems to us that Simon's postulates are little more than an ingenious attempt to provide a possible *rapprochement* between game theory and learning theory for one simple schedule of reinforcement.

Statistical decision theory (Blackwell and Girshick [1954]) offers yet another viewpoint concerning rational behavior. In this light the subject should not regard himself as playing a 2×2 game with a fixed payoff matrix, but should analyze the situation as a multi-stage decision problem against an unknown opponent. Unfortunately, selecting an optimal strategy for this multi-stage problem is far from a trivial task mathematically. It would take us too far afield to discuss the theory in adequate detail, for a considerably more elaborate set

[2] This remark also applies to Siegel [1959].

of concepts is needed than was required for the discussion of game theory in Sec. 1.6. Readers are referred to Appendix 8 of Luce and Raiffa [1957] and to the excellent bibliography there of the work of Blackwell, Hannan, Robbins, and others.

*3.7. Multi-Element Models

Despite the close correspondence between the predicted and observed asymptotic response probabilities, there are several disturbing features to consider about the results for this one-element model; in particular: (1) in the theoretical matrix the anti-diagonal is identically 0, whereas in the observed transition matrices several entries on the anti-diagonals are markedly different from 0; (2) the theoretical transition probabilities are constant, while in the observed matrices changes occur over trials; (3) the maximum-likelihood estimates of θ yield values that are too large to account for the observed rate of learning and too small to account for the observed asymptotic variance; (4) for the Mixed Group, there is evidence that the process is a second-order Markov chain in the response random variables.

Fortunately, these contradictions between theory and data are to a large extent peculiar to the one-element model. Before applying the methods of Sec. 2.5 to the detailed empirical analysis of the two element model, we consider some general aspects of multi-element models; some of the results are independent of the number of stimuli in the set available for sampling.

Concerning problem (1) above, consider the conditional probability

$$P(A_{2,n+1}B_{2,n+1} \mid A_{1,n}B_{1,n}) ,$$

which is 0 in the one-element model. In the two-element model

$$P(A_{2,n+1}B_{2,n+1}\, A_{1,n}B_{1,n}) = \tfrac{1}{8}\,\theta(1-a_1)u_{21}^{(n)} + \tfrac{1}{8}\,\theta a_1 u_{12}^{(n)} + \tfrac{1}{16}(1+\theta)u_{11}^{(n)} , \tag{3.7.1}$$

where $u_{ij}^{(n)}$ is the probability that on trial n Player A has i elements conditioned to the A_1 response and Player B has j elements conditioned to B_1; to fix $u_{ij}^{(n)}$ uniquely we assume that elements have an equal probability of being conditioned to either response on trial 1. The method of deriving this result is similar to the ones used in Sec. 2.5. And also by similar methods

$$\alpha_n = u_{22}^{(n)} + u_{21}^{(n)} + u_{20}^{(n)} + \tfrac{1}{2}(u_{12}^{(n)} + u_{11}^{(n)} + u_{10}^{(n)}) , \tag{3.7.2}$$

$$\gamma_n = P(A_{1,n}B_{1,n}) = u_{22}^{(n)} + \tfrac{1}{2}(u_{21}^{(n)} + u_{12}^{(n)}) + \tfrac{1}{4}u_{11}^{(n)} . \tag{3.7.3}$$

Dividing (3.7.1) by (3.7.3) yields the desired conditional probability

$$P(A_{2,n+1}B_{2,n+1} \mid A_{1,n}B_{1,n}) ,$$

which is not 0 in the two-element model. Also it is to be noted that in the one-element model this conditional probability is independent of n, while in the two element model it depends on n.

In Sec. 2.5 we have already indicated some of the mathematical difficulties in working with multi-element models for two-person situations. However, under certain limited conditions, simple results can be obtained for the zero-sum game. We shall now examine these to strengthen our understanding of the properties of the theory.

In terms of the axioms of Sec. 1.2, α_n has the following recursion, where N_A is the number of stimuli in Player A's stimulus set, and N_B is similarly defined for Player B:

$$\begin{aligned}\alpha_{n+1} = (1-\theta_A)\alpha_n + \theta_A\Big\{P(A_{1,n}B_{1,n})\Big[a_1\alpha_n + (1-a_1)\Big(\alpha_n - \frac{1}{N_A}\Big)\Big] \\ + P(A_{1,n}B_{2,n})\Big[a_2\alpha_n + (1-a_2)\Big(\alpha_n - \frac{1}{N_A}\Big)\Big] \\ + P(A_{2,n}B_{1,n})\Big[a_3\alpha_n + (1-a_3)\Big(\alpha_n + \frac{1}{N_A}\Big)\Big] \\ + P(A_{2,n}B_{2,n})\Big[a_4\alpha_n + (1-a_4)\Big(\alpha_n + \frac{1}{N_A}\Big)\Big]\Big\}.\end{aligned} \tag{3.7.4}$$

Note that

$$P(A_{1,n}B_{1,n}) = \gamma_n\,, \quad P(A_{1,n}B_{2,n}) = \alpha_n - \gamma_n\,, \quad P(A_{2,n}B_{1,n}) = \beta_n - \gamma_n\,.$$

Hence,

$$\begin{aligned}\alpha_{n+1} = \Big[1 - \frac{\theta_A}{N_A}(2 - a_2 - a_4)\Big]\alpha_n + \frac{\theta_A}{N_A}(a_4 - a_3)\beta_n \\ + \frac{\theta_A}{N_A}(a_1 + a_3 - a_2 - a_4)\gamma_n + \frac{\theta_A}{N_A}(1 - a_4)\,.\end{aligned} \tag{3.7.5}$$

Similarly, we obtain

$$\begin{aligned}\beta_{n+1} = \Big[1 - \frac{\theta_B}{N_B}(a_3 + a_4)\Big]\beta_n + \frac{\theta_B}{N_B}(a_2 - a_4)\alpha_n \\ + \frac{\theta_B}{N_B}(a_3 + a_4 - a_1 - a_2)\gamma_n + \frac{\theta_B}{N_B}a_4\,.\end{aligned} \tag{3.7.6}$$

It can be shown that the limits α, β, and γ exist; hence, by letting $\alpha_{n+1} = \alpha_n = \alpha$, $\beta_{n+1} = \beta_n = \beta$, and $\gamma_n = \gamma$, we obtain the following pair of relations:

$$\begin{aligned}\alpha(2 - a_2 - a_4) + \beta(a_3 - a_4) + \gamma(a_2 + a_4 - a_1 - a_3) = 1 - a_4\,, \\ \alpha(a_4 - a_2) + \beta(a_3 + a_4) + \gamma(a_1 + a_2 - a_3 - a_4) = a_4\,.\end{aligned} \tag{3.7.7}$$

By eliminating γ from these two equations we obtain the linear relation (1.4.12) in α and β; this relation is independent of θ_A, θ_B, N_A, and N_B.

Of course, when $a_1a_3 = a_2a_4$ (as was the case in the Mixed Group) the coefficient of β vanishes in (1.4.12), and thus for any θ_A, θ_B, N_A, and N_B

$$\alpha = \frac{\frac{1}{2}(a_3 + a_4 - a_1 - a_2) + a_4(a_2 - a_3)}{a_3 + a_4 - a_1 - a_2 + a_1a_2 - a_3a_4}\,. \tag{3.7.8}$$

Similarly, when $a_3 + a_4 + a_1a_2 = a_1 + a_2 + a_3a_4$, the coefficient of α vanishes in

(1.4.12) and

$$\beta = \frac{\frac{1}{2}(a_3 + a_4 - a_1 - a_2) + a_4(a_2 - a_3)}{a_2 a_4 - a_1 a_3} . \tag{3.7.9}$$

Naturally, we can go one step further and introduce a restriction such that the coefficient of γ_n vanishes in both (3.7.5) and (3.7.6): namely,

$$a_1 + a_3 = a_2 + a_4 , \qquad a_1 + a_2 = a_3 + a_4 . \tag{3.7.10}$$

Under these conditions a great deal can be said about α_n and β_n. To simplify notation, we shall rewrite (3.7.5) and (3.7.6) with the restrictions of (3.7.10)

$$\begin{aligned} \alpha_{n+1} &= (1 - a\varphi)\alpha_n - b\varphi\beta_n + c\varphi , \\ \beta_{n+1} &= (1 - a'\varphi')\beta_n - b'\varphi'\alpha_n + c'\varphi' , \end{aligned} \tag{3.7.11}$$

where $\varphi = \theta_A/N_A$, $\varphi' = \theta_B/N_B$, $a = 2 - a_2 + a_4$, $b = a_3 - a_4$, and so on.

Solutions for (3.7.11) are well known and can be obtained by a number of different techniques. Burke [1959], in applying the linear model described in Sec. 1.5 to two-person interactions, encountered a similar pair of equations. He investigated their properties by studying the characteristic equation of the difference equation (3.7.11). Of course, Burke's results are directly applicable to our work, since the difference equations for the linear model and our model are identical under the conditions of (3.7.10). That is, the two models predict the same mean functions for α_n and β_n.

We now present the solutions of (3.7.11) for the zero-sum situation. The reader is referred to Burke's paper for a detailed discussion of this material. Three cases are distinguished; except for Case III

$$\alpha = \frac{a'c + bc'}{aa' + bb'} , \qquad \beta = \frac{ac' - b'c}{aa' - bb'} . \tag{3.7.12}$$

Case I: If $4bb'\varphi\varphi' \neq -(a\varphi - a'\varphi')^2$, the solutions of (3.7.11) are

(3.7.13)

$$\begin{aligned} \alpha_n = \alpha &+ \frac{[a\varphi - a'\varphi' + \sqrt{(a\varphi - a'\varphi')^2 + 4bb'\varphi\varphi'}]\,\frac{1}{2}(\alpha_1 - \alpha) + b\varphi(\beta_1 - \beta)}{\sqrt{(a\varphi - a'\varphi')^2 + 4bb'\varphi\varphi'}} r_+^{n-1} \\ &+ \frac{[a'\varphi' - a\varphi + \sqrt{(a\varphi - a'\varphi')^2 + 4bb'\varphi\varphi'}]\,\frac{1}{2}(\alpha_1 - \alpha) - b\varphi(\beta_1 - \beta)}{\sqrt{(a\varphi - a'\varphi')^2 + 4bb'\varphi\varphi'}} r_-^{n-1} , \end{aligned}$$

(3.7.14)

$$\begin{aligned} \beta_n = \beta &+ \frac{[a'\varphi' - a\varphi + \sqrt{(a\varphi - a'\varphi')^2 + 4bb'\varphi\varphi'}]\,\frac{1}{2}(\beta_1 - \beta) + b'\varphi'(\alpha_1 - \alpha)}{\sqrt{(a\varphi - a'\varphi')^2 + 4bb'\varphi\varphi'}} r_+^{n-1} \\ &+ \frac{[a\varphi - a'\varphi' + \sqrt{(a\varphi - a'\varphi')^2 + 4bb'\varphi\varphi'}]\,\frac{1}{2}(\beta_1 - \beta) - b'\varphi'(\alpha_1 - \alpha)}{\sqrt{(a\varphi - a'\varphi')^2 + 4bb'\varphi\varphi'}} r_-^{n-1} , \end{aligned}$$

where $r_\pm = 1 - \frac{1}{2}[a\varphi + a'\varphi' \pm \sqrt{(a\varphi - a'\varphi')^2 + 4bb'\varphi\varphi'}]$.

Case II: If $4bb'\varphi\varphi' = -(a\varphi - a'\varphi')^2$ and $a\varphi + a'\varphi' \neq 2$, the solutions of (3.7.11) are

$$\alpha_n = \alpha + (\alpha_1 - \alpha)r^{n-1} + [(a'\varphi' - a\varphi)\tfrac{1}{2}(\alpha_1 - \alpha) - b\varphi(\beta_1 - \beta)](n-1)r^{n-2} ,$$
$$\beta_n = \beta + (\beta_1 - \beta)r^{n-1} + [(a\varphi - a'\varphi')\tfrac{1}{2}(\beta_1 - \beta) - b'\varphi'(\alpha_1 - \alpha)](n-1)r^{n-2} . \tag{3.7.15}$$

Case III: If $4bb'\varphi\varphi' = -(a\varphi - a'\varphi')^2$ and $a\varphi + a'\varphi' = 2$, then (3.7.11) becomes

$$\alpha_{n+1} = (1 - a\varphi)\alpha_n - b\varphi\beta_n + c\varphi ,$$
$$\beta_{n+1} = -(1 - a\varphi)\beta_n + \frac{(1 - a\varphi)^2}{b\varphi}\alpha_n + c'\varphi' , \tag{3.7.16}$$

and the asymptotic solutions reduce to

$$\alpha = \varphi(2 - a\varphi)c - bc'\varphi\varphi' ,$$
$$\beta = ac'\varphi\varphi' + \frac{(1 - a\varphi)^2}{b}c . \tag{3.7.17}$$

The solution of the difference equations is

$$\alpha_2 = (1 - a\varphi)\alpha_1 - b\varphi\beta_1 + c\varphi , \qquad \alpha_n = \alpha \quad (n \geqq 3) ,$$
$$\beta_2 = (1 - a'\varphi')\beta_1 - b'\varphi'\alpha_1 + c'\varphi' , \quad \beta_n = \beta \quad (n \geqq 3) . \tag{3.7.18}$$

Empirical analysis of the two-element model. In Sec. 2.5 procedures were presented for estimating the transition numbers associated with the unobservable states of conditioning in the multi-element model. The analysis was carried out in detail for the two-element model in the zero-sum, two-person situation. We now apply the methods of that section to the data of this chapter.

The transition matrix for the two-element model applied to the zero-sum situation is given in (2.5.21). The states are designated (ij), where $i, j = 0, 1, 2$; i indicates the number of elements conditioned to A_1 and j the number to B_1. As in Sec. 2.5, we use $m_{ij,i'j'}$ to denote the transition frequencies associated with the unobservable states of conditioning and n_{kl} $(k, l = X, Y, Z, W)$ to denote transition frequencies for observable response pairs. [See (2.5.22) for a definition of X, Y, Z, and W.]

Equations (2.5.23–26) give 40 linear equations in $m_{ij,i'j'}$ and n_{kl}. There are 33 $m_{ij,i'j'}$'s and 16 n_{kl}'s. However, to ensure that $\sum n = \sum m = N$, where N is the total number of observed transitions, we eliminate $m_{00,00}$ from the set of 40 equations and replace it by N minus the sum of the other $m_{ij,i'j'}$'s [see (2.5.15) if this point is not clear]. This may be accomplished most simply by discarding the last equation in (2.5.24). We then have 39 equations, with 32 $m_{ij,i'j'}$'s. Applying the least-squares method to these equations leads to a system of 32 linear nonhomogeneous equations in the 32 unknowns $m_{ij,i'j'}$ [see (2.5.17)].

In solving for $m_{ij,i'j'}$ we use n_{kl}'s for the last 100 trials aggregated over subject-pairs; that is, the n_{kl} defined at the outset of Sec. 3.4. It would have been preferable to estimate the $m_{ij,i'j'}$'s for each subject-pair in terms of their particular n_{kl}'s. However, the number of observations for each pair of subjects did not seem large enough to yield reliable estimates when so many quantities are involved.

The solution of the system of equations yielded the transition frequencies given in Table 3.6. The point about the estimated transition frequencies for the Mixed Group which most immediately demands comment is that four of them are negative. Certainly these negative entries have no empirical interpretation, although two of them ($m_{20.10}$ and $m_{10.00}$) are close enough to 0 to be interpreted as 0. On the other hand, there is nothing in the least-squares procedure to guarantee non-negativity. From a mathematical standpoint it is simple to add this constraint, but from a computational standpoint it is wholly impractical. Moreover, from the standpoint of testing the two-element model it is of interest to see if the fit is good enough to preclude negative entries.

TABLE 3.6
ESTIMATED VALUES OF $m_{ij,i'j'}$ FOR THE MIXED GROUP, THE PURE GROUP, AND THE SURE GROUP

Transition Frequency	Mixed	Pure	Sure	Transition Frequency	Mixed	Pure	Sure
$m_{22.22}$	61.6	125.5	71.6	$m_{11.10}$	48.9	65.3	19.0
$m_{22.21}$	48.0	93.4	43.0	$m_{11.01}$	39.5	33.1	12.7
$m_{22.12}$	92.6	96.2	42.5	$m_{10.20}$	27.3	27.4	2.5
$m_{21.22}$	92.7	138.5	40.6	$m_{10.11}$	37.5	46.4	18.2
$m_{21.21}$	208.6	40.2	230.7	$m_{10.10}$	−46.0	−0.2	−113.5
$m_{21.20}$	29.7	69.3	22.3	$m_{10.00}$	−0.4	3.6	−1.7
$m_{21.11}$	61.8	27.9	22.5	$m_{02.12}$	33.0	33.1	31.6
$m_{20.21}$	58.4	98.5	19.9	$m_{02.02}$	37.1	49.2	−40.6
$m_{20.20}$	0.9	69.4	−29.8	$m_{02.01}$	30.7	40.4	9.0
$m_{20.10}$	−0.6	−0.8	1.6	$m_{01.11}$	48.7	52.0	12.6
$m_{12.22}$	50.3	49.3	44.3	$m_{01.02}$	6.4	8.3	11.3
$m_{12.12}$	153.1	−7.3	587.1	$m_{01.01}$	−41.3	−59.7	−18.8
$m_{12.11}$	86.5	96.4	44.2	$m_{01.00}$	18.3	20.8	1.8
$m_{12.02}$	56.3	65.3	28.3	$m_{00.10}$	15.7	14.5	−1.0
$m_{11.21}$	78.1	82.9	24.0	$m_{00.01}$	2.9	7.4	2.9
$m_{11.12}$	67.9	81.4	41.8	$m_{00.00}$	23.0	11.4	17.2
$m_{11.11}$	536.9	456.0	803.3				

An analogous problem arises in making a maximum-likelihood estimate of the conditioning parameter θ. For instance, in the one-element, noncontingent model estimates of θ for individual subjects often yield $\hat{\theta} > 1$. Such estimates can be eliminated by adding the constraint $\theta \leqq 1$, but it is of interest in this situation to find the value of θ that best fits the data. When that θ is greater than 1 it provides another indication that the model is fitting the data poorly.

Using the estimated values of $m_{ij,i'j'}$ for the Mixed Group, we obtain $\hat{\theta} = .525$ by (2.5.27). We now wish to take this value of θ and test the goodness-of-fit of asymptotic transition probabilities in the two-element model to the observed data. The procedure is discussed in detail after (2.5.27); the particular transition probabilities of interest are

$$P_{\infty}(A_{i,n+1}B_{j,n+1} \mid E_{k,n}A_{i',n}B_{j',n}) .$$

Using $\hat{\theta}$, we obtain numerical values for these probabilities by means of (2.5.28). Table 3.7 presents the observed transition data for the last block of 100 trials aggregated over all subjects in the Mixed Group. The expected theoretical frequencies that also are given in the table were obtained in terms of the row sums of the observed frequencies and the predicted transition probabilities. In general, the correspondence between theoretical and observed quantities is not too bad considering the number of observations. However, the χ^2 of (2.4.4) yields a value of 99.3. This result is interpreted with $7 \times 3 - 1 = 20$ degrees of freedom (one degree of freedom is lost owing to the estimate of θ) and is highly significant.[3]

TABLE 3.7

OBSERVED TRANSITION FREQUENCIES OVER THE LAST 100 TRIALS FOR THE MIXED GROUP

(Theoretical frequencies for $\theta = .525$ are given in parentheses)

$n+1$ / n	X	Y	Z	W
E_1X	90 (110.6)	111 (92.8)	30 (28.5)	24 (23.1)
E_2X	190 (216.5)	60 (52.3)	184 (189.1)	68 (44.1)
E_1Y	247 (211.4)	112 (158.0)	68 (50.5)	29 (36.1)
E_2Y	0 (0)	0 (0)	0 (0)	0 (0)
E_1Z	49 (40.6)	43 (32.8)	78 (99.1)	81 (78.5)
E_2Z	126 (108.3)	35 (24.3)	70 (87.4)	8 (19.0)
E_1W	13 (7.6)	6 (5.4)	17 (17.7)	7 (12.3)
E_2W	59 (39.9)	85 (83.4)	28 (31.2)	46 (63.5)

In addition to the above computations, we also computed theoretical frequencies for an array of θ values ($\theta = .02i$, for $i = 1, \cdots, 50$), and then for each θ we computed the χ^2 discussed in the preceding paragraph. The obtained values of χ^2 as a function of θ for the Mixed Group are plotted in Fig. 3.5; of course, the minimum of this function represents another procedure for estimating θ in the two-element model. It is interesting that the value of θ estimated in terms of the $m_{ij,i'j'}$'s is not too close to the minimum of the χ^2 function, which occurs at about $\theta = .68$.

[3] It was noted in Sec. 2.4 that the χ^2 goodness-of-fit test does not have complete validity for the two-element model. The reason is that the sequence of pairs of response random variables in the two-element model is a chain of infinite order, not a first-order Markov chain.

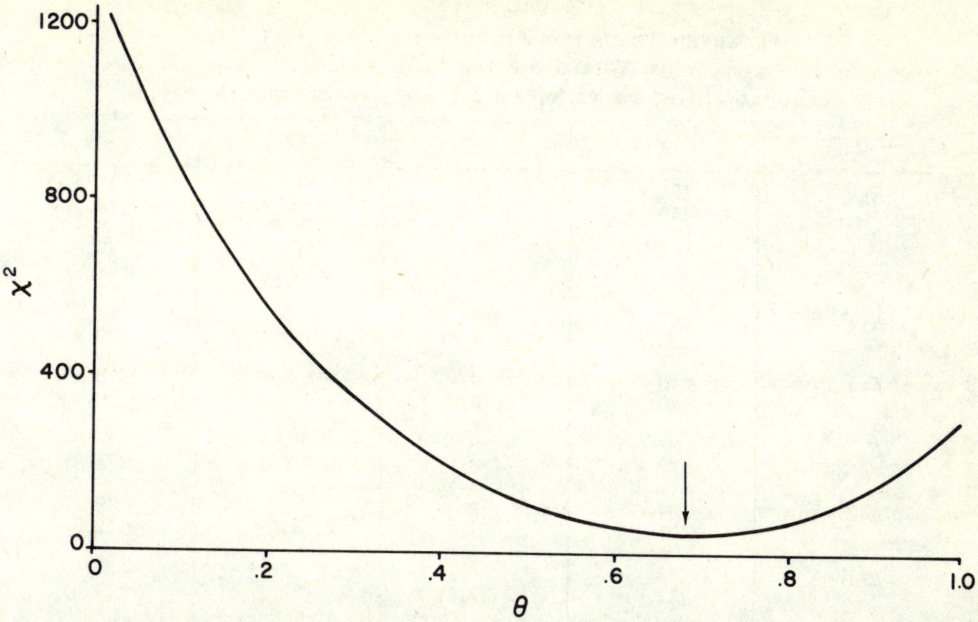

FIGURE 3.5. χ^2 as a function of θ for the Mixed Group.

The same procedures were also applied to the data of the Pure and Sure Groups (see Table 3.6). For the Pure Group the estimated value of θ in terms of (2.5.27) was .653. Table 3.8 presents the observed and predicted transition frequencies for the Pure Group with $\theta = .653$. The obtained value of χ^2 was 86.1; with 20 degrees of freedom the result is highly significant. Figure 3.6 presents χ^2 as a function of θ for the Pure Group; the minimum occurs at approximately $\theta = .64$, which is extremely close to the estimate of θ in terms of the $m_{ij,i'j'}$'s. For the Sure Group the estimated value of θ in terms of (2.5.27) was .245. Table 3.9 presents the observed and predicted transition frequencies for the Sure Group when $\theta = .245$. The value of χ^2 is 455.4, which is much larger than either of the other two χ^2's. Figure 3.7 presents χ^2 as a function of θ for the Sure Group; the minimum occurs at approximately $\theta = .64$.

The stability of the θ values as estimated by minimizing χ^2 (.68, .64, .64, for the three groups) suggests that in practice this may be a better method than the cumbersome procedure of estimating the $m_{ij,i'j'}$'s. As already remarked in Chapter 2, even prior to making the present computations we felt that this method of estimating the transition numbers $m_{ij,i'j'}$ was of more theoretical and methodological interest than of practical interest. The presence of some negative $m_{ij,i'j'}$'s in the estimates for each experimental group and the rather poor goodness-of-fit results reinforce this conclusion. Of course, a substantial part of the poor goodness-of-fit results are due not to this method of estimating $m_{ij,i'j'}$ but to the simple fact that the data are not fitting the two-element model in detail. Because the minimum-χ^2 method of estimating θ indicates the

TABLE 3.8

OBSERVED TRANSITION FREQUENCIES OVER THE LAST 100 TRIALS FOR THE PURE GROUP

(Theoretical frequencies for $\theta = .653$ are given in parentheses)

n \ $n+1$	X	Y	Z	W
E_1X	151 (148.1)	143 (169.5)	49 (31.3)	40 (34.1)
E_2X	135 (150.4)	57 (39.6)	141 (150.1)	45 (37.9)
E_1Y	268 (283.7)	166 (172.2)	61 (53.2)	44 (29.9)
E_2Y	0 (0)	0 (0)	0 (0)	0 (0)
E_1Z	38 (33.1)	38 (36.1)	64 (71.0)	75 (74.8)
E_2Z	84 (101.2)	42 (24.0)	51 (55.3)	16 (12.5)
E_1W	18 (13.9)	8 (7.8)	17 (27.5)	21 (14.8)
E_2W	67 (42.7)	87 (92.2)	24 (22.2)	25 (45.9)

TABLE 3.9

OBSERVED TRANSITION FREQUENCIES OVER THE LAST 100 TRIALS FOR THE SURE GROUP

(Theoretical frequencies for $\theta = .245$ are given in parentheses)

n \ $n+1$	X	Y	Z	W
E_1X	207 (272.2)	151 (110.9)	75 (63.5)	37 (23.4)
E_2X	175 (246.6)	24 (49.4)	190 (111.9)	39 (20.1)
E_1Y	169 (124.5)	63 (150.2)	72 (24.3)	21 (26.0)
E_2Y	0 (0)	0 (0)	0 (0)	0 (0)
E_1Z	37 (34.1)	26 (12.6)	48 (77.0)	39 (26.3)
E_2Z	236 (160.9)	46 (26.9)	130 (212.5)	20 (31.7)
E_1W	42 (25.4)	22 (27.2)	63 (49.9)	25 (49.5)
E_2W	12 (7.5)	10 (13.2)	9 (8.9)	13 (14.4)

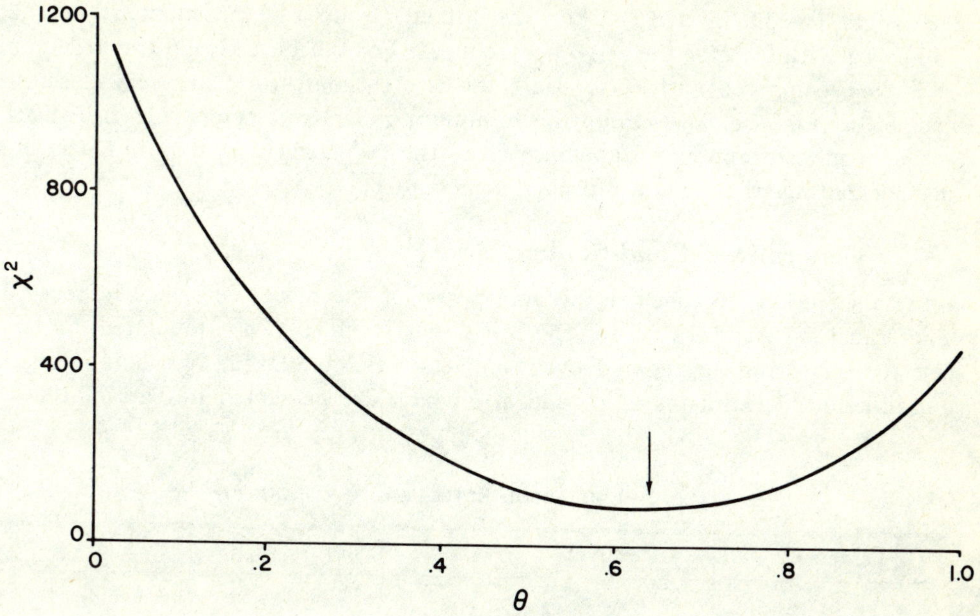

FIGURE 3.6. χ^2 as a function of θ for the Pure Group.

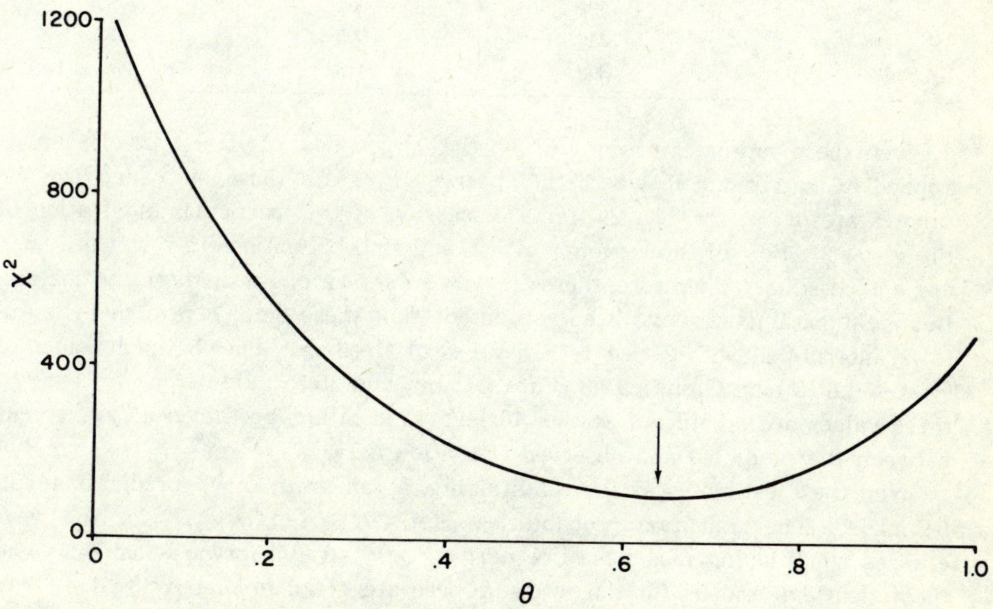

FIGURE 3.7. χ^2 as a function of θ for the Sure Group.

best possible fit of the data to the model in terms of a χ^2 value, we may directly compare these χ^2's with those obtained from the estimated $m_{ij,i'j'}$'s (see Figs. 3.5, 3.6, and 3.7). For the Mixed Group the minimum χ^2 is 47.2, compared with 99.3 for the $m_{ij,i'j'}$ procedure. For the Pure Group the minimum χ^2 is 85.6, compared with 86.1. Here the two estimates of θ are very close (.64 and .653). For the Sure Group the minimum χ^2 is 108.8, compared with 455.4.

The results of applying this approach to the two-element model are better for the noncontingent schedule reported in Sec. 10.6.

*3.8. Generalized Conditioning Model

In this section we consider the application of the model described in Sec. 2.6. The transition numbers n_{ij} for the 8×4 matrix (2.6.9) were tabulated for the last 100 trials and aggregated over subjects. Using (2.6.10), we obtain maximum-likelihood estimates of c_{ij} and d_{ij}, which are presented in Table 3.10.

TABLE 3.10

MAXIMUM-LIKELIHOOD ESTIMATES OF c_{ij} AND d_{ij}

Parameter	Mixed	Pure	Sure
c_{11}	.788	.790	.742
c_{21}	.667	.707	.639
c_{12}	.498	.508	.465
c_{22}	.378	.366	.421
d_{11}	.769	.720	.850
d_{21}	.691	.604	.725
d_{12}	.488	.505	.592
d_{22}	.399	.448	.477

Given these estimates of c_{ij} and d_{ij}, the χ^2 goodness-of-fit test (2.4.4) can be applied to matrix (2.6.9), where the observed n_{ij}'s are those on which the estimates are made (see Sec. 2.6 for a discussion of this particular application of the χ^2 test). For all three groups $a_2 = 1$, and thus transitions in row 4 of (2.6.9) are not possible. Therefore, there are 7×3 degrees of freedom for the matrix. But eight parameters have been estimated from these data; hence, the χ^2's are to be interpreted with $21 - 8 = 13$ degrees of freedom. The obtained values of χ^2 are 13.8 (Mixed Group), 19.3 (Pure Group), and 16.3 (Sure Group). None of these values are significant at the .10 level, indicating good over-all agreement between the predicted and observed transition data.

Given these estimates of the conditioning parameters, other predictions can be made. The transition probabilities $P(A_{i,n}B_{j,n} \mid A_{i',n-1}B_{j',n-1})$ for the generalized conditioning model may be derived in a straightforward fashion. The theoretical expressions for the zero-sum case are given in matrix (3.8.1). This matrix corresponds to (1.4.3) for the one-element stimulus model.

(3.8.1)

$$
\begin{array}{c|llll}
 & (1,1) & (1,2) & (2,1) & (2,2) \\
\hline
(1,1) & a_1c_{11}d_{12} & a_1c_{11}(1-d_{12}) & a_1(1-c_{11})d_{12} & a_1(1-c_{11})(1-d_{12}) \\
 & +(1-a_1)c_{12}d_{11} & +(1-a_1)c_{12}(1-d_{11}) & +(1-a_1)(1-c_{12})d_{11} & +(1-a_1)(1-c_{12})(1-d_{11}) \\
(1,2) & a_2c_{11}d_{21} & a_2c_{11}(1-d_{21}) & a_2(1-c_{11})d_{21} & a_2(1-c_{11})(1-d_{21}) \\
 & +(1-a_2)c_{12}d_{22} & +(1-a_2)c_{12}(1-d_{22}) & +(1-a_2)(1-c_{12})d_{22} & +(1-a_2)(1-c_{12})(1-d_{22}) \\
(2,1) & a_3c_{22}d_{12} & a_3c_{22}(1-d_{12}) & a_3(1-c_{22})d_{12} & a_3(1-c_{22})(1-d_{12}) \\
 & +(1-a_3)c_{21}d_{11} & +(1-a_3)c_{21}(1-d_{11}) & +(1-a_3)(1-c_{21})d_{11} & +(1-a_3)(1-c_{21})(1-d_{11}) \\
(2,2) & a_4c_{22}d_{21} & a_4c_{22}(1-d_{21}) & a_4(1-c_{22})d_{21} & a_4(1-c_{22})(1-d_{21}) \\
 & +(1-a_4)c_{21}d_{22} & +(1-a_4)c_{21}(1-d_{22}) & +(1-a_4)(1-c_{21})d_{22} & +(1-a_4)(1-c_{21})(1-d_{22})
\end{array}
$$

Table 3.11 presents the predicted transition probabilities based on the maximum-likelihood estimates of c_{ij} and d_{ij} and also the observed proportions computed over the last 100 trials. As would be expected from the results of the χ^2 tests reported above, the correspondence between theoretical and observed values is extremely good.

TABLE 3.11

PREDICTED TRANSITION MATRICES FOR THE GENERALIZED CONDITIONING MODEL

(For purposes of comparison the observed probabilities, originally presented in Table 3.3, are given in parentheses)

	Mixed Group				Pure Group				Sure Group			
n \ $n+1$	(1, 1)	(1, 2)	(2, 1)	(2, 2)	(1, 1)	(1, 2)	(2, 1)	(2, 2)	(1, 1)	(1, 2)	(2, 1)	(2, 2)
(1, 1)	.38 (.37)	.22 (.23)	.29 (.28)	.11 (.12)	.38 (.38)	.27 (.26)	.23 (.25)	.12 (.11)	.42 (.43)	.19 (.19)	.30 (.30)	.09 (.08)
(1, 2)	.54 (.54)	.24 (.25)	.15 (.15)	.07 (.06)	.48 (.50)	.31 (.31)	.13 (.11)	.08 (.08)	.54 (.52)	.20 (.19)	.19 (.22)	.07 (.07)
(2, 1)	.35 (.36)	.17 (.16)	.28 (.30)	.20 (.18)	.34 (.30)	.19 (.20)	.27 (.28)	.20 (.22)	.47 (.47)	.11 (.12)	.32 (.31)	.10 (.10)
(2, 2)	.27 (.28)	.35 (.35)	.18 (.17)	.20 (.20)	.29 (.32)	.34 (.36)	.19 (.15)	.18 (.17)	.30 (.28)	.17 (.16)	.36 (.37)	.17 (.19)

Given these theoretical transition matrices, we may compute u_i, α, and β for each of the groups by (1.3.17), (1.4.10), and (1.4.11). The results of these computations are presented in Table 3.12. A comparison of Table 3.12 with Table 3.2 and Fig. 3.4 indicates good agreement between predicted and observed asymptotic quantities.

On the basis of the results just reported, several comments may be made about the generalized conditioning model. Referring to Table 3.10, we note first that the expected monotonicity relations hold in each of the three columns; namely, an E_1 reinforcement and an A_1 response are more effective in con-

TABLE 3.12

PREDICTED ASYMPTOTIC QUANTITIES FOR THE GENERALIZED CONDITIONING MODEL

Quantity	Mixed	Pure	Sure
u_1	.40	.39	.44
u_2	.23	.27	.17
u_3	.24	.20	.29
u_4	.13	.14	.10
α	.62	.66	.61
β	.64	.59	.73

ditioning the stimulus to the A_1 response than are an E_1 and an A_2, which in turn are more effective than E_2 and A_1, which in turn are more effective than E_2 and A_2. In other words, we have

$$c_{11} > c_{21} > c_{12} > c_{22}\,, \qquad d_{11} > d_{21} > d_{12} > d_{22}\,.$$

The fact that $c_{21} > c_{12}$ and $d_{21} > d_{12}$ indicates that the actual response made on a trial does not have as strong a self-reinforcing effect as the experimenter-determined reinforcement E_i has. As should be evident from preceding discussions, it does not necessarily follow from our fundamental theory that the actual response has no effect on the state of conditioning at the beginning of the next trial. In the one-element and multi-element models

$$P(A_{1,n+1} \mid E_{1,n}A_{1,n}) > P(A_{1,n+1} \mid E_{1,n}A_{2,n})$$

if the latter probability is not 1. The generalized conditioning model does not have this inequality as a consequence of its assumptions; and it is, therefore, pleasing to find it empirically confirmed in Table 3.10. One of the main reasons for analyzing our data from the standpoint of the generalized model was to provide quantitative information of this sort.

In setting forth the theory of the generalized conditioning model in Sec. 2.6, we mentioned that we would be concerned to see if the conditioning of the stimulus is more affected by the occurrence on the preceding trial of the response or reinforcement that has on the average occurred more often. Table 3.2 shows that on the average both players in all groups made more A_1 and B_1 responses than A_2 and B_2; and Table 3.1 shows that this average effect almost holds for individual subjects. Also, on the last 100 trials the average probability of an E_1 reinforcement for Player A was .60, .67, and .64 in the Mixed, Pure, and Sure Groups, respectively; the average probability of an E_1 reinforcement for Player B was .63, .59, and .67 in the same groups, respectively. The effects of the more numerous A_1 and B_1 responses and E_1 reinforcements are clearly indicated in Table 3.10, because $c_{11} > 1 - c_{22}$ and $d_{11} > 1 - d_{22}$ for all three groups. In fact, the values of $1 - c_{22}$ and $1 - d_{22}$ in the Pure and Sure Groups are surprisingly small. The values of c_{22} and d_{22} represent the most striking deviation of the generalized conditioning model from the one-element stimulus-sampling model, in which, as stated in (2.6.2), $c_{22} = d_{22} = 0$.

The remarkably good results on the goodness-of-fit tests are worth emphasizing for, as noted earlier, this test is extremely sensitive when the number of observations is large.[4] Readers who feel that it is a simple matter to propound a theory with eight parameters that will fit an 8×4 stochastic matrix may find it interesting to try fitting an arbitrary stochastic matrix of this size with eight parameters. They will soon agree, we believe, that the close fit reported here is not simply an artifact of the model.

We now turn to the case $c_{ij} = d_{ij}$ and use (2.6.11) to estimate the conditioning parameters. The $\hat{c}_{ij}$'s are given in Table 3.13. The same χ^2 tests reported above were run for predicted values based on the $\hat{c}_{ij}$'s given in the table. However, in this case only four parameters are estimated from the data; consequently, the χ^2's are to be interpreted with 17 degrees of freedom. The obtained values of χ^2 are 16.0 (Mixed Group), 43.0 (Pure Group), and 72.8 (Sure Group). Except for the Mixed Group the χ^2 are highly significant and thus indicate substantial disagreement between predicted and observed quantities.

TABLE 3.13

MAXIMUM-LIKELIHOOD ESTIMATES OF $c_{ij} = d_{ij}$

Parameter	Mixed	Pure	Sure
c_{11}	.778	.763	.798
c_{21}	.680	.645	.682
c_{12}	.493	.506	.540
c_{22}	.387	.400	.428

The values of c_{ij} given in Table 3.13 have all the qualitative features mentioned in the discussion of Table 3.10. Also, careful inspection indicates that the quantitative values are close to those for c_{ij} and d_{ij} in the earlier table. What is disappointing is the relatively poor goodness-of-fit results in two of the three groups. It bears testimony to the sensitivity of the χ^2 goodness-of-fit test when the number of observations is at least 2000. One interpretation of the results is that the conditioning parameters are sufficiently affected by the detailed pattern of actual responses and reinforcements to require different parameters for A and B. On the other hand, results like the present ones for the completely symmetric situation of Chapter 5 show that this is not the whole story. We shall return to this point in later sections that deal with the generalized conditioning model.

[4] Because of these excellent results, we attempted to increase their significance by testing the first-order versus second-order Markov character of the observed data in the full response-reinforcement sequence. Unfortunately, with 64 columns for entries it turned out that many cells in the 64×8 matrix had theoretical expectations that were too small to make the test satisfactory.

3.9. Summary

In this chapter the mathematical models and methods developed in Chapters 1 and 2 are applied to data obtained in simple two-person interaction experiments. The experimental situation involved a series of 200 discrete trials. On a given trial each of the two players independently made one of two responses. The probability that a player was correct or incorrect in his response depended in part on the response of the other player. It was by virtue of this dependency that the game aspect of the experiment arose. The reinforcement schedule was specified by the following payoff matrix:

	B_1	B_2
A_1	a_1	a_2
A_2	a_3	a_4

With probability a_i Player A was correct and Player B incorrect, given the occurrence of the corresponding pair of responses; with probability $1 - a_i$ the situation was reversed. The game was zero-sum in that exactly one player was correct on each trial.

The interaction of the two players was limited in at least three ways: (1) neither subject was shown the payoff matrix; (2) neither subject was informed of the responses of the other; (3) subjects were in fact told that they were not interacting at all, but were being run in pairs only as a matter of experimental convenience.

Three experiments are reported in this chapter, each using a different set of a_i values; the selection of these values was dictated on the basis of game-theory considerations. Each group had available either a sure-thing strategy, a pure minimax strategy, or a mixed minimax strategy. The details of the experimental method are reported in Sec. 3.2. Mean learning curves and asymptotic results are presented in Sec. 3.3; the predictions of asymptotic response probabilities are well confirmed. Tests for stationarity and order of the Markov process, independent of a particular model, are given in Sec. 3.4. For two of the three groups the χ^2 test for second-order versus first-order Markov chain is not significant. Estimates of θ and predicted asymptotic variances are presented in Sec. 3.5; the maximum-likelihood estimates of θ are very close to .6. In Sec. 3.6 the observed data are compared with predictions derived from game theory. The data fail to support the hypothesis that players approach an optimal strategy, and various proposals are discussed for a *rapprochement* of game theory and learning theory. Multi-element stimulus-sampling models are discussed in Sec. 3.7. The goodness-of-fit tests for the two-element model are highly significant for all three groups. Finally, in Sec. 3.8 the data are analyzed in terms of the generalized conditioning model. Using maximum-likelihood estimates of the conditioning parameters, we find that the fit of the model to the data over the last 100 trials is extremely good.

4

Some Non-Zero-Sum, Two Person Situations

4.1. Introduction

The experiments in this chapter differ from those of the preceding chapter in two important respects. In the first place, the third experimental restriction mentioned at the beginning of Sec. 3.1 was removed. Subjects were told that they were interacting with each other and that the correctness of their prediction depended in part on the responses of the other person. They were in fact advised to think of the situation as a game of a rather complicated character between two players. In the second place, the reinforcement schedule was no longer zero-sum, but was such that on many trials both players could be correct in their responses or both incorrect. The payoff matrix (1.4.1) was replaced by

$$
\text{(4.1.1)} \qquad \begin{array}{c|cc} & B_1 & B_2 \\ \hline A_1 & (a_1, b_1, c_1, d_1) & (a_2, b_2, c_2, d_2) \\ A_2 & (a_3, b_3, c_3, d_3) & (a_4, b_4, c_4, d_4) \end{array}
$$

where a_i is the probability that both players are correct, b_i the probability that A is correct and B incorrect, c_i the probability that A is incorrect and B correct, and d_i the probability that both players are incorrect. Thus $a_i + b_i + c_i + d_i = 1$, for $i = 1, \cdots, 4$. In Sec. 4.7 we consider a related matrix that is more suitable for analyzing the situation as a non-zero-sum game. Optimal game-theory strategies for the parameters selected are discussed there.

Two groups of subjects were run. On the basis of the game-theory considerations to be set forth in Sec. 4.7, the groups were designated *Sure* and *Mixed*. Parameter values for the Sure Group were:

$$
\begin{array}{llll}
a_1 = 1/4, & a_2 = 3/4, & a_3 = 5/32, & a_4 = 5/16, \\
b_1 = 3/4, & b_2 = 0, & b_3 = 3/32, & b_4 = 3/16, \\
c_1 = 0, & c_2 = 1/4, & c_3 = 15/32, & c_4 = 5/16, \\
d_1 = 0, & d_2 = 0, & d_3 = 9/32, & d_4 = 3/16.
\end{array}
$$

For the Mixed Group they were

$$
\begin{array}{llll}
a_1 = 0, & a_2 = 3/8, & a_3 = 0, & a_4 = 15/64, \\
b_1 = 1, & b_2 = 0, & b_3 = 0, & b_4 = 25/64, \\
c_1 = 0, & c_2 = 5/8, & c_3 = 5/8, & c_4 = 9/64, \\
d_1 = 0, & d_2 = 0, & d_3 = 3/8, & d_4 = 15/64.
\end{array}
$$

The experimental data for these two groups are reported in Atkinson and Suppes [1959], but are here for the first time analyzed by the methods described in Chapter 2. The experiments reported in Chapters 5–11 have not been previously published in any form.

4.2. Model

If we restrict ourselves to the assumption of one stimulus element for each player, it is clear that the stipulation of the payoff matrix in (4.1.1) determines the derivation of the Markov process from the axioms of Sec. 1.2. On each trial the single stimulus element of Player A is conditioned to response A_1 or A_2, and the element of Player B to response B_1 or B_2. Therefore, as was the case for the zero-sum games, there are four possible states of conditioning that will be represented by the ordered pairs (1, 1), (1, 2), (2, 1), (2, 2). The first member

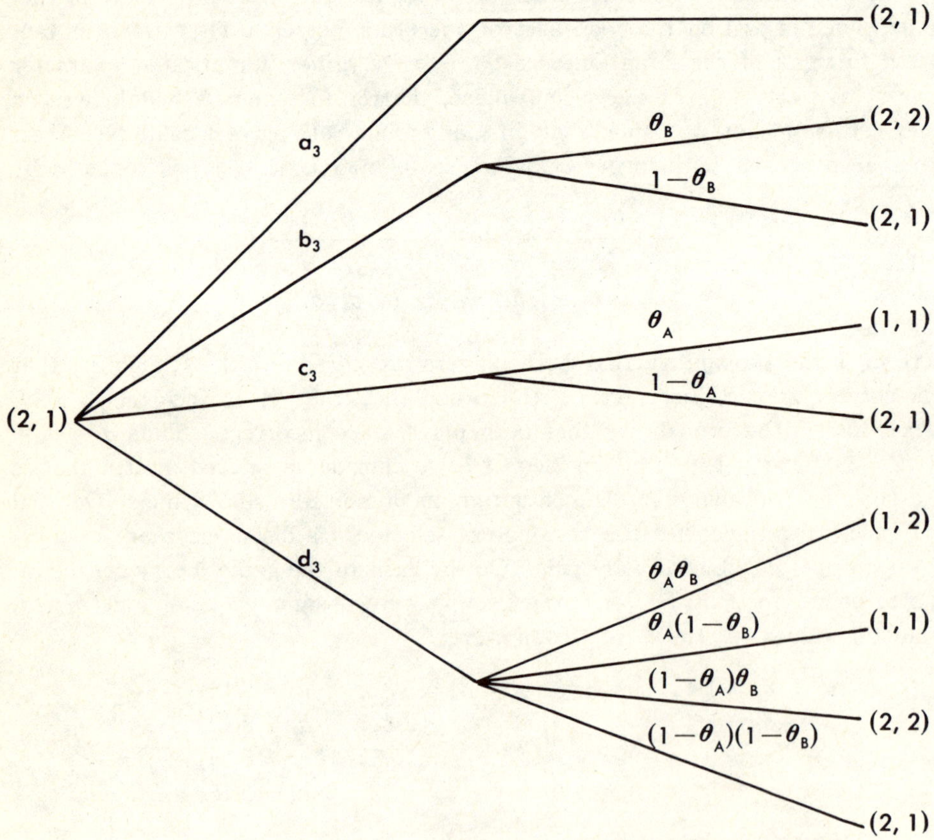

FIGURE 4.1

of each pair indicates the conditioning of Player A's stimulus element, and the second member the conditioning of Player B's element.

In view of the detailed derivations given in Chapter 1, we shall derive only one of the four trees for the Markov process, the tree starting from state (2, 1) at the start of trial n, which is given in Fig. 4.1. As noted in Sec. 1.3, the effectiveness of conditioning is irrelevant when the response made is reinforced. Consequently, on the upper branch of the tree θ_A and θ_B are omitted, for a_3 denotes the probability that both players are correct, given the response pair (A_2B_1). Similarly, θ_A is omitted from the second main branch, for with probability b_3 Player A is correct and Player B is incorrect; and so on.

Figure 4.1 and the three analogous trees yield the following transition matrix:

(4.2.1)

	(1, 1)	(1, 2)	(2, 1)	(2, 2)
(1,1)	$a_1+b_1(1-\theta_B)$ $+c_1(1-\theta_A)$ $+d_1(1-\theta_A)(1-\theta_B)$	$b_1\theta_B$ $+d_1\theta_B(1-\theta_A)$	$c_1\theta_A$ $+d_1\theta_A(1-\theta_B)$	$d_1\theta_A\theta_B$
(1,2)	$b_2\theta_B$ $+d_2\theta_B(1-\theta_A)$	$a_2+b_2(1-\theta_B)$ $+c_2(1-\theta_A)$ $+d_2(1-\theta_A)(1-\theta_B)$	$d_2\theta_A\theta_B$	$c_2\theta_A$ $+d_2\theta_A(1-\theta_B)$
(2,1)	$c_3\theta_A$ $+d_3\theta_A(1-\theta_B)$	$d_3\theta_A\theta_B$	$a_3+b_3(1-\theta_B)$ $+c_3(1-\theta_A)$ $+d_3(1-\theta_A)(1-\theta_B)$	$b_3\theta_B$ $+d_3\theta_B(1-\theta_A)$
(2,2)	$d_4\theta_A\theta_B$	$c_4\theta_A$ $+ d_4\theta_A(1-\theta_B)$	$b_4\theta_B$ $+d_4\theta_B(1-\theta_A)$	$a_4+b_4(1-\theta_B)$ $+c_4(1-\theta_A)$ $+d_4(1-\theta_A)(1-\theta_B)$

Again, to simplify notation the states will be numbered as follows: $1 = (1, 1)$, $2 = (1, 2)$, $3 = (2, 1)$, and $4 = (2, 2)$. The transition matrix in (4.2.1) will be denoted $[p_{ij}]$, for $i, j = 1, 2, 3, 4$.

The general expressions for u_i $(i = 1, 2, 3, 4)$, α, and β are too lengthy to reproduce here, but special forms of them will be presented later.

At the moment, it is advantageous to turn from the one-element model to the general case, in which N_A and N_B elements are posited for players A and B, respectively. Such an analysis permits us to establish several general results concerning the particular experimental parameter values used in this chapter.

As in (1.2.9) we define $\alpha_n = P(A_{1,n})$ and $\beta_n = P(B_{1,n})$. Then, in terms of the axioms of Sec. 1.2, for the recursion of α_n

(4.2.2)
$$\begin{aligned}\alpha_{n+1} = \alpha_n(1-\theta_A)+ \theta_A\{&P(A_{1,n}B_{1,n})[(a_1+b_1)\alpha_n + (c_1+d_1)(\alpha_n - 1/N_A)]\\ &+ P(A_{1,n}B_{2,n})[(a_2+b_2)\alpha_n+ (c_2+d_2)(\alpha_n - 1/N_A)]\\ &+ P(A_{2,n}B_{1,n})[(a_3+b_3)\alpha_n + (c_3+d_3)(\alpha_n + 1/N_A)]\\ &+ P(A_{2,n}B_{2,n})[(a_4+b_4)\alpha_n + (c_4+d_4)(\alpha_n + 1/N_A)]\}\,.\end{aligned}$$

In (1.2.9), $\gamma_n = P(A_{1,n}B_{1,n})$. Further, $P(A_{1,n}B_{1,n}) + P(A_{1,n}B_{2,n}) = \alpha_n$ and $P(A_{1,n}B_{1,n}) + P(A_{2,n}B_{1,n}) = \beta_n$; therefore, $P(A_{1,n}B_{2,n}) = \alpha_n - \gamma_n$, $P(A_{2,n}B_{1,n}) = \beta_n - \gamma_n$, and $P(A_{2,n}B_{2,n}) = 1 - \alpha_n - \beta_n + \gamma_n$. Substituting these results in

(4.2.2), we obtain

$$\text{(4.2.3)} \quad \alpha_{n+1} = \left[1 - \frac{\theta_A}{N_A}(c_2 + d_2 + c_4 + d_4)\right]\alpha_n + \frac{\theta_A}{N_A}(c_3 + d_3 - c_4 - d_4)\beta_n$$
$$+ \frac{\theta_A}{N_A}[(c_2 + d_2) + (c_4 + d_4) - (c_1 + d_1) - (c_3 + d_3)]\gamma_n + \frac{\theta_A}{N_A}(c_4 + d_4)\,.$$

Similarly, for the recursion of β_n

$$\text{(4.2.4)} \quad \beta_{n+1} = \left[1 - \frac{\theta_B}{N_B}(b_3 + d_3 + b_4 + d_4)\right]\beta_n + \frac{\theta_B}{N_B}(b_2 + d_2 - b_4 - d_4)\alpha_n$$
$$+ \frac{\theta_B}{N_B}[(b_3 + d_3) + (b_4 + d_4) - (b_1 + d_1) - (b_2 + d_2)]\gamma_n + \frac{\theta_B}{N_B}(b_4 + d_4)\,.$$

From these recursions results can be established for the non-zero-sum game similar to those presented in (3.7.8–9) for the zero-sum game. However, in this chapter we shall be interested only in the special case in which the coefficient of γ_n vanishes in both (4.2.3) and (4.2.4). To this end we require the following restrictions on our experimental parameters:

$$\text{(4.2.5)} \quad \begin{aligned} c_2 + d_2 + c_4 + d_4 &= c_1 + d_1 + c_3 + d_3\,, \\ b_1 + d_1 + b_2 + d_2 &= b_3 + d_3 + b_4 + d_4\,. \end{aligned}$$

These restrictions apply to the experimental parameters for the stuay reported in this chapter. With them in mind, we now rewrite (4.2.3) and (4.2.4) as

$$\text{(4.2.6)} \quad \begin{aligned} \alpha_{n+1} &= (1 - a\varphi)\alpha_n - b\varphi\beta_n + c\varphi\,, \\ \beta_{n+1} &= (1 - a'\varphi')\beta_n - b'\varphi'\alpha_n + c'\varphi'\,, \end{aligned}$$

where $\varphi = \theta_A/N_A$, $\varphi' = \theta_B/N_B$, $a = c_2 + d_2 + c_4 + d_4$, $b = c_4 + d_4 - c_3 - d_3$, and so on.

The solution of (4.2.6) is discussed in Sec. 3.7 for the zero-sum case. For the non-zero-sum game additional complications arise for very special sets of parameter values. Since these special cases do not occur in our experimental work, we shall not consider them here; for a discussion of them the reader is referred to Burke's article mentioned in Sec. 3.7. In general, however, the asymptotes for α_n and β_n exist and are independent of both the initial conditions and the parameters θ_A, θ_B, N_A, and N_B. Further, the solutions of (4.2.6) are given by (3.7.12–18). Employing 3.7.12, we obtain the following asymptotic results for an A_1 and a B_1 response, respectively:

(4.2.7)

$$\alpha = \frac{(b_3 + d_3)(c_4 + d_4) + (c_3 + d_3)(b_4 + d_4)}{(c_2 + d_2 + c_4 + d_4)(b_3 + d_3 + b_4 + d_4) - (a_4 + b_4 - a_3 - b_3)(a_4 + c_4 - a_2 - c_2)}\,,$$

$$\beta = \frac{(c_2 + d_2)(b_4 + d_4) + (c_4 + d_4)(b_2 + d_2)}{(c_2 + d_2 + c_4 + d_4)(b_3 + d_3 + b_4 + d_4) - (a_4 + b_4 - a_3 - b_3)(a_4 + c_4 - a_2 - c_2)}\,.$$

The fact that α and β are independent of θ_A, θ_B, N_A, and N_B under the restrictions of (4.2.5), in no way implies that γ_n is also independent of these quantities. As an example, u_1 for the one-element model presented in this section is not independent of θ_A and θ_B; nor for that matter is it independent of $\theta = \theta_A = \theta_B$. This can be seen by inspecting the linear equations (1.3.17) for the transition matrix (4.2.1).

4.3. Experimental Method

Subjects. The subjects were 88 undergraduates obtained from introductory courses in psychology at Stanford University. They were randomly assigned to the two groups, with the restriction that there were 24 pairs of subjects in the Sure Group and 20 in the Mixed Group.

Apparatus. The apparatus described in Chapter 3 was used.

Procedure. For each pair of subjects, one was randomly selected as Player A and the other as Player B. Further, for each subject one of the two response keys was randomly designated Response 1 and the other Response 2, with the restriction that the following four possible combinations occurred equally often in each of the experimental groups: (1) A_1 and B_1 on the right, (2) A_1 on the right and B_1 on the left, (3) A_1 on the left and B_1 on the right, and (4) A_1 and B_1 on the left.

When the subjects had been seated, they were read the following instructions:

"This experiment is analogous to a real-life situation where what you gain or lose depends not only on what you do but also on what someone else does. In fact, you should think of the situation as a game involving you and another player, the person at the other end of the table.

"The experiment for each of you consists of a series of trials. The top center lamp on your panel will light for about two seconds to indicate the start of each trial. Shortly thereafter one or the other of the two lower lamps will light up. Your job is to predict on each trial which one of the two lower lamps will light and to indicate your prediction by pressing the proper key. That is, if you expect the left lamp to light, press the left key; if you expect the right lamp to light, press the right key. On each trial press one or the other of the two keys, but never both. If you are not sure which key to press, then guess.

"Be sure to indicate your choice by pressing the proper key immediately after the onset of the signal light. That is, when the signal light goes on, press one or the other key down and release it. Then wait until one of the lower lights goes on. If the light above the key you pressed goes on, your prediction was correct; if the light above the key opposite from the one you pressed goes on, you were incorrect.

"Being correct or incorrect on a given trial depends on the key you press and also on the key the other player presses. With some combinations of your key choice with the other player's key choice, you may both be correct; with other combinations one player will be correct and the other incorrect; for still other combinations you may both be incorrect.

"As you have probably already guessed, the situation is fairly complicated. The object of the experiment is to see how many correct predictions you can get over a series of trials."

Questions were answered by paraphrasing the appropriate part of the instructions.

After the instructions were read, 210 trials were run in continuous sequence. For each pair of subjects, sequences of reinforcing lights were generated in accordance with assigned values of (a_i, b_i, c_i, d_i) and observed responses.

On all trials the signal light was lighted for 3.5 seconds; the time between successive signal exposures was 10 seconds. The reinforcing light followed the cessation of the signal light by 1.5 seconds and remained on for 2 seconds.

4.4. Mean Learning Curves and Asymptotic Results

Figure 4.2 presents the observed mean proportions of A_1 and B_1 responses in successive 30-trial blocks for the entire sequence of 210 trials. An inspection of this figure indicates that response proportions are very stable

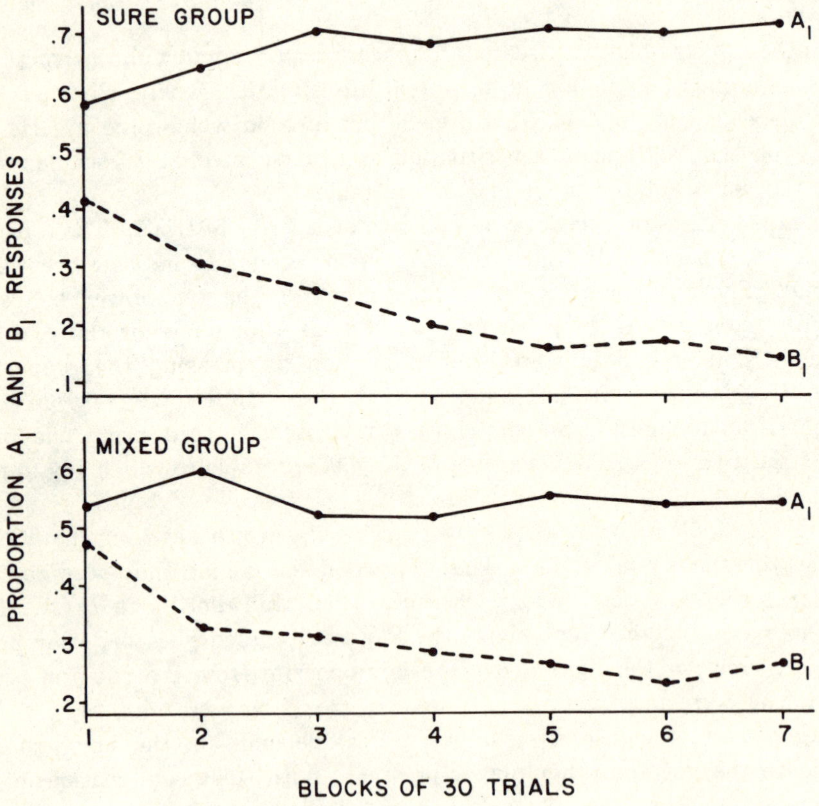

FIGURE 4.2. Observed mean proportions of A_1 and B_1 responses in successive 30-trial blocks.

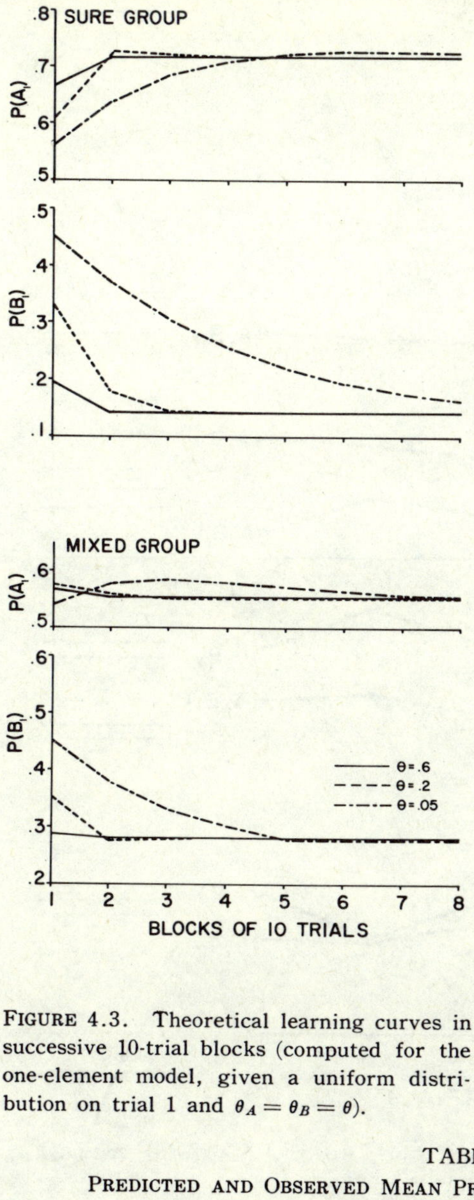

FIGURE 4.3. Theoretical learning curves in successive 10-trial blocks (computed for the one-element model, given a uniform distribution on trial 1 and $\theta_A = \theta_B = \theta$).

over the final block of 90 trials. Consequently, the proportions computed over the last 90 trials are used as estimates of the asymptotic probabilities of A_1 and B_1 responses. Table 4.1 presents the observed mean proportions in the final 90-trial block and the observed standard deviations. Entries for the Sure Group are based on $N = 24$; for the Mixed Group, $N = 20$. The values predicted by the model as given in (4.2.7) are also presented in the table. It should be emphasized that the predictions in Table 4.1 are based solely on experimentally assigned values of a_i, b_i, c_i, and d_i. Thus they are entirely *a priori* and do not make use of any estimated parameters. Consequently, differences between subjects, which can be represented by inter-subject variability in the number of stimulus elements and the conditioning parameter, do not affect these asymptotic quantities. An inspection of Table 4.1 indicates remarkably good agreement between observed and predicted values.

Figure 4.3 presents theoretical learning functions for the one-element model when $\theta_A = \theta_B = \theta$. The curves give the proportions of A_1 and B_1 responses in successive 10-trial blocks for initial values of $u_1^{(1)} = u_2^{(1)} = u_3^{(1)} = u_4^{(1)} = \frac{1}{4}$. For all groups the curves are virtually

TABLE 4.1

PREDICTED AND OBSERVED MEAN PROPORTIONS OF A_1 AND B_1 RESPONSES OVER THE LAST 90 TRIALS

Group	A_1			B_1		
	Predicted	Observed	SD	Predicted	Observed	SD
Sure	.71	.72	.090	.14	.16	.099
Mixed	.55	.55	.056	.28	.25	.079

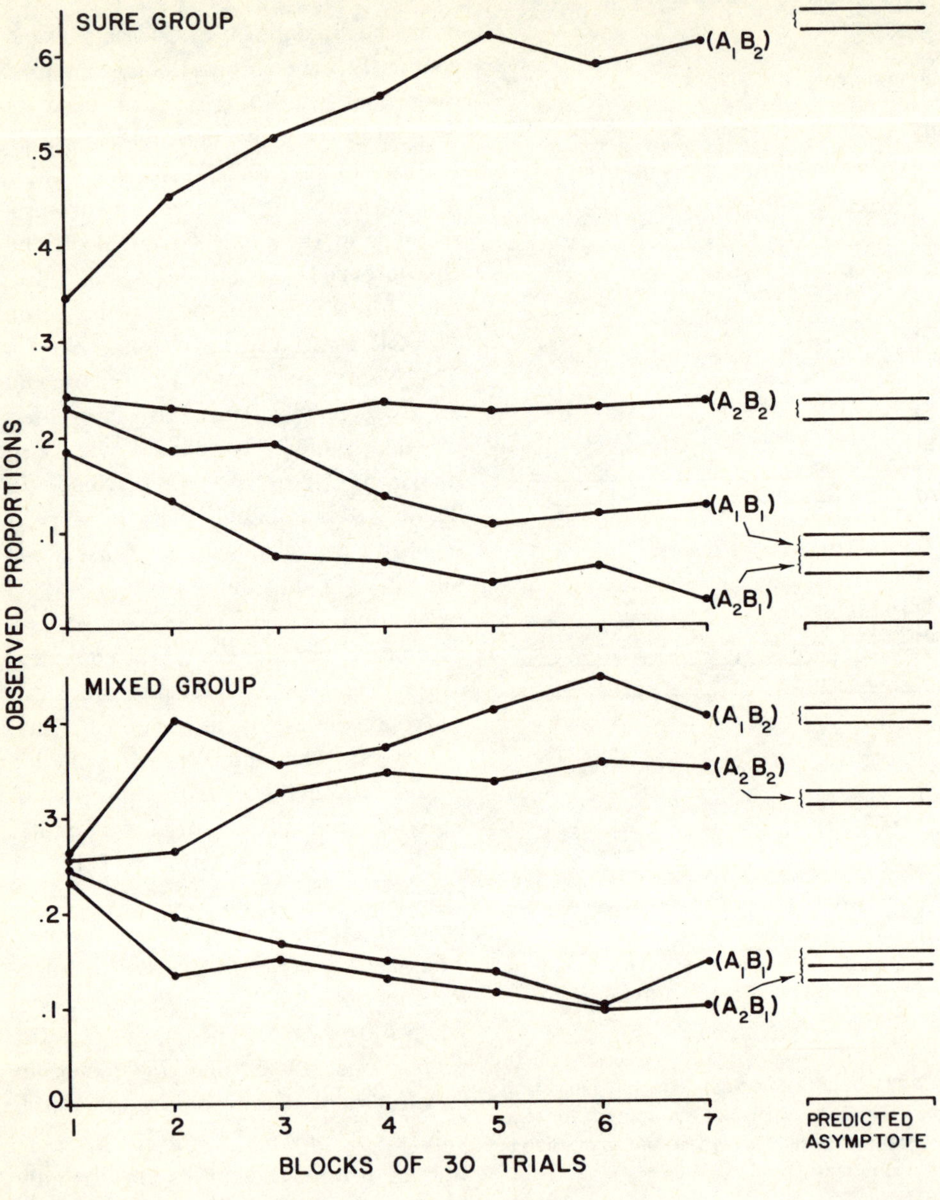

FIGURE 4.4. Observed proportions of the response pairs (A_1B_1), (A_1B_2), (A_2B_1), and (A_2B_2) in successive 30–trial blocks and predicted asymptotic values.

at asymptote within the first or second block of trials for $\theta = .2$ and $\theta = .6$. The curves for $\theta = .05$ compare more favorably with the observed data in Fig. 4.2 when appropriately scaled. However, as was the case in Chapter 3, the theoretical functions for $\theta = .05$ approach the asymptote slightly more rapidly than do the observed curves. Thus a value of θ that would provide a close fit to this aspect of the data would be below .05.

Figure 4.4 presents the observed mean proportions of the response pairs (A_1B_1), (A_1B_2), (A_2B_1), and (A_2B_2) in successive blocks of 30 trials. As remarked in Sec. 4.2, theoretical expressions for these quantities are a function of the conditioning parameters and the number of stimulus elements, and appear to have no simple form. Consequently, we only present the asymptotic predictions for these quantities in the one-element model when $\theta_A = \theta_B = \theta$. They are, of course, u_1, u_2, u_3, and u_4 as defined by (1.3.17) and (4.2.1). Specifically, for the Sure Group they are

$$(4.4.1)\qquad \begin{aligned} &.071 < u_1 = \frac{8-\theta}{7(16-5\theta)} \leqq .091\,, &&.052 \leqq u_3 = \frac{4(2-\theta)}{7(16-5\theta)} < .071\,,\\ &.623 \leqq u_2 = \frac{24(3-\theta)}{7(16-5\theta)} < .643\,, &&.214 < u_4 = \frac{6(4-\theta)}{7(16-5\theta)} \leqq .234\,; \end{aligned}$$

and for the Mixed Group they are

$$(4.4.2)\qquad \begin{aligned} &.140 < u_1 = \frac{25(64-15\theta)}{89(128-39\theta)} \leqq .154\,, &&.126 \leqq u_3 = \frac{100(16-3\theta)}{89(128-39\theta)} < .140\,,\\ &.396 \leqq u_2 = \frac{64(73-24\theta)}{89(128-39\theta)} < .410\,, &&.309 < u_4 = \frac{320(11-3\theta)}{89(128-39\theta)} \leqq .323\,. \end{aligned}$$

The theoretical bounds on the u_j's are indicated on the far right of Fig. 4.4. An inspection of the figure indicates fairly good agreement between predicted asymptotes and observed values in the last few blocks of trials. The narrowness of these bounds means that the theory is in effect predicting the u_i to very close approximation, independent of the estimate of the conditioning parameter θ. The maximum-likelihood estimate $\hat{\theta}$ given in the next section can be used to yield an exact numerical prediction.

4.5. Estimates of θ_A and θ_B

Explicit expressions can be obtained for $\partial L(\theta_A, \theta_B)/\partial\theta_A$ and $\partial L(\theta_A, \theta_B)/\partial\theta_B$ in the non-zero-sum case. Setting these expressions equal to 0 and simultaneously solving for θ_A and θ_B yields maximum-likelihood estimates of the conditioning parameters. Unfortunately, the equations are cumbersome to work with; we have accordingly resorted to computing the function

$$(4.5.1)\qquad L(\theta_A, \theta_B) = \sum_{i,j} n_{ij} \log p_{ij}(\theta_A, \theta_B)$$

for an array of values on θ_A and θ_B and then using the results to approximate its maximum.

All computations were carried out on n_{ij}'s for the last 100 trials aggregated over subject-pairs. Figure 4.5 presents $L(\theta_A, \theta_B)$ for the special case, in which $\theta_A = \theta_B = \theta$. Computations were carried out for $\theta = .01i$ $(i = 1, \cdots, 100)$. As can be seen from the figure, the maximum value of the function occurred at $\theta = .91$ for the Sure Group and at $\theta = .72$ for the Mixed Group.

Similar computations were performed for various pairs of θ_A and θ_B. Specifi-

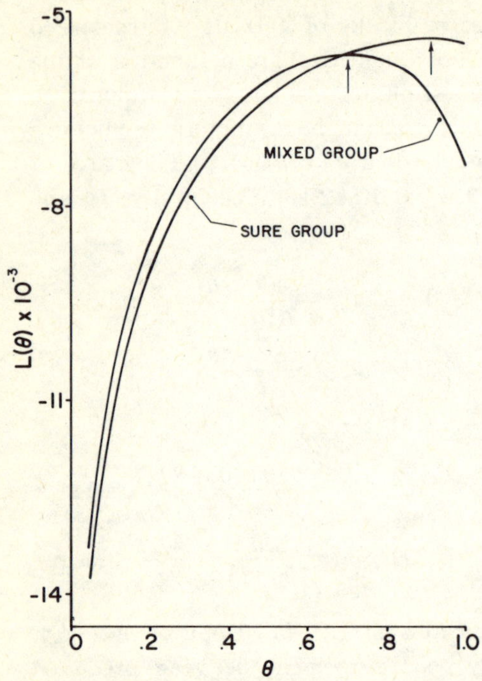

FIGURE 4.5. $L(\theta_A, \theta_B)$ for the special case $\theta_A = \theta_B = \theta$.

cally, $\theta_A = .05i$, $\theta_B = .05j$ $(i, j = 1, 2, \cdots, 20)$, and $L(\theta_A, \theta_B)$ were computed for all 400 combinations of θ_A and θ_B. The maximum of $L(\theta_A, \theta_B)$ was $-5{,}358.9$ for the Sure Group and occurred at $\theta_A = .80$, $\theta_B = .95$. For the Mixed Group the maximum was $-5{,}536.0$ and occurred at $\theta_A = .65$, $\theta_B = .90$.

It was noted in the preceding section that the observed mean learning curves are best fit by $\theta < .05$. This same disparity between the maximum-likelihood estimate $\hat{\theta}$ and the value of θ that fits the mean learning curve was observed in Chapter 3. As the data of subsequent chapters will indicate, this discrepancy recurs and is one of the most difficult problems to solve in any adequate fashion. As the recursions (4.2.3) and (4.2.4) indicate, what is estimated from the learning curves are the ratios θ_A/N_A and θ_B/N_B. If there were some reasonable way to compute the conditional probabilities $P(A_{i,n+1}B_{j,n} \mid A_{i',n}B_{j',n})$ in the N-element model, we could estimate θ and N by the pseudomaximum-likelihood methods of Sec. 2.3 and compare the ratio of these estimated values with that required for the learning curves. Unfortunately, however, numerical computation of these conditional probabilities for even a small spread of θ and N values is a staggering task, one that was far too great for the facilities available to us.

Moreover, we have some doubts whether we would solve the problem even if we were able to carry through this approach. The agreement of $\hat{\theta}$ estimates for the first and last halves of the trials, the particular shape of the learning curves in several experiments, and some other related facts suggest that the models we are considering fit the data much better after the first 40 or 50 trials. During the initial trials the organism may be rapidly sampling a large number of stimuli before settling on the few that are relevant to the experimental situation.

This type of hypothesis has been used for somewhat different purposes by Restle [1955*a*, *b*] and LaBerge [1959*a*, *b*]. Certain other hypotheses seem as likely to be correct on *a priori* grounds. But the experimental situation used for our studies is too complicated to pursue this problem in detail. We merely want to call attention to the ubiquitous character of this discrepancy. Results of the application of our estimation procedures to other studies are not available; but it is fairly certain, on the basis of our own results for noncontingent reinforcement schedules reported in Chapter 10, that this same discrepancy would appear in the Estes-Straughan [1954] data. Using the kind of linear model described in Sec. 1.5, Estes and Straughan obtained estimates for θ of the order of .05 from the mean learning curves. We are convinced that maximum-likelihood estimates for the one-element stimulus-sampling model, or pseudomaximum-likelihood estimates for the multi-element stimulus-sampling model and for the linear model, would have yielded radically different values for θ. From a statistical standpoint there is little doubt that the maximum-likelihood methods are better than the one based on the mean learning curve.

4.6. Tests for Markov Properties

Table 4.2 presents values for the observed transition matrix $[\hat{p}_{ij}]$ computed over the last 100 trials. The maximum-likelihood estimate $\hat{p}_{ij}$ was obtained as

TABLE 4.2
OBSERVED TRANSITION MATRICES CORRESPONDING TO THE THEORETICAL TRANSITION MATRIX
(Computed over the last 100 trials)

	Sure Group				Mixed Group			
n \ $n+1$	(1, 1)	(1, 2)	(2, 1)	(2, 2)	(1, 1)	(1, 2)	(2, 1)	(2, 2)
(1, 1)	.25	.55	.07*	.13*	.18	.53	.05*	.24*
(1, 2)	.07*	.70	.02*	.21	.07*	.45	.06*	.42
(2, 1)	.21	.43	.16	.20	.25	.35	.21	.19
(2, 2)	.14	.43	.07	.36	.12	.40	.14	.34

* These entries are theoretically 0.

follows: (1) Let $n_{ij}(t)$ denote the number of subject-pairs in state i on trial $t-1$ and in state j on trial n; and (2) let

$$n_{ij} = \sum_{t=110}^{210} n_{ij}(t) .$$

Then $\hat{p}_{ij} = n_{ij}/\sum_j n_{ij}$. No statistical test is required to see that some of the $\hat{p}_{ij}$'s differ from the theoretical values for the one-element model. It suffices to observe that in the theoretical matrix (4.2.1), for the set of experimental parameter values employed in both the Mixed and Sure Groups, the last two entries in row 1 and the first and third entries in row 2 are identically 0; in the observed matrices, entries in these cells are in some cases markedly different from 0.

The hypothesis that the transition probabilities are constant over trials was tested by computing (2.4.1) for $t = 1, 2$; $t = 1$ refers to the block of trials from 11 to 110, and $t = 2$ refers to the 111–210 block of trials. For both the Mixed and Sure Groups the obtained values of χ^2 were significant at the .05 level.

The same test was also run for the last 50-trial block and next-to-last 50-trial block. In this case neither χ^2 was significant at the .05 level, indicating that transition probabilities approached stationarity in the last 100 trials of the experiment.

The tests of hypotheses that the chain is of a given order deal only with transition data obtained over the final block of 100 trials. The zero-order versus first-order hypothesis was tested by calculating (2.4.2). The χ^2's for both groups were highly significant. The first-order versus second-order hypothesis was tested by computing the χ^2 of (2.4.3). The obtained values were 50.3 for the Sure Group and 47.6 for the Mixed Group. With 36 degrees of freedom neither result is significant at the .05 level. Thus in this particular set of data there is no statistically significant improvement in prediction if one knows the response history of the pair of subjects on the previous two trials as against only the preceding trial. Of course, the one-element model is a one-stage process, and this finding supports the model. However, as noted earlier, this assumption is not necessary for our general theoretical approach. For example, with more than one element for each player the observable sequence of response random variables is a chain of infinite order.

4.7. Game-Theory Comparisons

In Sec. 1.6 the basic classification of non-zero-sum games as either cooperative or noncooperative was described; it is apparent that from a game standpoint the experiments reported in this chapter were of the noncooperative type. It was also mentioned in Sec. 1.6 that there is at present no completely satisfactory theory of optimality for noncooperative non-zero-sum games, but that Nash's concept of an equilibrium point is probably the best one yet introduced. Recall that an equilibrium point is a set of strategies, one for each player, such that no player can do better in the game by changing his strategy if all other players remain at the equilibrium point.

Methods for computing an equilibrium point of even a very simple game with a 2×2 matrix were not given in Sec. 1.6. We shall briefly consider the problem at this point. (For a more extensive discussion the reader is referred to Nash [1950], [1951], or McKinsey [1952].) To begin with, the payoff matrices (4.1.1) are not suitable from a game standpoint. We need to have them in the form (a_{ij}, b_{ij}), where now a_{ij} is the probability of Player A's being correct independent of B, and b_{ij} is the probability of Player B's being correct independent of A. The numbers a_{ij} and b_{ij} $(i, j = 1, 2)$ are then the expected outcomes for A and B, respectively. In these terms the general 2×2 matrix has the form

(4.7.1)

	B_1	B_2
A_1	(a_{11}, b_{11})	(a_{12}, b_{12})
A_2	(a_{21}, b_{21})	(a_{22}, b_{22})

Now equilibrium-point strategies x^* and y^*, with x_i^* the probability that A chooses A_i and y_i^* the probability that B chooses B_i, have the property that for all strategies x

$$\sum_{i,j} a_{ij}x_i^* y_j^* \geqq \sum_{i,j} a_{ij}x_i y_j^* , \tag{4.7.2}$$

and for all strategies y

$$\sum_{i,j} b_{ij}x_i^* y_j^* \geqq \sum_{i,j} b_{ij}x_i^* y_j . \tag{4.7.3}$$

For only two responses available to each player let $x_1 = x$, $x_2 = 1 - x$, $y_1 = y$, and $y_2 = 1 - y$. Whether setting x^* or y^* equal to 0 or 1 yields an equilibrium-point strategy may be checked by direct inspection. If not, then to compute x^*, we may simplify (4.7.2) to

$$[a_{11}y^* + a_{12}(1 - y^*) - a_{21}y^* - a_{22}(1 - y^*)]x^* \geqq [a_{11}y^* + a_{12}(1 - y^*) - a_{21}y^* - a_{22}(1 - y^*)]x .$$

Now if the expression in the square brackets on both sides of this inequality is not equal to 0, it must be positive or negative. If it is positive, then $x^* = 1$ does not satisfy the hypothesis, for by supposition $x^* < 1$. Similarly, if the expression is negative, then $x^* = 0$ will not satisfy the inequality, for our supposition is also that $x^* > 0$. Thus we must have

$$a_{11}y^* + a_{12}(1 - y^*) - a_{21}y^* - a_{22}(1 - y^*) = 0 ,$$

whence

$$y^* = \frac{a_{12} - a_{22}}{a_{21} - a_{11} + a_{12} - a_{22}} . \tag{4.7.4}$$

By a similar argument, when $0 < y^* < 1$, we find from (4.7.3) that

$$x^* = \frac{b_{21} - b_{22}}{b_{12} - b_{11} + b_{21} - b_{22}} . \tag{4.7.5}$$

For the Sure Group the matrix (4.7.1) has the form

(4.7.6)

	B_1	B_2
A_1	$(1, \frac{1}{4})$	$(\frac{3}{4}, 1)$
A_2	$(\frac{1}{4}, \frac{5}{8})$	$(\frac{1}{2}, \frac{5}{8})$

Direct inspection indicates that the set of strategies $\{A_1, B_2\}$ is an equilibrium point of pure strategies. Moreover, A_1 and B_2 both satisfy the still stronger sure-thing principle, which accounts for the labeling of this group. As in the case of the Pure and Sure Groups of Chapter 3, the observed asymptotic mean response probabilities do not even roughly approximate the sure-thing strategies. Examination of individual responses over the last 90 trials yields $\max \alpha = .867$ and $\min \beta = .033$.

For the Mixed Group the game matrix is

(4.7.7)

	B_1	B_2
A_1	$(1, 0)$	$(\frac{3}{8}, 1)$
A_2	$(0, \frac{5}{8})$	$(\frac{5}{8}, \frac{3}{8})$

Applying (4.7.4) and (4.7.5), we obtain unique equilibrium-point strategies $x^* = \frac{1}{5}$ for A and $y^* = \frac{1}{5}$ for B. The results of t tests indicate that the observed asymptotic probability of an A_1 response differs significantly from x^* at beyond the .001 level, and the probability of a B_1 response differs significantly at the .02 level. The weaker result for B may be accounted for by the fact that the learning-theory prediction is $\beta = .28$, which is exceedingly close to .20; whereas $\alpha = .55$. Because of the closeness of β and the equilibrium-point strategy, we did not run the sign test used in Chapter 3 to see if either player was moving significantly toward the game-theory strategy relative to the other player.

TABLE 4.3
GAME-THEORY, LEARNING-THEORY, AND OBSERVED VALUES OF THE NON-ZERO-SUM GAMES

Type of Value	Sure Group		Mixed Group	
	A	B	A	B
Game Theory	.750	1.000	.500	.460
Learning Theory	.687	.827	.499	.618
Observed	.708	.808	.517	.597

The three values for the game are given in Table 4.3. The game-theory value is computed in terms of the equilibrium-point strategies. Because the games are non-zero-sum, we have computed separately the value for Player A and the value for Player B. The use of the term "value" in this sense for non-zero-sum games is not standard, but the intended definition is obvious: the game value for A is $\sum_{ij} x_i^* y_j^* a_{ij}$, and for B it is $\sum_{ij} x_i^* y_j^* b_{ij}$. The learning-theory values were based on $\hat{\theta} = .91$ for the Sure Group and $\hat{\theta} = .72$ for the Mixed Group. The observed values are computed on the basis of estimating the u_i from the last 30 trials. In the Sure Group for both Player A and Player B the observed values are decisively closer to the learning-theory values than to the game-theory values. This is also true for Player B in the Mixed Group. Because the game-theory and learning-theory values are essentially identical for Player A in the Mixed Group, no real comparison for this case is possible. In Chapter 3 we noted that the observed value in all cases was between the game and learning values, which might be interpreted as a slight trend to move toward the game-theory solution. In the present study this is true for Player A in the Sure Group and Player B in the Mixed Group, but not for Player B in the Sure Group (we again rule out Player A in the Mixed Group). When we refer to Player A or Player B in this discussion, we are of course referring to the central tendencies of the A or B subjects.

*4.8. Generalized Conditioning Model

In this section we apply the model discussed in Sec. 2.6 to the non-zero-sum case. The transition matrix corresponding to (2.6.10) for the zero-sum case is given in (4.8.1). The matrix presents $P(A_{i,n+1}B_{j,n+1} \mid A_{i',n}E_{k,n}A_{j',n}E_{k',n})$, where $E_{k,n}$ indicates the reinforcing event for Player A, and $E_{k',n}$ the reinforcing event for Player B. Note that the eight rows of (2.6.10) are replaced by sixteen in (4.8.1) because here the reinforcing events for the two players are statistically independent, given the responses of the players; whereas in the zero-sum case they are deterministically dependent.

(4.8.1)

	A_1B_1	A_1B_2	A_2B_1	A_2B_2
$A_1E_1B_1E_1$	$c_{11}d_{11}$	$c_{11}(1-d_{11})$	$(1-c_{11})d_{11}$	$(1-c_{11})(1-d_{11})$
$A_1E_1B_1E_2$	$c_{11}d_{12}$	$c_{11}(1-d_{12})$	$(1-c_{11})d_{12}$	$(1-c_{11})(1-d_{12})$
$A_1E_2B_1E_1$	$c_{12}d_{11}$	$c_{12}(1-d_{11})$	$(1-c_{12})d_{11}$	$(1-c_{12})(1-d_{11})$
$A_1E_2B_1E_2$	$c_{12}d_{12}$	$c_{12}(1-d_{12})$	$(1-c_{12})d_{12}$	$(1-c_{12})(1-d_{12})$
$A_1E_1B_2E_1$	$c_{11}d_{21}$	$c_{11}(1-d_{21})$	$(1-c_{11})d_{21}$	$(1-c_{11})(1-d_{21})$
$A_1E_1B_2E_2$	$c_{11}d_{22}$	$c_{11}(1-d_{22})$	$(1-c_{11})d_{22}$	$(1-c_{11})(1-d_{22})$
$A_1E_2B_2E_1$	$c_{12}d_{21}$	$c_{12}(1-d_{21})$	$(1-c_{12})d_{21}$	$(1-c_{12})(1-d_{21})$
$A_1E_2B_2E_2$	$c_{12}d_{22}$	$c_{12}(1-d_{22})$	$(1-c_{12})d_{22}$	$(1-c_{12})(1-d_{22})$
$A_2E_1B_1E_1$	$c_{21}d_{11}$	$c_{21}(1-d_{11})$	$(1-c_{21})d_{11}$	$(1-c_{21})(1-d_{11})$
$A_2E_1B_1E_2$	$c_{21}d_{12}$	$c_{21}(1-d_{12})$	$(1-c_{21})d_{12}$	$(1-c_{21})(1-d_{12})$
$A_2E_2B_1E_1$	$c_{22}d_{11}$	$c_{22}(1-d_{11})$	$(1-c_{22})d_{11}$	$(1-c_{22})(1-d_{11})$
$A_2E_2B_1E_2$	$c_{22}d_{12}$	$c_{22}(1-d_{12})$	$(1-c_{22})d_{12}$	$(1-c_{22})(1-d_{12})$
$A_2E_1B_2E_1$	$c_{21}d_{21}$	$c_{21}(1-d_{21})$	$(1-c_{21})d_{21}$	$(1-c_{21})(1-d_{21})$
$A_2E_1B_2E_2$	$c_{21}d_{22}$	$c_{21}(1-d_{22})$	$(1-c_{21})d_{22}$	$(1-c_{21})(1-d_{22})$
$A_2E_2B_2E_1$	$c_{22}d_{21}$	$c_{22}(1-d_{21})$	$(1-c_{22})d_{21}$	$(1-c_{22})(1-d_{21})$
$A_2E_2B_2E_2$	$c_{22}d_{22}$	$c_{22}(1-d_{22})$	$(1-c_{22})d_{22}$	$(1-c_{22})(1-d_{22})$

In terms of (4.8.1) maximum-likelihood estimates c_{ij} and d_{ij} can be obtained. Let the transition numbers n_{ij} be specified for the 16×4 matrix. Further, let $h_i = n_{i1} + n_{i2}$, $k_i = n_{i1} + n_{i3}$, and $t_i = n_{i1} + n_{i2} + n_{i3} + n_{i4}$. Then the maximum-likelihood estimates of c_{ij} and d_{ij} are

$$
\begin{aligned}
\hat{c}_{11} &= \frac{h_1 + h_2 + h_5 + h_6}{t_1 + t_2 + t_5 + t_6}, & \hat{d}_{11} &= \frac{k_1 + k_3 + k_9 + k_{11}}{t_1 + t_3 + t_9 + t_{11}}, \\
\hat{c}_{12} &= \frac{h_3 + h_4 + h_7 + h_8}{t_3 + t_4 + t_7 + t_8}, & \hat{d}_{12} &= \frac{k_2 + k_4 + k_{10} + k_{12}}{t_2 + t_4 + t_{10} + t_{12}}, \\
\hat{c}_{21} &= \frac{h_9 + h_{10} + h_{13} + h_{14}}{t_9 + t_{10} + t_{13} + t_{14}}, & \hat{d}_{21} &= \frac{k_5 + k_7 + k_{13} + k_{15}}{t_5 + t_7 + t_{13} + t_{15}}, \\
\hat{c}_{22} &= \frac{h_{11} + h_{12} + h_{15} + h_{16}}{t_{11} + t_{12} + t_{15} + t_{16}}, & \hat{d}_{22} &= \frac{k_6 + k_8 + k_{14} + k_{16}}{t_6 + t_8 + t_{14} + t_{16}}.
\end{aligned}
\tag{4.8.2}
$$

Similar estimates can be obtained for the case in which $c_{ij} = d_{ij}$. They are rather lengthy, but we present them here since they will be used again in later chapters. Define $e_i = n_{i3} + n_{i4}$, $f_i = n_{i2} + n_{i4}$, $g_i = n_{i2} + n_{i3} + n_{i4}$, and $s_i = n_{i2} + n_{i3}$. Further, let

$$\begin{aligned}
x_1 &= 2n_{11} + h_2 + h_5 + h_6 + k_3 + k_9 + k_{11}\,, \\
x_2 &= 2n_{41} + h_3 + h_7 + h_8 + k_2 + k_{10} + k_{12}\,, \\
x_3 &= 2n_{13,1} + h_9 + h_{10} + h_{14} + k_5 + k_7 + k_{15}\,, \\
x_4 &= 2n_{16,1} + h_{11} + h_{12} + h_{15} + k_6 + k_8 + k_{14}\,,
\end{aligned}$$

$$\begin{aligned}
y_1 &= 2g_1 + e_2 + e_5 + e_6 + f_3 + f_9 + f_{11}\,, \\
y_2 &= 2g_4 + e_3 + e_7 + e_8 + f_2 + f_{10} + f_{12}\,, \\
y_3 &= 2g_{13} + e_9 + e_{10} + e_{14} + f_5 + f_7 + f_{15}\,, \\
y_4 &= 2g_{16} + e_{11} + e_{12} + e_{15} + f_6 + f_8 + f_{14}\,.
\end{aligned}$$

Then the maximum-likelihood estimates of $c_{ij} = d_{ij}$ are

$$\hat{c}_{11} = \frac{x_1 + s_1}{x_1 + y_1}\,, \quad \hat{c}_{12} = \frac{x_2 + s_4}{x_2 + y_2}\,, \quad \hat{c}_{21} = \frac{x_3 + s_{13}}{x_3 + y_3}\,, \quad \hat{c}_{22} = \frac{x_4 + s_{16}}{x_4 + y_4}\,. \tag{4.8.3}$$

The transition numbers n_{ij} for the 16×4 matrix given in (4.8.1) were tabulated for the last 100 trials and aggregated over subjects. Using (4.8.2), we obtained the $\hat{c}_{ij}$'s and $\hat{d}_{ij}$'s given in Table 4.4.

TABLE 4.4

MAXIMUM-LIKELIHOOD ESTIMATES OF c_{ij} AND d_{ij}

Parameter	Sure	Mixed	Parameter	Sure	Mixed
c_{11}	.807	.672	d_{11}	.431	.588
c_{21}	.619	.601	d_{21}	.335	.352
c_{12}	.656	.436	d_{12}	.278	.240
c_{22}	.534	.472	d_{22}	.102	.131

Given these estimates of c_{ij} and d_{ij}, the χ^2 goodness-of-fit test (2.4.4) can be applied to matrix (4.8.1), where the observed n_{ij}'s are those on which the estimates are based (see Sec. 2.6 for a discussion of this particular application of the χ^2 test). Since $c_1 = d_1 = b_2 = d_2 = 0$ for the Sure Group, transitions in rows 3, 4, 5, and 7 are not possible; therefore, there are $12 \times 3 = 36$ degrees of freedom for the matrix. For the Mixed Group, $a_1 = c_1 = d_1 = b_2 = d_2 = a_3 = b_3 = 0$, and transitions in rows 1, 3, 4, 5, 7, 11, and 12 are not possible. Thus, there are $9 \times 3 = 27$ degrees of freedom. But the eight parameters c_{ij} and d_{ij} have been estimated from these data; therefore, the χ^2 are to be interpreted with $36 - 8 = 28$ and $27 - 8 = 19$ degrees of freedom for the Sure and Mixed Groups, respectively. The obtained values of χ^2 are 37.7 for the Sure Group and 19.1 for the Mixed Group. Neither is significant at the .10 level, indicating substantial agreement between predicted and observed transition data.

Given these estimates of the conditioning parameters, predictions can be made for the transition probabilities $P(A_{i,n+1}B_{j,n+1} \mid A_{i',n}B_{j',n})$. The derivation of these quantities is straightforward; they are given in (4.8.4). This matrix

(4.8.4)

$$
\begin{array}{c|llll}
 & (1,1) & (1,2) & (2,1) & (2,2) \\
\hline
(1,1) & \begin{array}{l} a_1c_{11}d_{11} \\ +d_1c_{12}d_{12} \\ +b_1c_{11}d_{12} \\ +c_1c_{12}d_{11} \end{array} & \begin{array}{l} a_1c_{11}(1-d_{11}) \\ +d_1c_{12}(1-d_{12}) \\ +b_1c_{11}(1-d_{12}) \\ +c_1c_{12}(1-d_{11}) \end{array} & \begin{array}{l} a_1(1-c_{11})d_{11} \\ +d_1(1-c_{12})d_{12} \\ +b_1(1-c_{11})d_{12} \\ +c_1(1-c_{12})d_{11} \end{array} & \begin{array}{l} a_1(1-c_{11})(1-d_{11}) \\ +d_1(1-c_{12})(1-d_{12}) \\ +b_1(1-c_{11})(1-d_{12}) \\ +c_1(1-c_{12})(1-d_{11}) \end{array} \\
(1,2) & \begin{array}{l} a_2c_{11}d_{22} \\ +d_2c_{12}d_{21} \\ +b_2c_{11}d_{21} \\ +c_2c_{12}d_{22} \end{array} & \begin{array}{l} a_2c_{11}(1-d_{22}) \\ +d_2c_{12}(1-d_{21}) \\ +b_2c_{11}(1-d_{21}) \\ +c_2c_{12}(1-d_{22}) \end{array} & \begin{array}{l} a_2(1-c_{11})d_{22} \\ +d_2(1-c_{12})d_{21} \\ +b_2(1-c_{11})d_{21} \\ +c_2(1-c_{12})d_{22} \end{array} & \begin{array}{l} a_2(1-c_{11})(1-d_{22}) \\ +d_2(1-c_{12})(1-d_{21}) \\ +b_2(1-c_{11})(1-d_{21}) \\ +c_2(1-c_{12})(1-d_{22}) \end{array} \\
(2,1) & \begin{array}{l} a_3c_{22}d_{11} \\ +d_3c_{21}d_{12} \\ +b_3c_{22}d_{12} \\ +c_3c_{21}d_{11} \end{array} & \begin{array}{l} a_3c_{22}(1-d_{11}) \\ +d_3c_{21}(1-d_{12}) \\ +b_3c_{22}(1-d_{12}) \\ +c_3c_{21}(1-d_{11}) \end{array} & \begin{array}{l} a_3(1-c_{22})d_{11} \\ +d_3(1-c_{21})d_{12} \\ +b_3(1-c_{22})d_{12} \\ +c_3(1-c_{21})d_{11} \end{array} & \begin{array}{l} a_3(1-c_{22})(1-d_{11}) \\ +d_3(1-c_{21})(1-d_{12}) \\ +b_3(1-c_{22})(1-d_{12}) \\ +c_3(1-c_{21})(1-d_{11}) \end{array} \\
(2,2) & \begin{array}{l} a_4c_{22}d_{22} \\ +d_4c_{21}d_{21} \\ +b_4c_{22}d_{21} \\ +c_4c_{21}d_{22} \end{array} & \begin{array}{l} a_4c_{22}(1-d_{22}) \\ +d_4c_{21}(1-d_{21}) \\ +b_4c_{22}(1-d_{21}) \\ +c_4c_{21}(1-d_{22}) \end{array} & \begin{array}{l} a_4(1-c_{22})d_{22} \\ +d_4(1-c_{21})d_{21} \\ +b_4(1-c_{22})d_{21} \\ +c_4(1-c_{21})d_{22} \end{array} & \begin{array}{l} a_4(1-c_{22})(1-d_{22}) \\ +d_4(1-c_{21})(1-d_{21}) \\ +b_4(1-c_{22})(1-d_{21}) \\ +c_4(1-c_{21})(1-d_{22}) \end{array}
\end{array}
$$

corresponds to (4.2.1) for the one-element stimulus-sampling model. Table 4.5 presents the predicted transition probabilities based on the maximum-likelihood estimates of c_{ij} and d_{ij}. As would be expected from the values of the χ^2's reported above, the correspondence between these values and the observed values is excellent.

Next, we compute u_i, α, and β from the theoretical transition matrices by (1.3.17), (1.4.10), and (1.4.11). The results of these computations are given in Table 4.6. A comparison of Table 4.6 with Table 4.1 and Fig. 4.4 indicates excellent agreement between the predicted and observed quantities.

TABLE 4.5

PREDICTED TRANSITION MATRICES FOR THE GENERALIZED CONDITIONING MODEL
(For purposes of comparison the observed proportions, originally given in Table 4.2, are presented in parentheses)

	Sure Group				Mixed Group			
n \ $n+1$	(1, 1)	(1, 2)	(2, 1)	(2, 2)	(1, 1)	(1, 2)	(2, 1)	(2, 2)
(1, 1)	.26 (.25)	.55 (.55)	.06 (.07)	.13 (.13)	.16 (.18)	.51 (.53)	.08 (.05)	.25 (.24)
(1, 2)	.08 (.07)	.69 (.70)	.02 (.02)	.21 (.21)	.07 (.07)	.46 (.45)	.06 (.06)	.41 (.42)
(2, 1)	.22 (.21)	.38 (.43)	.15 (.16)	.25 (.20)	.27 (.25)	.33 (.35)	.18 (.21)	.22 (.19)
(2, 2)	.11 (.14)	.47 (.43)	.08 (.07)	.34 (.36)	.14 (.12)	.38 (.40)	.13 (.14)	.35 (.34)

The qualitative features of the generalized conditioning model with parameters c_{ij} and d_{ij} are not quite as satisfactory for the data of this chapter as they were for Chapter 3. There is, for example, in Table 4.4 one reversal of the monotonic relations noted in Table 3.10. This occurs in the Sure Group where $c_{12} > c_{21}$, suggesting in this case that subjects who were in the Player A position found the emitted response slightly more self-reinforcing than the objective reinforcement. (This same effect is found in Table 4.7 for the Sure Group when $c_{ij} = d_{ij}$.) It will be noticed that the values of the d_{ij} parameters are smaller in Table 4.4 than in Table 3.10. Undoubtedly this difference reflects the more frequent occurrence on the average of B_2 responses and E_2 reinforcements for Player B (see Table 4.1). This effect is also reflected by the fact that here $1 - d_{22} > d_{11}$ for both groups. On the other hand, since A_1 responses and E_1 reinforcements were more frequent for Player A, the relation $c_{11} > 1 - c_{22}$, which held in Chapter 3, also holds here.

TABLE 4.6

PREDICTED ASYMPTOTIC QUANTITIES FOR THE GENERALIZED CONDITIONING MODEL

Quantity	Sure Group	Mixed Group
u_1	.11	.13
u_2	.61	.42
u_3	.05	.10
u_4	.23	.35
α	.72	.55
β	.16	.23

TABLE 4.7

MAXIMUM-LIKELIHOOD ESTIMATES OF $c_{ij} = d_{ij}$

Parameter	Sure Group	Mixed Group
c_{11}	.753	.657
c_{21}	.482	.458
c_{12}	.501	.363
c_{22}	.172	.235

Finally, the excellent goodness-of-fit results corroborate those of Sec. 3.8. Moreover, in the present case the transition matrix is larger, and the number of cells to be fitted relative to the number of parameters is more like that for the restraint $c_{ij} = d_{ij}$ in Sec. 3.8.

Similar computations were carried out for the case $c_{ij} = d_{ij}$. The maximum-likelihood estimates of c_{ij} are given in Table 4.7. The same χ^2 tests reported above were run for predicted values based on the $\hat{c}_{ij}$'s given in Table 4.7. The degrees of freedom were $36 - 4 = 32$ and $27 - 4 = 23$ for the Sure and Mixed Groups, respectively. The obtained values of χ^2 were 72.2 (Sure Group) and 33.3 (Mixed Group). Both are significant at the .10 level.

In view of the significance level obtained for this same model with $c_{ij} = d_{ij}$ in Sec. 3.8, it is not surprising that the fit is no better. For in this case there is greater asymmetry between the c_{ij} and d_{ij} of Table 4.4 than between those of Table 3.10. A comparison of Table 4.7 with Table 3.13 bears out once again the general conclusion that the values of the conditioning parameters are affected by the average frequency of responses and reinforcements. As a single example, in Table 4.7 c_{22} is .172 in one group and .235 in the other, whereas $\min c_{22} = .387$ in Table 3.13.

From the standpoint of a viable general theory of behavior it is unfortunate that the conditioning parameters in the generalized model appear to be so sensitive to the actual frequencies of responses and reinforcements. This fact implies that the model is not providing a satisfactory theory of the elementary process of conditioning, which should be essentially independent of a particular reinforcement schedule. Unfortunately, not until Chapter 10 are we in a position to make an adequately detailed comparison of the generalized model and the multi-element θ model; but it should be kept in mind that the latter model always provides, in principle, an alternative approach to the search for a model that can satisfy a stringent goodness-of-fit test.

4.9. Summary

The interaction experiments in this chapter differ from those of the preceding chapter in two important ways. In the first place, subjects were informed that they were involved in a two-person game and that whether they won or lost depended in part on the responses of the other subject. Second, the reinforcement schedule was no longer zero-sum, but was such that on many trials both players could be correct in their responses or both incorrect. Two experiments are reported, each using a different payoff matrix; the selection of the payoff matrices was made on the basis of game-theory considerations. Subjects were run for 210 trials.

The one-element model is presented in Sec. 4.2, and details of the experimental method are reported in Sec. 4.3. Mean learning curves and asymptotic results are given in Sec. 4.4. As in the case of Chapter 3 the predictions of asymptotic response probabilities are very well confirmed. Estimates of θ are given in Sec. 4.5; the values are .72 for one group and .91 for the other. The general tests for stationarity and order of the Markov process are presented in Sec. 4.6. Stationarity is approached in the last 100 trials; for neither group is the test of second-order versus first-order chain significant at the .05 level. In Sec. 4.7 the data are compared with game-theory predictions based on Nash's concept of an equilibrium point. Again there is no evidence to support the hypothesis that the players are approaching optimal game strategies. Finally, in Sec. 4.8 the data are analyzed in terms of the generalized conditioning model. Excellent goodness-of-fit results are obtained; the χ^2 test is not significant at the .10 level for either group.

5

A Symmetric Non-Zero-Sum Situation

5.1. Theoretical Description

The experiment reported in this chapter involves a non-zero-sum situation of the general sort described in Chapter 4. However, it differs from the experiment reported in that chapter in two important respects.

First, the situation is completely symmetric between players A and B. If responses A_1 and B_1 occur, both players are correct with independent probabilities w_1; and if responses A_2 and B_2 occur, both players are correct with independent probabilities w_2. If responses A_1 and B_2, or A_2 and B_1, occur then both players are incorrect with probability 1. Thus the game-theory matrix corresponding to (4.7.1) has the form

(5.1.1)
$$\begin{array}{c|cc} & B_1 & B_2 \\ \hline A_1 & (w_1, w_1) & (0, 0) \\ A_2 & (0, 0) & (w_2, w_2) \end{array}$$

The game-theory aspects of the situation are explored further below.

Second, the transition matrix derived from the one-element θ model for response pairs has no zeros in it. In the experiments already reported, there were at least four zeros in every such transition matrix derived from the one-element model; thus a satisfactory goodness-of-fit result was virtually precluded from the start.

For subsequent use we need certain detailed relations with the model of Chapter 4. To begin with, the experimental parameters (a_i, b_i, c_i, d_i) of (4.1.1) may be expressed as follows in terms of w_1 and w_2:

[1] This study was run and the data were in part analyzed by Professor William A. Wilson, Jr., of the University of Colorado while he was at Stanford University for the 1957 Summer Institute on the Application of Mathematics in Social Science Research, sponsored by the Social Science Research Council. We are grateful to him for granting us permission to present the results here.

$$
\begin{aligned}
&a_1 = w_1^2\,, && a_2 = a_3 = 0\,, && a_4 = w_2^2\,,\\
&b_1 = w_1(1-w_1)\,, && b_2 = b_3 = 0\,, && b_4 = w_2(1-w_2)\,,\\
&c_1 = w_1(1-w_1)\,, && c_2 = c_3 = 0\,, && c_4 = w_2(1-w_2)\,,\\
&d_1 = (1-w_1)^2\,, && d_2 = d_3 = 1\,, && d_4 = (1-w_2)^2\,.
\end{aligned}
\tag{5.1.2}
$$

The transition matrix (4.2.1) of the one-element model then assumes the following form upon substitution of the relations given by (5.1.2):

(5.1.3)

	(1, 1)	(1, 2)	(2, 1)	(2, 2)
(1, 1)	$[1-\theta(1-w_1)]^2$	$\theta(1-w_1) - \theta^2(1-w_1)^2$	$\theta(1-w_1) - \theta^2(1-w_1)^2$	$\theta^2(1-w_1)^2$
(1, 2)	$\theta(1-\theta)$	$(1-\theta)^2$	θ^2	$\theta(1-\theta)$
(2, 1)	$\theta(1-\theta)$	θ^2	$(1-\theta)^2$	$\theta(1-\theta)$
(2, 2)	$\theta^2(1-w_2)^2$	$\theta(1-w_2) - \theta^2(1-w_2)^2$	$\theta(1-w_2) - \theta^2(1-w_2)^2$	$[1-\theta(1-w_2)]^2$

In (5.1.3) we have set $\theta_A = \theta_B = \theta$, and with better reason than in any of the previous experiments, because of the completely symmetric reinforcement schedule.

Solving (5.1.3) for the asymptotic response probabilities, we obtain

$$
\begin{aligned}
u_1 &= (1-w_2)(1-\theta)/K\,,\\
u_2 = u_3 &= (1-w_1)(1-w_2)[1-\tfrac{1}{2}\theta(2-w_1-w_2)]/K\,,\\
u_4 &= (1-w_1)(1-\theta)/K\,,
\end{aligned}
\tag{5.1.4}
$$

where $K = 2(1-w_1)(1-w_2) + (2-w_1-w_2)\{1-\theta[1+(1-w_1)(1-w_2)]\}$.

Further, from (1.4.10)

$$
\alpha = \beta = (1-w_2)\{1-\theta+(1-w_1)[1-\tfrac{1}{2}\theta(2-w_1-w_2)]\}/K\,. \tag{5.1.5}
$$

Only one experimental group was run, for which the parameter values were $w_1 = .8$ and $w_2 = .4$. Thus, as functions of θ we have from (5.1.4) and (5.1.5)

$$
\begin{aligned}
u_1 &= .60(1-\theta)/(1.04-.90\theta)\,,\\
u_2 = u_3 &= (.12-.05\theta)/(1.04-.90\theta)\,,\\
u_4 &= .20(1-\theta)/(1.04-.90\theta) = \tfrac{1}{3}u_1\,,\\
\alpha = \beta &= (.72-.65\theta)/(1.04-.90\theta)\,.
\end{aligned}
\tag{5.1.6}
$$

5.2. Experimental Method

Subjects. The subjects were 48 undergraduate students hired through the Stanford University Student Employment Service. They were paid \$1.25 for participating.

Apparatus and procedure. The apparatus described in Chapter 3 was used. The procedure was identical to that described in Chapter 4. The only difference was that 220 trials were run, compared with 210 for the experiments in Chapter 4.

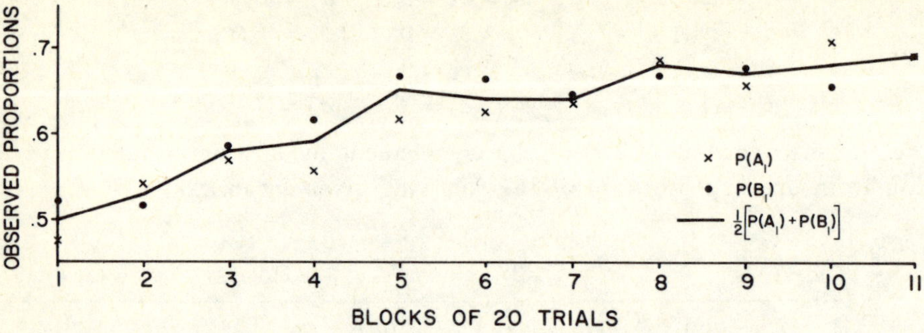

FIGURE 5.1. Observed proportions of A_1 and B_1 responses in successive 20-trial blocks.

5.3. Mean Learning Curves

Figure 5.1 presents the observed average proportion of A_1 and B_1 responses in successive 20-trial blocks for the entire sequence of 220 trials. It can easily be shown that if the states of the Markov chain are equiprobable on the initial trial and if $\theta_A = \theta_B$, then $P(A_{1,n}) = P(B_{1,n})$ for all n. Consequently, the mean of the A_1 and B_1 proportions is also presented.

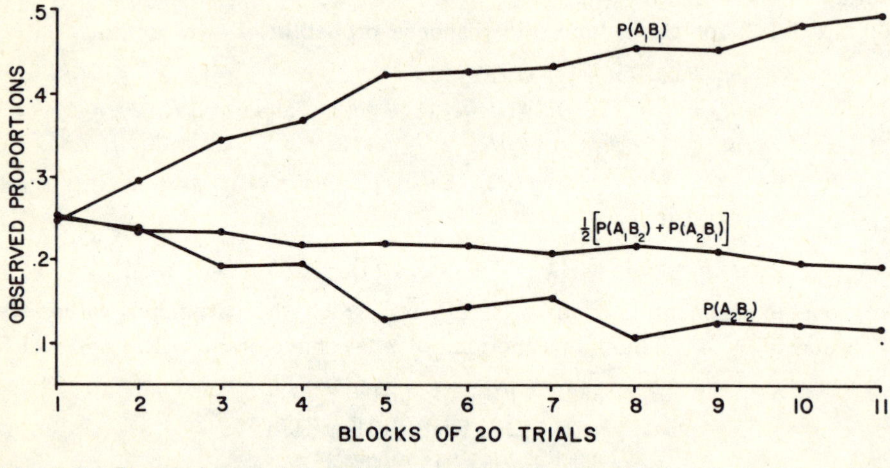

FIGURE 5.2. Observed proportions of the response pairs (A_1B_1), (A_1B_2), (A_2B_1), and (A_2B_2) in successive 20-trial blocks.

Figure 5.2 presents the observed proportions of the response pairs (A_1B_1), (A_1B_2), (A_2B_1), and (A_2B_2). Again, it can be shown that $P(A_{1,n}B_{2,n}) = P(A_{2,n}B_{1,n})$, if $\theta_A = \theta_B$ and the states are equiprobable on the initial trial. Individual curves for the proportion of (A_1B_2)'s and (A_2B_1)'s were virtually identical, and consequently only the mean of the two proportions is presented in Fig. 5.2.

From (5.1.6), by taking appropriate limits, it follows that

$$
\begin{aligned}
&0 \leqq u_1 < .577\,, && 0 \leqq u_4 < .192\,, \\
&0 < u_2 = u_3 \leqq .500\,, && .500 \leqq \alpha = \beta < .692\,.
\end{aligned}
\tag{5.3.1}
$$

An inspection of Figs. 5.1 and 5.2 indicates that the observed proportions over the last few blocks of trials were in the appropriate intervals. We shall return to this point later.

5.4. Maximum-Likelihood Estimate of θ and Related Predictions

The likelihood function described in (4.5.1) was computed for $\theta = .01i$ $(i = 1, \cdots, 100)$. The computations were carried out on the n_{ij}'s for the last 120 trials aggregated over all subject-pairs. The maximum value of the function occurred at $\hat{\theta} = .59$. Employing this estimated value of θ, we obtain the following predictions from (5.1.6):

$$u_1 = .481\,, \qquad u_2 = u_3 = .179\,, \qquad u_4 = .160\,, \qquad \alpha = \beta = .660\,. \tag{5.4.1}$$

If we use proportions computed over the last 80 trials as estimates of the asymptotic probabilities, the average over subjects is as follows:

$$\hat{u}_1 = .474\,, \qquad \tfrac{1}{2}(\hat{u}_2 + \hat{u}_3) = .204\,, \qquad \hat{u}_4 = .117\,, \qquad \tfrac{1}{2}(\hat{\alpha} + \hat{\beta}) = .678\,. \tag{5.4.2}$$

The observed standard deviations for these quantities are .161, .071, .054, and .122, respectively. Over all, the predicted and observed values are in fairly close correspondence. However, there is one disturbing feature about these results, as we can see from an inspection of Fig. 5.2: it appears that the curve for $P(A_{1,n}B_{1,n})$ may still be rising, and that an asymptotic level of responding has not yet been attained. If this is the case, it is of course possible that the true limiting value of $P(A_{1,n}B_{1,n})$ is greater than predicted by the theory.

We now turn to the one-stage transition data. Using $\hat{\theta} = .59$, we may compute the theoretical matrix (5.1.3) and compare these quantities with the observed transition matrix. Table 5.1 presents the observed transition probabilities computed over the last 100 trials. Also, the theoretical values are given in parentheses for $\hat{\theta} = .59$.

The over-all agreement between these two arrays of numbers is not very good. To test the hypothesis that $p_{ij}(\hat{\theta}) = \hat{p}_{ij}$ $(i, j = 1, \cdots, 4)$, we compute the

TABLE 5.1
THE OBSERVED TRANSITION MATRIX COMPUTED OVER THE LAST 100 TRIALS
(The theoretical predictions for $\theta = .59$ are in parentheses)

n \ $n+1$	(1, 1)	(1, 2)	(2, 1)	(2, 2)
(1, 1)	.649 (.778)	.151 (.104)	.152 (.104)	.048 (.014)
(1, 2)	.323 (.242)	.258 (.168)	.240 (.348)	.179 (.242)
(2, 1)	.365 (.242)	.253 (.348)	.221 (.168)	.161 (.242)
(2, 2)	.231 (.125)	.328 (.229)	.227 (.229)	.214 (.417)

goodness-of-fit χ^2 given by (2.4.4). There are $12 - 1 = 11$ degrees of freedom, since θ was estimated from the same set of n_{ij}'s as the $\hat{p}_{ij}$'s. The obtained value of χ^2 is 458.4 and indicates a highly significant difference between predicted and observed transition probabilities. Thus, the hope that in this experiment the one-element model would provide a better fit to the transition matrix is not borne out. Such a decisive result for this experiment rather conclusively establishes the inability of the one-element model to yield a good prediction of the transition matrix in the response pairs. It is unfortunate indeed that the multi-element models have proved so difficult to test in any detail in these two-person situations. In the one set of experiments for which this has been possible (Chapter 10), the over-all fit has been nearly as good as that of the generalized conditioning model.

5.5. Tests for Markov Properties

The hypothesis that the transition probabilities are constant over trials was tested by computing (2.4.1) for $t = 1, 2$; $t = 1$ refers to the block of trials from 1 to 110, and $t = 2$ refers to the 111–220 trial block. The obtained value of χ^2 was 109.3 (d.f. = 12), and is highly significant.

The test of the hypothesis that the chain is of a given order deals only with transition data over the final block of 120 trials. The zero-order versus first-order hypothesis was tested by calculating (2.4.2); the obtained value of χ^2 was 355.4 (d.f. = 12), and is highly significant. The first-order versus second-order hypothesis was tested by calculating (2.4.3); χ^2 was 117.7 (d.f. = 36), also highly significant.

These results indicate that the response data cannot be adequately explained by any stationary Markov model whose states are simply the possible response pairs. It is to be noted that this conclusion does not necessarily apply to the generalized conditioning model, for its states are the response-reinforcement quadruples; and it is perfectly feasible to have a Markov chain.

5.6. Game-Theory Comparisons

For the experimental parameters used in this study the game matrix (5.1.1) is as follows:

(5.6.1)

	B_1	B_2
A_1	(.8, .8)	(0, 0)
A_2	(0, 0)	(.4, .4)

From the standpoint of game theory, this matrix differs from the two of Chapter 4 in the following important respect: the sets $\{A_1, B_1\}$ and $\{A_2, B_2\}$ are both equilibrium points, whereas the matrices of Chapter 4 each had a unique equilibrium point. It is easy to give simple examples of zero-sum games that do not have unique minimax strategies. However, the value of the game is the same for the different strategies in the zero-sum case, whereas for the non-zero-sum case distinct equilibrium points do not necessarily yield the same expected outcomes. The various values of the game are given in Table 5.2, with

TABLE 5.2
GAME-THEORY, LEARNING-THEORY, AND OBSERVED VALUES OF THE NON-ZERO-SUM GAME

Type of Value	A	B
Game Theory $\{A_1, B_1\}$	.800	.800
Game Theory $\{A_2, B_2\}$	.400	.400
Learning Theory	.449	.449
Observed	.426	.426

the game-theory "value" stated for both equilibrium points. The learning-theory value is based on $\hat{\theta} = .59$ and the observed value on the last 80 trials, with the $\hat{u}_i$'s as given by (5.4.2). It is interesting to note that the observed value falls between the learning-theory value and the less favorable game-theory value. Consequently there is no evidence that subjects tended to move toward the intuitively optimal game strategy of responding A_1 and B_1 with probability 1. This tendency was substantially reflected in individual response sequences for only two of the 48 subjects, and then not perfectly. In particular, max $\alpha = .98$ and max $\beta = .93$, and these two players did not belong to the same pair. The next highest individual values were .88 for α and .90 for β. Moreover, the standard deviations for the means of α and β are not unduly high.

*5.7. Generalized Conditioning Model

In this section we again consider the application of the model described in Sec. 2.6. The transition numbers n_{ij} for the 16×4 matrix (4.8.1) were tabulated over the last 100 trials and aggregated over subjects. Using (4.8.2), we obtained the following maximum-likelihood estimates of c_{ij} and d_{ij}:

$$\begin{aligned} &\hat{c}_{11} = .853\,, \quad \hat{c}_{12} = .584\,, \quad \hat{d}_{11} = .834\,, \quad \hat{d}_{12} = .610\,, \\ &\hat{c}_{21} = .621\,, \quad \hat{c}_{22} = .465\,, \quad \hat{d}_{21} = .555\,, \quad \hat{d}_{22} = .357\,. \end{aligned} \tag{5.7.1}$$

Given these estimates of c_{ij} and d_{ij}, the χ^2 goodness-of-fit test (2.4.4) may be applied to matrix (4.8.1), where the observed n_{ij}'s are those on which the estimates are based (see Sec. 2.6 for a discussion of this particular application of the χ^2 test). The experimental parameters are such that transitions are not possible in rows 5, 6, 8, 9, 11, and 12; hence, there are $10 \times 3 = 30$ degrees of freedom for the matrix. However, since eight parameters have been estimated from these data, the test is to be interpreted with $30 - 8 = 22$ degrees of freedom. The obtained value of χ^2 is 24.3 and, of course, is not significant.

TABLE 5.3

THE THEORETICAL TRANSITION MATRIX FOR THE GENERALIZED CONDITIONING MODEL

(The observed quantities over the last 100 trials are given in parentheses)

n \ $n+1$	(1, 1)	(1, 2)	(2, 1)	(2, 2)
(1, 1)	.631 (.649)	.169 (.151)	.158 (.152)	.042 (.048)
(1, 2)	.324 (.323)	.260 (.258)	.231 (.240)	.185 (.179)
(2, 1)	.379 (.365)	.242 (.253)	.231 (.221)	.148 (.161)
(2, 2)	.266 (.231)	.294 (.328)	.210 (.227)	.230 (.214)

Using the estimates given in (5.7.1), we can obtain predicted transition probabilities for $P(A_{i,n+1}B_{j,n+1} \mid A_{i',n}B_{j'n})$ in terms of (4.8.4). Table 5.3 presents the theoretical values and, in parentheses, the observed values computed over the last 100 trials (as presented earlier in Table 5.1). As would be expected from the results of the χ^2 test, the correspondence between predicted and observed quantities is fairly good.

The theoretical quantities u_i, α, and β have also been computed in terms of the estimates of c_{ij} and d_{ij} by (1.3.17), (1.4.10), and (1.4.11). The values are given below:

$$u_1 = .473 \,, \quad \tfrac{1}{2}(u_2 + u_3) = .205 \,, \quad u_4 = .116 \,, \quad \tfrac{1}{2}(\alpha + \beta) = .678 \,. \tag{5.7.2}$$

The results correspond almost perfectly with the observed quantities (5.4.2).

Although the goodness-of-fit result for the generalized model is very good, as it was in Chapters 3 and 4, there are two respects in which the estimated values of the parameters c_{ij} and d_{ij} are unsatisfactory. In the first place, $d_{21} < d_{12}$, which is contrary to the expected monotonic relation $d_{11} > d_{21} > d_{12} > d_{22}$. A similar reversal was observed in Sec. 4.8. Second, the estimates of c_{ij} and the corresponding d_{ij} are not as close as one would expect given the complete symmetry of the reinforcement conditions. For instance, $\hat{c}_{21} - \hat{d}_{21} = .066$ and $\hat{c}_{22} - \hat{d}_{22} = .108$. These discrepancies seem to reflect the extreme sensitivity of the generalized conditioning model to the detailed pattern of responses and reinforcements. Unfortunately, the labor involved was too great to compute individual c_{ij}'s and d_{ij}'s in order to test the null hypothesis that $c_{ij} = d_{ij}$.

As indicated in Sec. 5.1, the experimental situation is perfectly symmetric for both players; that is, the payoff probabilities are the same for Player A and Player B. Therefore, we should not expect systematic differences between the group of subjects who were run as Player A and the group run as Player B. This argument implies that $c_{ij} = d_{ij}$.

With this in mind, we computed the maximum-likelihood estimates of $c_{ij} = d_{ij}$

by (4.8.3). They are as follows:

$$\hat{c}_{11} = .805\,, \qquad \hat{c}_{21} = .598\,, \qquad \hat{c}_{12} = .575\,, \qquad \hat{c}_{22} = .439\,.$$

When these estimates were used, the χ^2 goodness-of-fit test, with $30 - 4 = 26$ degrees of freedom, yielded a value of 53.8, which is highly significant. This kind of significance result for $c_{ij} = d_{ij}$ agrees with those obtained in Secs. 3.8 and 4.8.

5.8. Summary

The interaction experiment in this chapter involves a non-zero-sum situation of the sort described in Chapter 4. However, it differs from the earlier experiments in that the situation was perfectly symmetric between the two players. Specifically, if (A_1B_1) occurred, both players were correct with independent probabilities of .8; if (A_2B_2) occurred, both players were correct with independent probabilities of .4. However, if either (A_1B_2) or (A_2B_1) occurred, then both players were incorrect with probability 1.

The theoretical predictions based on the one-element stimulus-sampling model are derived in Sec. 5.1. The asymptotic mean response probabilities $P_\infty(A_1)$ and $P_\infty(B_1)$ are functions of θ, which was not the case in Chapters 3 and 4. The details of the experimental method are given in Sec. 5.2. Mean learning curves and asymptotic results are presented in Sec. 5.3. The maximum-likelihood estimate of θ and related predictions are given in Sec. 5.4. As in Chapter 3, the estimate of θ is very close to .6 ($\hat{\theta} = .59$). Predictions of asymptotic response probabilities are reasonably well confirmed in terms of this estimated θ. On the other hand, the over-all goodness-of-fit test for the one-element model is highly significant. The tests for stationarity and order of the Markov chain are presented in Sec. 5.5. In Sec. 5.6 the data are compared with game-theory predictions based on Nash's concept of an equilibrium point. There is no evidence of a tendency to move toward the pure equilibrium points (A_1B_1) or (A_2B_2). In Sec. 5.7 the data are analyzed in terms of the generalized conditioning model. Excellent goodness-of-fit results are again obtained.

6

Communication and Discrimination in Two-Person Situations

6.1. Introduction

In the experimental situations described in the preceding three chapters the subjects were permitted to interact only by means of their mutual control of the reinforcement schedule. In this and the next chapter we shall consider situations in which the extent of interaction is increased by informing one player about the choices of the other. Specifically, the situation is as follows: At the start of each trial Player A makes either an A_1 or an A_2 response; Player B is then informed of Player A's response and, in turn, chooses either B_1 or B_2. A payoff matrix of the type used in Chapter 3 represents the contingent probabilities of each player's being correct:

	A_1	A_2
B_1	a_1	a_2
B_2	a_3	a_4

With probability a_i Player A is correct and Player B incorrect, given the occurrence of the corresponding pair of responses; with probability $1-a_i$ the situation is reversed. Of course, as in the previous studies, the payoff matrix remains fixed over all trials of the experiment.

The behavioral consequence of Player B's receiving this additional information may be analyzed from at least three different general theoretical positions within the behavioral sciences. One approach is by way of the communication models of Bales [1950], Bavelas [1950], Christie, Luce, and Macy [1953], and others. Their models, however, are oriented toward interaction situations that are less structured than ours; in general they do not provide methods for making quantitative predictions of the kind we have considered in preceding chapters.

A second approach is provided by the theory of games. From the standpoint of game theory, Player B is playing a game of perfect information. That is, he has full knowledge of his opponent's moves, and thus possesses a pure optimal strategy. Moreover, the game aspects of the situation were stressed in the instructions to subjects (see Sec. 6.3). Discrepancies between the theory and experimental results are qualitatively comparable to those obtained in the experiments already reported. Details are given in Sec. 6.6.

The third orientation is provided by the stimulus-sampling theory of learning that we have been applying. The fundamental postulates of Sec. 1.2 remain unchanged. What is required is a way of bringing Player B's perception of Player A's responses within the conceptual framework of the theory. Some indication of how this is to be done has been given in Sec. 1.7 in our discussion of hypothesis models; but we may profitably review the approach here. The central idea is very simple. The communication of A's responses to B amounts to presenting B with two stimuli (A_1 and A_2) between which he learns to discriminate. Such an identification is not in any sense unique to the present experiment. Our viewpoint is that any social interaction situation may, *in principle*, be analyzed entirely as a situation in which each member of the group provides stimuli and reinforcements for the other member; the behavior of any individual in the group may be explained in terms of the sequence of stimuli and reinforcements he receives. At this stage of our knowledge, the phrase "in principle" is all important, since we cannot actually analyze in quantitative detail all interaction situations in terms of subjects discriminating in their responses between different stimuli.[1]

What we can do is offer some concrete, constructive examples of such an analysis in relatively simple, highly structured social situations. Such examples are provided in the present chapter and the next, as well as in some of the material in Chapter 12. We emphasize the importance of these examples for indicating how stimulus-sampling theory may be applied to increasingly complicated group interaction situations.

Prior to developing the two-person models that incorporate the passing of information from A to B, it will be instructive to consider a simple one-person discrimination model as a prototype of the method of analysis. There are two stimuli, s_1 and s_2. On each trial the experimenter presents exactly one of them to the subject. For simplicity, we consider only the case of a noncontingent presentation schedule. Let τ $(0 < \tau < 1)$ be the probability that s_1 is presented and $1 - \tau$ the probability that s_2 is presented. The probability of an E_1 reinforcement depends on the stimulus presented at the beginning of the trial: i.e.,

[1] Viewed from the standpoint of the history of science, it would be surprising if the situation were otherwise. Most physical theories can explain only in principle many phenomena that fall within their scope, even though these theories have been intensely developed over a period of many years. A classic example, intuitively nearly equal in complexity to many unstructured social situations, is the problem of predicting the weather. There is good evidence that the advancement of meteorology is not primarily dependent upon the elaboration of new fundamental laws of physics, but upon a deeper and more elaborate application of known physical principles.

(6.1.1) $$P(E_1 \mid s_1) = \pi_1 \,, \qquad P(E_1 \mid s_2) = \pi_2 \,.$$

Because there are two stimuli, there are four states of conditioning, which we indicate by the ordered pair (i, j). The first member of the pair is 1 or 2, depending on whether s_1 is conditioned to A_1 or A_2; the second member is also 1 or 2, depending on whether s_2 is conditioned to A_1 or A_2. This kind of notation for states is more convenient for later use in this chapter than the set notation introduced in Sec. 1.3.

A typical tree for the process (the tree starting from state (2, 1), i.e., the state with s_1 conditioned to A_2, and s_2 conditioned to A_1) is given in Fig. 6.1.

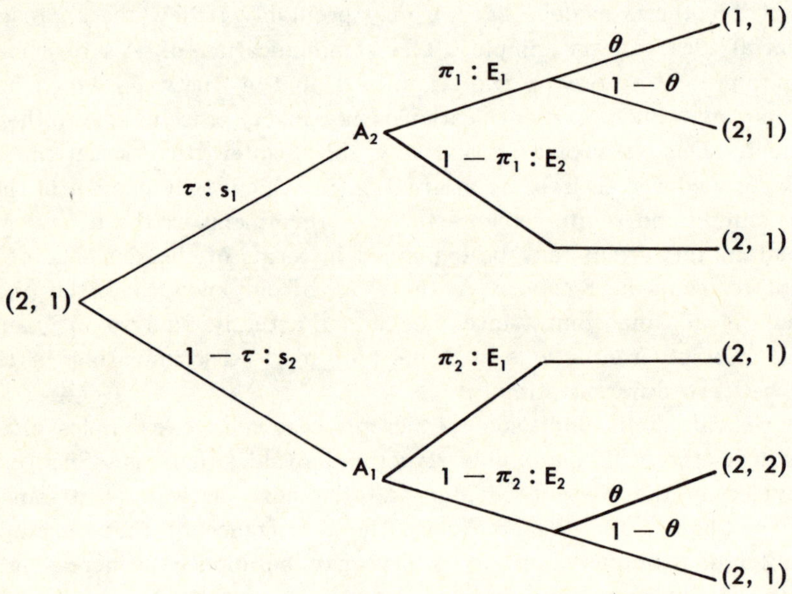

FIGURE 6.1

At the beginning of the trial, s_1 is presented with probability τ, and s_2 with probability $1 - \tau$. Whichever stimulus is presented by the experimenter is sampled with probability 1. When s_1 is sampled, response A_2 occurs with probability 1, for s_1 is conditioned to A_2; similarly when s_2 is sampled, A_1 occurs with probability 1. Also, when s_1 is sampled, E_1 occurs with probability π_1, and E_2 with probability $1 - \pi_1$. In the former case, if conditioning is effective, s_1 becomes conditioned to the reinforced response A_1, and the new state is (1, 1). If conditioning is not effective, the state remains unchanged. The state also remains unchanged on the upper branch when E_2 occurs, for the existing conditioning of s_1 to A_2 is then simply further reinforced. The analysis of the remainder of the lower branch of the tree proceeds in like fashion. From the construction of similar trees for the other three states we may derive the following transition matrix, which resembles both (1.3.12) and (1.7.3):

(6.1.2)

$$\begin{array}{c|cccc} & (1, 1) & (1, 2) & (2, 1) & (2, 2) \\ \hline (1, 1) & \begin{array}{c}1-(1-\tau)\theta(1-\pi_2)\\ -\tau\theta(1-\pi_1)\end{array} & (1-\tau)\theta(1-\pi_2) & \tau\theta(1-\pi_1) & 0 \\ (1, 2) & (1-\tau)\theta\pi_2 & \begin{array}{c}1-(1-\tau)\theta\pi_2\\ -\tau\theta(1-\pi_1)\end{array} & 0 & \tau\theta(1-\pi_1) \\ (2, 1) & \tau\theta\pi_1 & 0 & \begin{array}{c}1-\tau\theta\pi_1\\ -(1-\tau)\theta(1-\pi_2)\end{array} & (1-\tau)\theta(1-\pi_2) \\ (2, 2) & 0 & \tau\theta\pi_1 & (1-\tau)\theta\pi_2 & \begin{array}{c}1-\tau\theta\pi_1\\ -(1-\tau)\theta\pi_2\end{array} \end{array}$$

The methods of Sec. 1.3 may be applied directly to derive various asymptotic probabilities from this transition matrix. However, there is a simpler method, which may be applied rather widely and is also of some theoretical interest. The essential idea is that the conditioning of a stimulus can change only when it is sampled. If a stimulus is the only one sampled on certain trials, then the subsequence of trials on which it is sampled forms a learning process that may be studied independently of what happens on other trials (see Axiom C4 of Sec. 1.2). We then have, as the transition matrix for trials on which s_1 is sampled,

(6.1.3)

$$\begin{array}{c|cc} & A_1 & A_2 \\ \hline A_1 & 1-\theta(1-\pi_1) & \theta(1-\pi_1) \\ A_2 & \theta\pi_1 & 1-\theta\pi_1 \end{array}$$

which is just (1.3.4) with π replaced by π_1. Note that the states are designated A_1 and A_2 to indicate the conditioning of s_1. We have at once

(6.1.4)
$$P_\infty(A_1 \mid s_1) = \pi_1 ,$$

and by similar considerations for s_2,

(6.1.5)
$$P_\infty(A_1 \mid s_2) = \pi_2 .$$

By elementary probability we then have

(6.1.6)
$$\begin{aligned} P_\infty(A_1) &= P_\infty(A_1 \mid s_1)P(s_1) + P_\infty(A_1 \mid s_2)P(s_2) \\ &= \pi_1\tau + \pi_2(1-\tau) . \end{aligned}$$

This derivation of $P_\infty(A_1)$ is simpler than a direct analysis of the matrix (6.1.2). The gain in simplicity here emphasizes the importance of keeping the fundamental theory constantly in mind and not letting a particular mathematical representation like (6.1.2) become too dominant.

On the other hand, there are definite limitations to this approach of considering a separate learning or conditioning process for each stimulus. For many experimental situations it is reasonable to postulate the presence of additional stimulus elements that occur on all trials and are not subject to direct manipulation by the experimenter. Typically, two sources of common elements are identified in a discrimination problem. One subset of common elements may be associated with the actual stimuli to be discriminated; the other may be associated with background stimuli, such as characteristics of the experimental chamber, sounds from the apparatus, and proprioceptive stimulation. These common stimuli would be sampled on both s_1 and s_2 trials, and thus their state of conditioning would be changing on both types of trials. Consequently, when their presence is postulated, the simplified approach is not theoretically justified and yields incorrect results for asymptotic probabilities like $P_\infty(A_1)$. In the two-person discrimination studies considered in this chapter and the next we exclude the possibility of common stimuli, mainly because of the greater complexity of the resulting models. Even so, the simplified approach of separate processes cannot be used without restriction when the presentation of a stimulus to one subject is contingent upon the response of another, as we shall see in the next chapter.[2]

Concerning parameters controlled by the experimenter, the situation for the discrimination studies of this chapter is the same as that for the studies already reported. For the group labeled M, we have

(6.1.7)

	B_1	B_2
A_1	$\frac{1}{3}$	1
A_2	$\frac{1}{2}$	$\frac{1}{6}$

(The label M is used because (6.1.7) represents the payoff matrix of a 2×2 game for which both players have mixed minimax strategy.) For the group labeled S, we have

(6.1.8)

	B_1	B_2
A_1	$\frac{1}{2}$	1
A_2	$\frac{1}{4}$	$\frac{3}{4}$

(The label S was chosen because in the 2×2 game both players have strategies satisfying the sure-thing principle.) These two payoff matrices were also used for two of the three studies reported in Chapter 3.

6.2. Model

We begin, as before, by assuming that Player A has a single stimulus s available for sampling on each trial. However, Player B's behavior may be influenced by Player A's response. To permit this possibility we postulate two elements, s_1 and s_2, for Player B. We assume that s_1 will be sampled with

[2] The reader who is interested in recent quantitative theories of discrimination learning that incorporate the notion of common stimulus elements is referred to the following: Bush and Mosteller [1951]; Wyckoff [1952]; Restle [1955*a*], [1955*b*]; Burke and Estes [1957]; Atkinson [1958], [1960*a*]; and Green [1958]. For related experimental results see Estes and Burke [1955]; Estes, Burke, Atkinson, and Frankmann [1957]; Popper and Atkinson [1958]; Atkinson, Bogartz, and Turner [1959]; and Atkinson [1959].

probability 1 if Player A makes an A_1 response, and that s_2 will be sampled with probability 1 if an A_2 response occurs.

Restricting ourselves to this description of the stimuli, the stipulation of a payoff matrix like (6.1.7) determines the Markov process in the states of conditioning. On each trial the element s for Player A is conditioned to A_1 or A_2. Similarly, for Player B on each trial s_1 is conditioned to either B_1 or B_2, and s_2 is conditioned to either B_1 or B_2. Consequently, there are eight possible states of conditioning, represented by the ordered triples (1;11), (1;12), (1;21), (1;22), (2;11), (2;12), (2;21), and (2;22). The first member of each triple indicates the conditioning of Player A's element s. The second and third members indicate the conditioning of Player B's s_1 and s_2 elements, respectively. Thus (2;21) represents the conditioning state in which s is conditioned to A_2, s_1 is conditioned to B_2, and s_2 is conditioned to B_1. If we are in state (2;21) on trial n, Player A will make response A_2, and, necessarily, s_2 will be sampled by Player B. The sampling of s_2, in turn, elicits the B_1 response.

One of the trees for the Markov process, the tree starting from state (1;21), is given in Fig. 6.2. The response pair A_1B_2 occurs with probability 1, given

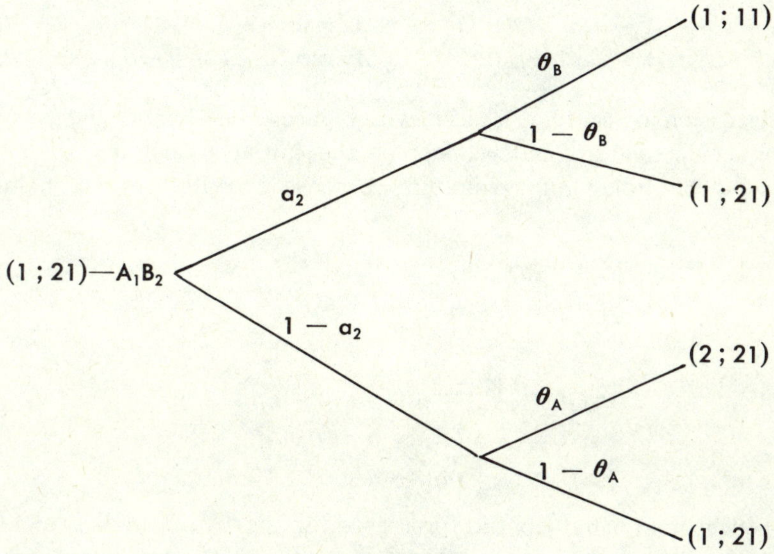

FIGURE 6.2

state (1;21), and the s_1 element is sampled by Player B. On the upper branch, a_2 indicates that Player A is correct and Player B incorrect; i.e., both players receive E_1 events. Only θ_B is indicated on the upper branch, since effectiveness of conditioning is irrelevant when the response is reinforced (see Sec. 1.3). With probability θ_B the element s_1 is conditioned to B_1; with probability $1-\theta_B$ its conditioning state remains unchanged. On the lower branch Player B is correct and Player A incorrect. Thus, with probability θ_A the element s is conditioned to A_2; with probability $1 - \theta_A$ its conditioning state remains unchanged. Of course, no change can occur for s_2, since it is not sampled.

Construction of the analogous trees yields the following transition matrix:

(6.2.1)

	1;11	1;12	1;21	1;22	2;11	2;12	2;21	2;22
1;11	A_1		$a_1\theta_B$		$(1-a_1)\theta_A$			
1;12		A_1		$a_1\theta_B$		$(1-a_1)\theta_A$		
1;21	$a_2\theta_B$		A_2				$(1-a_2)\theta_A$	
1;22		$a_2\theta_B$		A_2				$(1-a_2)\theta_A$
2;11	$(1-a_3)\theta_A$				A_3	$a_3\theta_B$		
2;12		$(1-a_4)\theta_A$				A_4		
2;21			$(1-a_3)\theta_A$				A_3	$a_3\theta_B$
2;22				$(1-a_4)\theta_A$			$a_4\theta_B$	A_4

$$A_1 = a_1(1-\theta_B) + (1-a_1)(1-\theta_A)\,, \qquad A_3 = a_3(1-\theta_B) + (1-a_3)(1-\theta_A)\,,$$
$$A_2 = a_2(1-\theta_B) + (1-a_2)(1-\theta_A)\,, \qquad A_4 = a_4(1-\theta_B) + (1-a_4)(1-\theta_A)\,.$$

From the discussion in Sec. 1.3 it is clear, with regard to (6.2.1), that the asymptotic quantities u_{ijk} exist and are independent of the initial probability distribution of the states. The following asymptotic response probabilities are defined in terms of u_{ijk}:

$$P_\infty(A_1B_1) = u_{111} + u_{112}\,, \qquad P_\infty(A_2B_1) = u_{211} + u_{221}\,,$$
$$P_\infty(A_1B_2) = u_{121} + u_{122}\,, \qquad P_\infty(A_2B_2) = u_{212} + u_{222}\,. \tag{6.2.2}$$

Also

$$P_\infty(A_1) = u_{111} + u_{112} + u_{121} + u_{122}\,,$$
$$P_\infty(B_1 \mid A_1) = (u_{111} + u_{112})/P_\infty(A_1)\,, \tag{6.2.3}$$
$$P_\infty(B_1 \mid A_2) = (u_{211} + u_{221})/P_\infty(A_2)\,.$$

The method of analyzing separate processes for the two stimuli s_1 and s_2 will not yield results for the mean asymptotic probabilities $P_\infty(A_iB_j)$ or $P_\infty(B_i)$, because the probability of an s_1 or s_2 is just the probability of an A_1 or A_2 response. And the mean probability of an A_1 or A_2 response is a function of what happens on all trials, not just those trials on which that response occurs. However, we can compute by the separation approach the asymptotic conditional probabilities

$$P_\infty(B_1 \mid s_1) = P_\infty(B_1 \mid A_1)\,, \qquad P_\infty(B_1 \mid s_2) = P_\infty(B_1 \mid A_2)\,.$$

The transition matrix for $s_1 = A_1$ trials is

$$\text{(6.2.4)} \qquad \begin{array}{c|cc} & B_1 & B_2 \\ \hline B_1 & 1-\theta_B a_1 & \theta_B a_1 \\ B_2 & \theta_B a_2 & 1-\theta_B a_2 \end{array}$$

whence

$$\text{(6.2.5)} \qquad P_\infty(B_1 \mid s_1) = \frac{a_2}{a_1 + a_2} \,.$$

By a similar argument,

$$\text{(6.2.6)} \qquad P_\infty(B_1 \mid s_2) = \frac{a_4}{a_3 + a_4} \,.$$

It may be noted that (6.2.5) and (6.2.6) hold independent of the values of θ_A and θ_B.[3]

When $\theta_A = \theta_B = \theta$, the u_{ijk}'s also are independent of θ. This follows from the observation that the terms on the main diagonal of (6.2.1) are of the form $(1-\theta) + \theta f_{ii}(a_k)$, whereas all other non-zero terms are of the form $\theta f_{ij}(a_k)$; consequently, θ cancels in the solution for the u's by (1.3.17). Hence, for this case the asymptotic predictions are entirely *a priori* and do not make use of parameters estimated from the data. Specifically, when $\theta_A = \theta_B$, the predicted u_{ijk}'s for Group M are

$$\text{(6.2.7)} \qquad \begin{array}{llll} u_{111} = 3/40 \,, & u_{121} = 1/40 \,, & u_{211} = 4/40 \,, & u_{221} = 0 \,, \\ u_{112} = 15/40 \,, & u_{122} = 5/40 \,, & u_{212} = 12/40 \,, & u_{222} = 0 \,. \end{array}$$

For Group S, the predicted values are

$$\text{(6.2.8)} \qquad \begin{array}{llll} u_{111} = 18/46 \,, & u_{121} = 9/46 \,, & u_{211} = 12/46 \,, & u_{221} = 0 \,, \\ u_{112} = 2/46 \,, & u_{122} = 1/46 \,, & u_{212} = 4/46 \,, & u_{222} = 0 \,. \end{array}$$

In a later section of this chapter, a pseudomaximum-likelihood estimate of the conditioning parameters is presented. Because of the greater complexity of both the theory and the experimental situation considered in this chapter, the related multi-element and generalized conditioning models are not considered.

6.3. Experimental Method

Subjects. The subjects were 96 undergraduates obtained from introductory courses in psychology at Stanford University. They were randomly assigned to the two groups, with the restriction of 24 pairs in each group.

Apparatus. The apparatus described in Chapter 3 was used, with the addition of green panel lights mounted 1 inch above each of Player B's keys. On every trial one of the green lights went on simultaneously with Player A's response and went off with the termination of the reinforcing light. When Player A's right key was pressed, Player B's right green light went on; when Player A's left key was pressed, Player B's left green light went on.

[3] These results also hold for the multi-element model.

The time intervals were identical to those given in Chapter 3. On all trials the signal light was lighted for 3.5 seconds. The time between successive signal exposures was 10 seconds. The reinforcing light followed the cessation of the signal light by 1.5 seconds and remained on for 2 seconds.

Procedure. For each pair of subjects, one was randomly selected as Player A and the other as Player B. Further, within each of the two experimental groups, four subgroups of 6 subject-pairs were formed by counterbalancing right and left positions of the A_i and B_j response keys.

The subjects were read the following instructions:

"This experiment is similar to a game situation involving two players, where on each play of the game one wins and the other loses. To keep things straight, I will call the person on my left Player A and the person on my right Player B.

"The experiment for each of you consists of a series of trials. The top center light on your panel will light to indicate the start of each trial. When this light goes on, Player A will press one or the other of the two keys in front of him. When he does this, one of Player B's green lights will go on, telling Player B what key Player A pressed. That is, if Player A pressed his left key, then the left green light will go on; if Player A pressed his right key, the right green light will go on. Knowing what key Player A has pressed, Player B will now press one of his two keys.

"Shortly thereafter one or the other of the two lower white lights will light on both your panels. If the white light above the key you pressed goes on, your key choice was correct and the other player's choice was incorrect. If the white light above the key opposite the one you pressed goes on, you were incorrect and the other player was correct.

"Being correct on a given trial depends on the combination of your key press with the other player's key press. For example, it could be that on some trials if you press your left key and the other player also presses his left key, you will be correct and he will be incorrect; whereas if he had pressed his right key, you would be incorrect and he would be correct.

"However, the combinations of your key press with his key press that were correct on one trial will not necessarily be correct on some later trial.

"Your job is to obtain as many correct key presses as you possibly can; that is, to maximize the number of trials on which the lower white light goes on above the key you press. At the start you will not know what keys to press; consequently, just guess. But as time goes on you will be able to improve on these guesses. Player B, in particular, will be able to take advantage of the fact that he knows what key Player A pressed before he must make his key choice.

"The trials move along rapidly and you must each make your key choice quickly. Player A, make your key choice immediately after the onset of the top center light. Player B, make your key choice immediately after the onset of one of the green lights."

After the instructions were read, 200 trials were run in continuous sequence. For each pair of subjects, sequences of reinforcing lights were randomly generated in accordance with the assigned values of a_i and the observed response pairs.

6.4. Mean Learning Curves and Asymptotic Results

For each pair of subjects a count was made of the number of (A_1B_1), (A_1B_2), (A_2B_1), and (A_2B_2) response pairs in successive blocks of 40 trials. From this data three proportions were computed for individual subjects. (1) The proportion of A_1 responses, designated $\hat{P}(A_1)$, was computed for each 40-trial block by dividing the number of (A_1B_1)'s plus (A_1B_2)'s by 40. (2) The proportion of B_1 responses given an A_1 response, designated $\hat{P}(B_1 \mid A_1)$, was computed for each 40-trial block by dividing the number of (A_1B_1)'s by the number of (A_1B_1)'s plus (A_1B_2)'s. (3) The proportion of B_1 responses given an A_2 response, designated $\hat{P}(B_1 \mid A_2)$, was computed by dividing the number of (A_2B_1)'s by the number of (A_2B_1)'s plus (A_2B_2)'s.

Figure 6.3 presents the average values for each of the quantities. An inspection of this figure indicates that response proportions were very stable over the last 100 trials. To check the stability of response proportions for individual data, t's for paired measures were computed between proportions for the first and last halves of the final block of 100 trials; in all cases, the obtained values of t were not significant at the .10 level. In view of these results, we use the proportions computed over the last 100 trials as estimates of asymptotic proportions.

TABLE 6.1

PREDICTED AND OBSERVED ASYMPTOTIC VALUES

(Computed over the last block of 100 trials)

Asymptotic Value	Group M			Group S		
	Predicted	Observed	SD	Predicted	Observed	SD
$P_\infty(A_1)$	.600	.611	.085	.652	.663	.127
$P_\infty(B_1 \mid A_1)$	.750	.751	.128	.667	.706	.126
$P_\infty(B_1 \mid A_2)$	.250	.268	.181	.750	.738	.181

Table 6.1 presents the means and observed standard deviations computed over the last 100 trials for $\hat{P}(A_1)$, $\hat{P}(B_1 \mid A_1)$, and $\hat{P}(B_1 \mid A_2)$. Each entry is based on $N = 24$. The predicted values for $\theta_A = \theta_B$ are also presented in the table and are obtained by substituting (6.2.7) and (6.2.8) into (6.2.3). Inspection of Table 6.1 indicates good agreement between predicted and observed quantities. Certainly from a statistical viewpoint there is no evidence in these measures for rejecting the hypothesis that no differences exist between theory and observation.

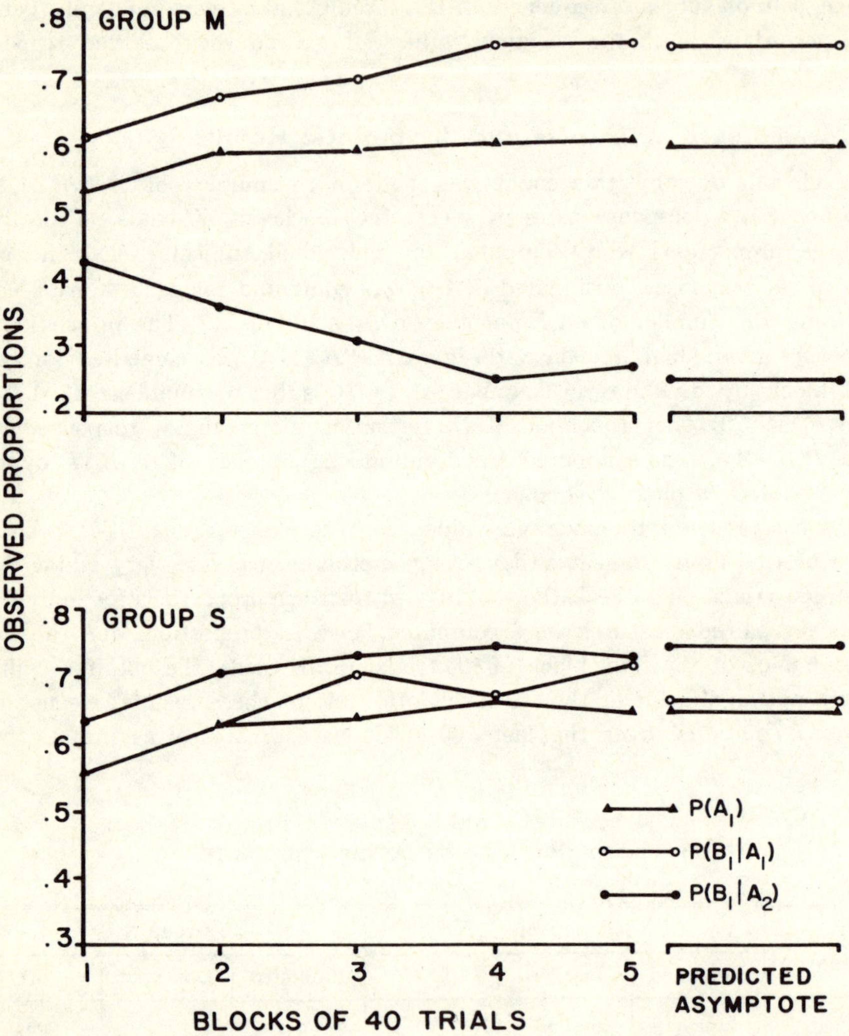

FIGURE 6.3. Observed means for $P(A_1)$, $P(B_1 | A_1)$, and $P(B_1 | A_2)$ in successive 40-trial blocks and predicted asymptotic values.

The largest difference between the predicted and empirical values is .039 for $P_\infty(B_1 | A_1)$ in Group S. Dividing this quantity by the standard error of the mean (in this case $.126/\sqrt{24}$) yields a t of 1.5. It is worthy of note that in the case of Group M the striking numerical differences in the predicted probabilities $P_\infty(B_1 | A_1)$ and $P_\infty(B_1 | A_2)$ are well substantiated experimentally. The interpretation of these results is that the kind of discrimination process postulated at the beginning of the chapter for explaining B's pattern of responses to the

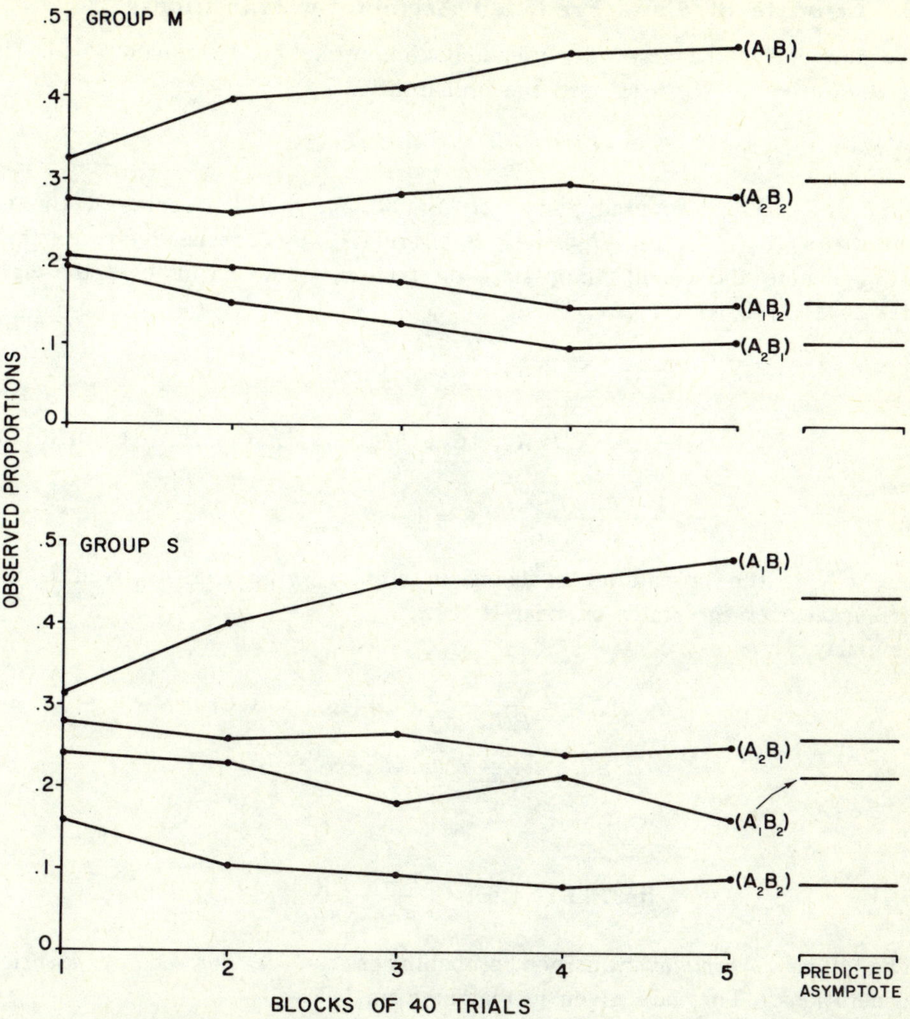

FIGURE 6.4. Observed proportions of the response pairs (A_1B_1), (A_1B_2), (A_2B_1), and (A_2B_2) in successive 40-trial blocks and predicted asymptotic values.

information given him about A's actions does provide a good first approximation to observed behavior. We say "first approximation," for, as in the tests of the one-element model reported in Chapters 3–5, the detailed analysis of transition data given in the next section is not as favorable to the model.

Figure 6.4 presents the mean proportions of the response pairs (A_1B_1), (A_1B_2), (A_2B_1), and (A_2B_2) in 40-trial blocks. The asymptotic predictions for these quantities are given by (6.2.2), (6.2.7), and (6.2.8) for $\theta_A = \theta_B$, and are indicated on the far left of the figure. Over-all agreement between the predicted values and the observed values in the last few blocks of trials is fairly good.

6.5. Estimate of θ and Predicted Asymptotic Transitions

We now derive the first-order pseudomaximum-likelihood estimate θ^* for the case $\theta_A = \theta_B = \theta$. We first need the probabilities

$$\lim_{n\to\infty} P(A_{i,n+1}B_{j,n+1} \mid A_{i',n}B_{j',n}) , \tag{6.5.1}$$

for $i, i', j, j' = 1, 2$. To abbreviate, let X_n, Y_n, Z_n, and W_n represent the response pairs $(A_{1,n}B_{1,n})$, $(A_{1,n}B_{2,n})$, $(A_{2,n}B_{1,n})$, and $(A_{2,n}B_{2,n})$, respectively. Further, let $C_{i,n}$ denote the conditioning state on trial n, where i runs over the eight states 1;11, 1;12, $\cdots$, 2;22. Then

$$\begin{aligned} P(X_{n+1} \mid X_n) &= \frac{1}{P(X_n)} \sum_{i,j} P(X_{n+1}C_{j,n+1}X_nC_{i,n}) \\ &= \frac{1}{P(X_n)} \sum_{i,j} P(X_{n+1} \mid C_{j,n+1})P(C_{j,n+1} \mid X_nC_{i,n})P(X_n \mid C_{i,n})P(C_{i,n}) \\ &= \frac{(1-\theta)(u_{111}^{(n)} + u_{112}^{(n)})}{u_{111}^{(n)} + u_{112}^{(n)}} = 1 - \theta , \end{aligned} \tag{6.5.2}$$

where $u_{ijk}^{(n)}$ is the probability of being in state $i;jk$ on trial n, given a fixed distribution over the states on trial 1.

Similarly,

$$\begin{aligned} P(W_{n+1} \mid X_n) &= \frac{1}{P(X_n)} \sum_{i,j} P(W_{n+1}C_{j,n+1}X_nC_{i,n}) \\ &= \frac{(1-a_1)\theta u_{112}^{(n)}}{u_{111}^{(n)} + u_{112}^{(n)}} , \end{aligned} \tag{6.5.3}$$

and

$$\lim P(W_{n+1} \mid X_n) = \frac{(1-a_1)\theta u_{112}}{u_{111} + u_{112}} . \tag{6.5.4}$$

The other *asymptotic* transition probabilities can be derived by identical considerations. They are given in matrix form below:

(6.5.5)

	X_{n+1}	Y_{n+1}	Z_{n+1}	W_{n+1}
X_n	$(1-\theta)$	$a_1\theta$	$\frac{(1-a_1)\theta u_{111}}{u_{111}+u_{112}}$	$\frac{(1-a_1)\theta u_{112}}{u_{111}+u_{112}}$
Y_n	$a_2\theta$	$(1-\theta)$	$\frac{(1-a_2)\theta u_{121}}{u_{121}+u_{122}}$	$\frac{(1-a_2)\theta u_{122}}{u_{121}+u_{122}}$
Z_n	$\frac{(1-a_3)\theta u_{211}}{u_{211}+u_{221}}$	$\frac{(1-a_3)\theta u_{221}}{u_{211}+u_{221}}$	$(1-\theta)$	$a_3\theta$
W_n	$\frac{(1-a_4)\theta u_{212}}{u_{212}+u_{222}}$	$\frac{(1-a_4)\theta u_{222}}{u_{212}+u_{222}}$	$a_4\theta$	$(1-\theta)$

Let $n_{XX}, n_{XY}, \cdots, n_{ZZ}$ be the observed transition numbers when the process is at asymptote. The function we want to maximize is

$$L^*(\theta) = n_{XX} \log (1 - \theta) + n_{XY} \log a_1\theta + n_{XZ} \log \frac{(1 - a_1)\theta u_{111}}{u_{111} + u_{112}} + \cdots + n_{ZZ} \log (1 - \theta) . \tag{6.5.6}$$

The first-order pseudomaximum-likelihood estimate θ^* is a solution of the equation

$$\frac{dL^*(\theta)}{d\theta} = - \frac{n_{XX}}{1 - \theta} + \frac{n_{XY}}{\theta} + \frac{n_{XZ}}{\theta} \cdots - \frac{n_{ZZ}}{1 - \theta} = 0 , \tag{6.5.7}$$

which simplifies to

$$\theta^* = \frac{\delta}{\delta + \delta'} , \tag{6.5.8}$$

where $\delta' = n_{XX} + n_{YY} + n_{ZZ} + n_{WW}$, $\delta = n_{XY} + n_{XZ} + n_{XW} + n_{YX} + n_{YZ} + n_{YW} + n_{ZX} + n_{ZY} + n_{ZW} + n_{WX} + n_{WY} + n_{WZ}$.

For each pair of subjects, n_{ij}'s $(i, j = X, Y, Z, W)$ were tabulated over the last block of 100 trials. Table 6.2 presents the observed transition probabilities over the last 100 trials aggregated over subject pairs; for example,

$$\hat{P}_\infty(X_{n+1} \mid X_n) = \frac{n_{XX}}{n_{XX} + n_{XY} + n_{XZ} + n_{XW}} .$$

TABLE 6.2

Observed Transition Probabilities for the Last 100 Trials Aggregated over Subject-Pairs

(Also, in parentheses, asymptotic theoretical probabilities based on $\theta^* = .60$)

	Group M				Group S			
n \ $n+1$	X	Y	Z	W	X	Y	Z	W
X	.456 (.395)	.138 (.202)	.098 (.067)	.308 (.336)	.477 (.396)	.222 (.302)	.218 (.272)	.083 (.030)
Y	.523 (.605)	.285 (.395)	.071 (0)	.121 (0)	.566 (.604)	.217 (.396)	.162 (0)	.055 (0)
Z	.345 (.302)	.147 (0)	.198 (.395)	.310 (.303)	.414 (.453)	.135 (0)	.351 (.396)	.100 (.151)
W	.452 (.504)	.119 (0)	.097 (.101)	.332 (.395)	.352 (.151)	.176 (0)	.314 (.453)	.158 (.396)

These probabilities, of course, correspond to the matrix (6.5.5). No statistical test is necessary to see that some of the observed values differ from the theoretical predictions. It suffices to observe that for $a_2 = 1$ and $u_{221} = u_{222} = 0$ (as is the case in both of our groups) the third and fourth entries in row two and the second entry in rows three and four are identically 0 in (6.5.5), whereas in the observed matrices entries in these cells are in some cases markedly different from 0.

Despite this discrepancy, we still are interested in examining the over-all fit between theoretical and observed transitions. Consequently, we estimate θ by (6.5.8), omitting n_{YZ}, n_{YW}, n_{ZY}, and n_{WY} from the computation, since the corresponding theoretical transition probabilities are equal to 0. The results of these estimates for individual pairs of subjects are given in Table 6.3. As shown in the table, the mean $\theta^* = .60$ for both groups. If these transition data are first aggregated over all subjects and θ^* is estimated, the result is .605 for Group M and .593 for Group S. The theoretical transitions given by (6.5.5) for $\theta^* = .60$ are presented in Table 6.2.

To test the correspondence between the observed and theoretical results we have computed the χ^2 given by (2.4.4), modified so that the theoretical quantity in the denominator is replaced by the observed quantity. The obtained values of χ^2 were 330.7 (Group M) and 1,961.0 (Group S), which are both highly significant with $12 - 1 = 11$ degrees of freedom.

TABLE 6.3

THE PSEUDOMAXIMUM-LIKELIHOOD ESTIMATE θ^* FOR INDIVIDUAL SUBJECT PAIRS

Subject Pairs	θ^*		Subject Pairs	θ^*	
	Group M	Group S		Group M	Group S
1	.53	.62	13	.64	.66
2	.56	.70	14	.77	.79
3	.64	.65	15	.70	.50
4	.69	.63	16	.59	.53
5	.36	.74	17	.70	.75
6	.69	.73	18	.69	.69
7	.46	.60	19	.64	.44
8	.67	.48	20	.59	.47
9	.49	.53	21	.70	.63
10	.56	.58	22	.53	.44
11	.50	.57	23	.63	.53
12	.57	.70	24	.62	.53

Mean θ^*: Group M, .60; Group S, .60.

6.6. Game-Theory Comparisons

The game-theory payoff matrix is 2×4, rather than 2×2, for both groups; a strategy for Player B does not consist simply of choosing B_1 or B_2, but of specifying what response is made after Player A's choice is known. We denote a strategy for B by (B_i, B_j), where B_i is the response to A_1 and B_j is the response to A_2. The game matrix in terms of the parameters a_i is then as follows:

(6.6.1)

	(B_1, B_1)	(B_1, B_2)	(B_2, B_1)	(B_2, B_2)
A_1	a_1	a_1	a_2	a_2
A_2	a_3	a_4	a_3	a_4

For Group M, the numerical matrix is

$$\begin{array}{c|cccc} & (B_1, B_1) & (B_1, B_2) & (B_2, B_1) & (B_2, B_2) \\ \hline A_1 & \frac{1}{3} & \frac{1}{3} & 1 & 1 \\ A_2 & \frac{1}{2} & \frac{1}{6} & \frac{1}{2} & \frac{1}{6} \end{array} \tag{6.6.2}$$

The outcome $\frac{1}{3}$ for the pair of strategies A_1 and (B_1, B_2) is a saddle point (and the only one) for this matrix; consequently, these strategies are pure minimax strategies for A and B, respectively. In fact, (B_1, B_2) satisfies the sure-thing principle for B.

For Group S, the numerical payoff matrix is

$$\begin{array}{c|cccc} & (B_1, B_1) & (B_1, B_2) & (B_2, B_1) & (B_2, B_2) \\ \hline A_1 & \frac{1}{2} & \frac{1}{2} & 1 & 1 \\ A_2 & \frac{1}{4} & \frac{3}{4} & \frac{1}{4} & \frac{3}{4} \end{array} \tag{6.6.3}$$

In this case, the outcome $\frac{1}{2}$ for the strategies A_1 and (B_1, B_1) is a saddle point (and the only one) for this payoff matrix; here also these strategies are pure minimax strategies, and again the minimax strategy for B satisfies the stronger sure-thing principle.

The observed asymptotic response probabilities given in Table 6.1 clearly show that the pure game-theory strategies are not even roughly approximated. There is not even any appreciable tendency for the observed probabilities to move away from the learning-theory predictions and toward the optimal game-theory strategies. In terms of individual subjects over the last 100 trials, for Group M, $\max \alpha = .83$, $\max P_\infty(B_1 \mid A_1) = .94$, and $\max P_\infty(B_2 \mid A_2) = .98$. For Group S, $\max \alpha = .87$, $\max P_\infty(B_1 \mid A_1) = .92$, and $\max P_\infty(B_1 \mid A_2) = .92$.

TABLE 6.4

GAME-THEORY, LEARNING-THEORY, AND OBSERVED VALUES OF THE TWO ZERO-SUM GAMES

Type of Value	Group M	Group S
Game Theory	−.167	.000
Learning Theory	−.100	.065
Observed	−.097	.044

The various values of the two games are given in Table 6.4. The learning-theory value is based on the u_{ijk} given in (6.2.7) and (6.2.8). It is interesting to note that each u_{ijk} specifies a pair of strategies, one for each player, among the eight possible pairs given in (6.6.1). The observed value is based on the proportion of A_iB_j response pairs in the last 40 trials. As in the case of Table 3.5, in computing these values we have subtracted $\frac{1}{2}$ from each a_i in order to obtain the zero-sum game that best represents the actual situation studied. An inspection of the table indicates that there is no well-defined tendency for the observed value to move away from the learning-theory value toward the game-theory value.

6.7. Summary

In the experiments of the previous chapters subjects interacted only by means of their mutual control of the reinforcement schedule. In this chapter the degree of interaction was extended by informing one player of the other player's responses. On each trial Player A made one of two possible responses; Player B was then informed of Player A's choice and, in turn, made one of two responses. The trial was concluded when one player was informed that his response was correct and the other player that his response was incorrect. A payoff matrix of the sort used in Chapter 3 represents the contingent probabilities of each player's being correct. Two experiments are reported, each using different payoff matrices; the selection of these matrices was dictated by game theory. Subjects were run for 200 trials.

The theoretical analysis of this extended game situation is in terms of discrimination learning; consequently, Sec. 6.1 is devoted to a discussion of a stimulus-sampling analysis of discrimination experiments. Details of how the theory is applied to our two-person situation are given in Sec. 6.2. The experimental method is described in Sec. 6.3. Mean learning curves and asymptotic results are presented in Sec. 6.4. Good agreement is obtained between predicted and observed asymptotic response probabilities. The pseudomaximum-likelihood estimate of θ and related predictions are given in Sec. 6.5. The estimate $\theta^* = .60$ agrees well with earlier results. The prediction of transition probabilities is rather poor, as is usual with the one-element model, partly because certain transitions are predicted to be identically zero. In Sec. 6.6 the data are analyzed in terms of game theory, and again there is no evidence that the observed values are moving away from learning-theory predictions and toward optimal game strategies. Because of the complexity of the experimental situation the multi-element and generalized conditioning models are not considered.

7

Another Communication Study

7.1. Introduction

In the preceding chapter we studied the effects of communicating Player A's response to Player B before B makes his choice. From a theoretical standpoint we analyzed the situation in terms of Player B's learning to discriminate between two conditioned stimuli. The present chapter is concerned with a similar study. The difference is that Player B is now shown Player A's response at the end of the trial, after B has already made his own response.

Viewed as games, the situations of this and the preceding chapter are quite different. As we have seen, the procedures of Chapter 6 correspond to a game of perfect information, and from the classical theorem of Zermelo [1912] there is a pure optimal strategy for Player B. This is not true for the procedures described in this chapter. In fact, the experimental situation does not have a simple game interpretation. If, as usual, we interpret each trial as one play of the game, the termination of the trial consists of showing Player B Player A's move. The standard concepts of game theory do not yield an analysis of the situation in which information is disclosed after the play of the game is complete; hence, there is no section on game-theory comparisons in the present chapter. It is important that there be no misunderstanding on this point. In principle it is possible to apply statistical decision theory, notably the theory of sequential decisions, to determine an optimal strategy for Player B; but this analysis is difficult and would take us outside the standard framework applied in previous chapters, in which each trial is considered as a complete play of the game.

On the other hand, stimulus-sampling theory can be easily extended to accommodate the new conditions. Player A's two possible responses are still the two stimuli that for Player B become conditioned in different ways. Details of the model are given in the next section.

For comparative purposes the reinforcement parameters a_i are exactly those used in Chapter 6. Group M has the payoff matrix

(7.1.1)

	B_1	B_2
A_1	$\frac{1}{3}$	1
A_2	$\frac{1}{2}$	$\frac{1}{6}$

and Group S has the payoff matrix

(7.1.2)

	B_1	B_2
A_1	$\frac{1}{2}$	1
A_2	$\frac{1}{4}$	$\frac{3}{4}$

7.2. Model

As in Chapter 6, we assume that Player A has a single stimulus element s available for sampling on each trial. However, Player B is shown, and thus may be affected by, the response of Player A on the previous trial; therefore we posit two stimulus elements, s_1 and s_2, for Player B. If Player A makes the A_1 response on trial $n-1$, the element s_1 will be sampled on trial n with probability 1; if the A_2 response is made on trial $n-1$, the element s_2 will be sampled on trial n with probability 1.

This description of the stimulus elements and the specification of (7.1.1) or (7.1.2) determine the derivation of the Markov process from the axioms of Sec. 1.2. On each trial Player A's element s is conditioned to either A_1 or A_2; similarly, Player B's elements s_1 and s_2 are each conditioned to either B_1 or B_2. In addition to these conditioning relations, it is necessary to specify which stimulus element is sampled on trial n by Player B. Therefore, there are 16 possible states to be represented. We use the triplet notation of Chapter 6 to indicate the conditioning of s, s_1, and s_2. In addition, an asterisk is appended

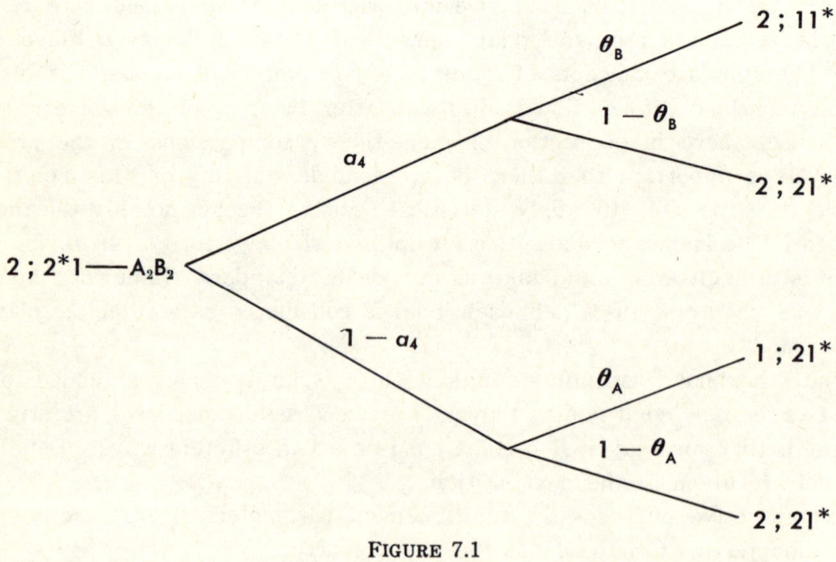

FIGURE 7.1

to either the second or the third member of the triple to indicate the stimulus element that is sampled. Thus if we are in state 1 ; 12* on trial n, the s_2 element is sampled and an $(A_{1,n}B_{2,n})$ event occurs. Of course, when we know that s_2 is sampled on trial n, we know that Player A made the A_2 response on trial $n-1$.

One of the trees for the Markov process, the tree starting from state 2 ; 2*1, is given in Fig. 7.1. The states on trial $n+1$ have the asterisk on the third position, since an A_2 response occurred on trial n.

Construction of the analogous trees yields the transition matrix (7.2.1), where $\varphi_i = a_i(1-\theta_B) + (1-a_i)(1-\theta_A)$ and $\delta_i = (1-a_i)\theta_A$.

(7.2.1)

	1 ; 1*1	1 ; 1*2	1 ; 2*1	1 ; 2*2	2 ; 1*1	2 ; 1*2	2 ; 2*1	2 ; 2*2	1 ; 11*	1 ; 12*	1 ; 21*	1 ; 22*	2 ; 11*	2 ; 12*	2 ; 21*	2 ; 22*
1 ; 1*1	φ_1		$a_1\theta_B$		δ_1											
1 ; 1*2		φ_1		$a_1\theta_B$		δ_1										
1 ; 2*1	$a_2\theta_B$		φ_2				δ_2									
1 ; 2*2		$a_2\theta_B$		φ_2				δ_2								
2 ; 1*1									δ_3				φ_3		$a_3\theta_B$	
2 ; 1*2										δ_3				φ_3		$a_3\theta_B$
2 ; 2*1											δ_4		$a_4\theta_B$		φ_4	
2 ; 2*2												δ_4		$a_4\theta_B$		φ_4
1 ; 11*	φ_1	$a_1\theta_B$			δ_1											
1 ; 12*	$a_2\theta_B$	φ_2				δ_2										
1 ; 21*			φ_1	$a_1\theta_B$			δ_1									
1 ; 22*			$a_2\theta_B$	φ_2				δ_2								
2 ; 11*									δ_3				φ_3	$a_3\theta_B$		
2 ; 12*										δ_4			$a_4\theta_B$	φ_4		
2 ; 21*											δ_3				φ_3	$a_3\theta_B$
2 ; 22*												δ_4			$a_4\theta_B$	φ_4

From the discussion in Sec. 1.3 it is clear that the limiting quantities u_{ij^*k} and u_{ijk^*} for the states exist and are independent of the initial probability distribution. In terms of these quantities the observable asymptotic response probabilities are defined as follows:

(7.2.2)
$$\begin{aligned}
P_\infty(A_1B_1) &= u_{11^*1} + u_{11^*2} + u_{111^*} + u_{121^*}\,,\\
P_\infty(A_1B_2) &= u_{12^*1} + u_{12^*2} + u_{112^*} + u_{122^*}\,,\\
P_\infty(A_2B_1) &= u_{21^*1} + u_{21^*2} + u_{211^*} + u_{221^*}\,,\\
P_\infty(A_2B_2) &= u_{22^*1} + u_{22^*2} + u_{212^*} + u_{222^*}\,.
\end{aligned}$$

In this experiment we shall also be interested in the additional quantities

$$P_\infty(A_1) = P_\infty(A_1B_1) + P_\infty(A_1B_2)\,,$$
$$P_\infty(B_1) = P_\infty(A_1B_1) + P_\infty(A_2B_1)\,,$$
(7.2.3)
$$\lim_{n\to\infty} P(B_{1,n} \mid A_{1,n-1}) = \frac{u_{11*1} + u_{11*2} + u_{21*1} + u_{21*2}}{P_\infty(A_1)}\,,$$
$$\lim_{n\to\infty} P(B_{1,n} \mid A_{2,n-1}) = \frac{u_{111*} + u_{121*} + u_{211*} + u_{221*}}{1 - P_\infty(A_1)}\,.$$

With regard to these asymptotic quantities, it should be pointed out that they all depend on θ_A and θ_B. Even when $\theta_A = \theta_B = \theta$, the u's are not independent of θ. This, of course, can be seen by inspecting (7.2.1). However, as will be noted later, this dependency has little effect, since the bounds on these quantities (i.e., for $\theta_A, \theta_B \to 0, 1$) are quite close.

Unfortunately, the technique used in Sec. 6.2 for deriving $\lim P(B_{1,n} \mid A_{1,n})$ and $\lim P(B_{1,n} \mid A_{2,n})$ is not applicable here for deriving $\lim P(B_{1,n} \mid A_{1,n-1})$ and $\lim P(B_{1,n} \mid A_{2,n-1})$. The reason is fairly obvious. In the earlier case we could study the subsequence of B's responses on those trials for which Player A made an A_1 response, since the behavior of Player B was determined by the stimulus sampled and the probability of reinforcement. However, in the present situation, if we consider the subsequence of trials immediately following trials on which an A_1 response occurred, the probability of an E_1 or E_2 reinforcing event for Player B is not determined; for this probability also depends on Player A's response on the trial under consideration. On the other hand, if we consider the subsequence of trials on which an A_1 response occurred, then the stimulus sampled by Player B is not specified.

Superficially, one might assume that this impasse could be overcome by considering the four subsequences of trials that take account of Player A's responses on the given trial and the preceding one. However, this will not do because the conditioning of s_1 in the subsequence defined by $A_{1,n}A_{1,n-1}$, say, is not independent of conditioning in the subsequence $A_{2,n}A_{1,n-1}$. For example, suppose we have the following event: $A_{1,n+2}A_{1,n+1}A_{2,n}A_{1,n-1}A_{1,n-2}$. Further, assume that $s_1 = A_{1,n-2}$ is conditioned to B_2 at the end of trial $n-1$, and that $s_1 = A_{1,n-1}$ is conditioned to B_1 at the end of trial n. When s_1 is sampled again, namely on trial $n+2$, response B_1 will be made. The important observation is that the subsequence defined by $A_{1,n}A_{1,n-1}$ is not self-contained with respect to the conditioning of s_1. As the example illustrates, what happens in the $A_{2,n}A_{1,n-1}$ subsequence affects conditioning in the first subsequence, and vice versa. An identical argument holds for the conditioning of $s_2 = A_2$ on a preceding trial.

In principle, the model considered in this section has an interesting application to the theory of coalitions. Von Neumann and Morgenstern, in their introductory remarks on coalitions [1947, p. 224], mention an alternative to their view that coalitions arise on the basis of explicit agreements between the players on a given play of the game: "If our theory were applied as a statistical analysis of a long series of plays of the same game—and not as the analysis of one isolated play—an alternative interpretation would suggest itself. We should then view agreements and all forms of cooperation as establishing themselves by repetition in such a long series of plays."

Their alternative could be realized experimentally by rather simple modifications of the setup used to test the model of this section. We could increase the number of players to three, and permit each player to see the response choices of the other two at the end of each trial, i.e., at the end of each play of the game. Stimulus-sampling theory yields specific predictions concerning the statistical tendency for the formation of coalitions under any given reinforcement schedule for the three players.

Unhappily, however, the practical difficulties of computing these predictions are staggering. The model already considered in this section has 16 states. In a three-person setup, if each player sees the responses of the other two, each player is sampling stimuli from a set of four; and, of course, for each player there would be 16 possible ways for his four stimuli to be conditioned. In addition, it would be necessary to specify the actual response for each player on the previous trial, in the same way as in the model of this section we needed to specify the response of Player A. For three players, there are eight possible patterns of actual responses. Putting these possibilities together, we see that there are $8 \times 16 \times 16 \times 16 = 32{,}768$ states in the Markov chain; and this represents the simplest three-person game situation in which each player is informed of the actions of the other two. The most discouraging thing about this large number of states is that there are no obvious principles of continuity that would permit us to convert the theory into a continuous one and apply more powerful mathematical methods of analysis. There are some possibilities of drastic simplification, but as yet they are too tentative to report here.

7.3. Experimental Method

Subjects. The subjects were 112 undergraduates obtained from introductory courses in psychology at Stanford University. They were randomly assigned to two groups with the restriction of 28 pairs in each group.

Apparatus. The apparatus described in Chapter 3 was used. In addition (as described in Chapter 6), green lights were mounted 1 inch above each of Player B's keys. At the end of each trial Player B was informed of the response made by Player A at the start of the trial. That is, if at the start of the trial Player A pressed his right key, then at the end of the trial Player B's right green light would go on; if Player A pressed his left key, Player B's left green light would go on.

The time intervals were identical to those given in Chapter 3. On all trials the signal light was lighted for 3.5 seconds; the time between successive signal exposures was 10 seconds. The reinforcing light followed the cessation of the signal light by 1.5 seconds and remained on for 2 seconds. Player B's green light went on 1 second after the onset of the reinforcing light and remained on for 2 seconds.

Procedure. One member of each pair was randomly selected as Player A and the other as Player B. Further, within each of the two experimental groups four subgroups of seven subject-pairs were formed by counterbalancing right and left positions of the A_i and B_i keys.

The subjects were read the following instructions:

"This experiment is similar to a game situation involving two players; on each play of the game one wins and the other loses. To keep things straight, I will call the person on my left Player A and the person on my right Player B.

"The experiment for each of you consists of a series of trials. The top center light on your panel will light to indicate the start of each trial. When this light goes on, you will each press one or the other of the two keys in front of you. Shortly thereafter, one or the other of the two lower white lights will light on each of your panels. If the white light above the key you pressed goes on, your key choice was correct and the other player's choice was incorrect. If, however, the white light above the key opposite from the one you pressed goes on, you were incorrect and the other player was correct. Thus, on each trial one player is correct and the other incorrect.

"Being correct on a given trial depends on the combination of your key press with the other player's key press. For example, it could be that on some trials if you press your left key and the other player also presses his left key, then you will be correct and he will be incorrect; whereas if he had pressed his right key, you would be incorrect and he would be correct. However, combinations of your key press with his key press that were correct on one trial will not necessarily be correct on some later trial.

"For Player A the above instructions provide all the information he has about the situation; however, for Player B we are going to present additional information. On each trial after you have both made a key press, Player B not only will be informed whether he was correct or incorrect but also will be told what key Player A pressed on the trial. That is, if Player A has pressed his right key, Player B's right green light will go on; whereas if Player A has pressed his left key, Player B's left green light will go on. Thus at the end of each trial Player B will know not only whether he was correct or incorrect but also what key Player A has pressed.

"For both of you, your job is to obtain as many correct key choices as you possibly can. At the start you will not know what keys to press; consequently just guess. But as time goes on, you will be able to improve this guessing behavior; Player B in particular will be able to take advantage of the fact that he knows what keys Player A has pressed on previous trials.

"The trials move along rapidly, and you must both make your key press immediately after the onset of the top center white light. That is, as soon as the signal light goes on, press one or the other key down and release it. Then wait until one of the lower white lights goes on. If the white light above the key you pressed goes on, your key choice was correct and the other player's choice was incorrect. If the white light above the key opposite from the one you pressed goes on, you were incorrect and the other player was correct."

After the instructions were read, 220 trials were run in continuous sequence. For each pair of subjects sequences of reinforcing lights were generated in accordance with the assigned values of a_i and the observed response pairs.

7.4. Mean Learning Curves and Asymptotic Results

Figure 7.2 presents the observed mean proportions for the response pair (A_1B_1), (A_1B_2), (A_2B_1), and (A_2B_2) in successive blocks of 20 trials. On the far right of the figure are the predicted asymptotic values given by (7.2.2) for $\theta_A = \theta_B = .6$.

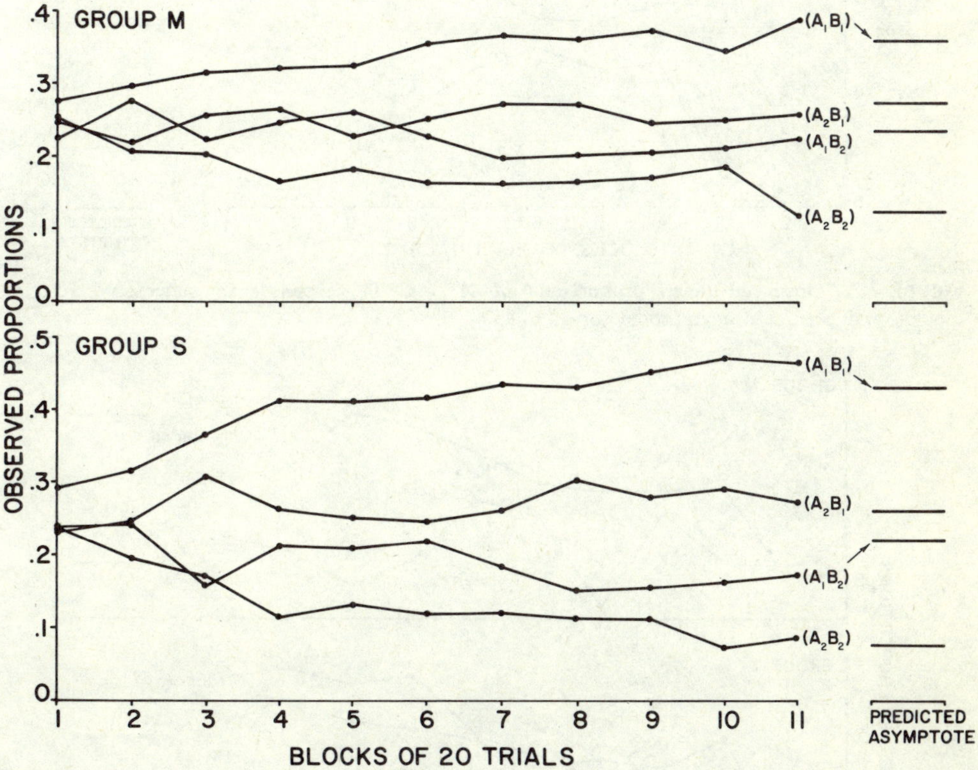

FIGURE 7.2. Observed mean proportions of the response pairs (A_1B_1), (A_1B_2), (A_2B_1), and (A_2B_2) in successive 20-trial blocks, and predicted asymptotes for $\theta_A = \theta_B = .6$.

Asymptotic predictions for other values of θ_A and θ_B will be presented later, and it will be evident that these predictions are relatively insensitive to variations in the values of θ_A and θ_B. Consequently, to provide a preliminary perusal of the data we use the predictions based on $\theta_A = \theta_B = .6$, since in the previous studies the estimated values have been of this magnitude.

Figure 7.3 presents the mean proportion of A_1 and B_1 responses in successive 20-trial blocks. On the far right of the figure are the predicted values for

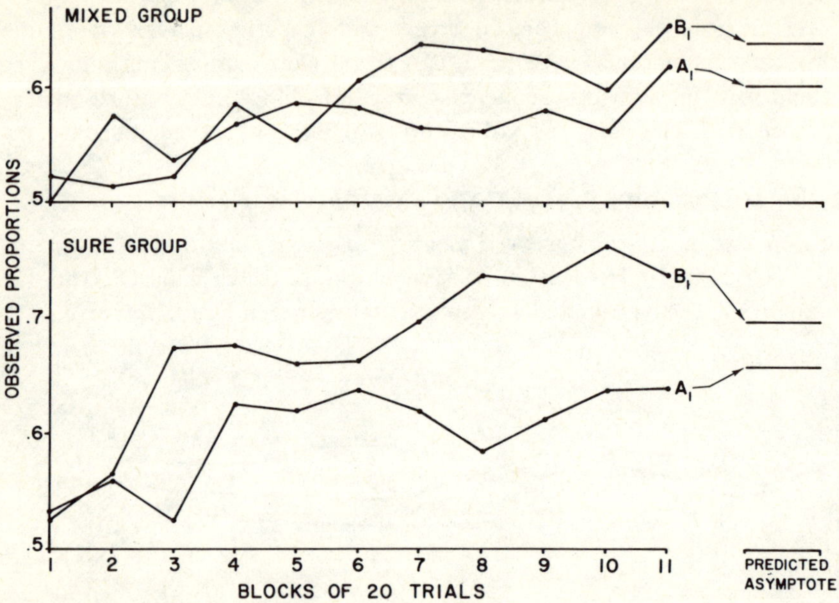

FIGURE 7.3. Observed mean proportions of A_1 and B_1 responses in successive 20-trial blocks, and predicted asymptotes for $\theta_A = \theta_B = .6$.

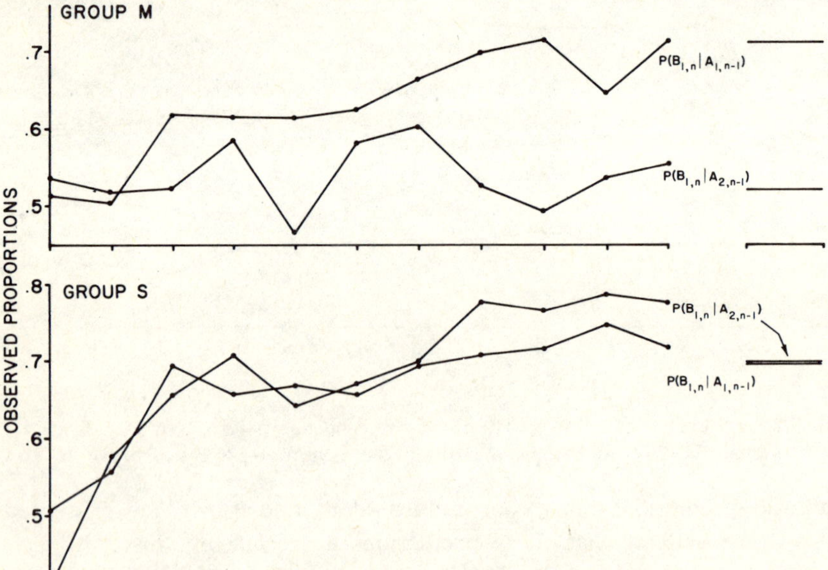

FIGURE 7.4. Observed means for $P(B_{1,n} | A_{1,n-1})$ and $P(B_{1,n} | A_{2,n-1})$ in successive 20-trial blocks, and predicted asymptotes for $\theta_A = \theta_B = .6$.

TABLE 7.1

ASYMPTOTIC PREDICTIONS FOR VARIOUS VALUES OF θ_A and θ_B

Theta Value		Asymptotic Prediction							
θ_A	θ_B	$P(A_1B_1)$	$P(A_1B_2)$	$P(A_2B_1)$	$P(A_2B_2)$	$P(A_1)$	$P(B_1)$	$P(B_{1,n} \mid A_{1,n-1})$	$P(B_{1,n} \mid A_{2,n-1})$
Group M									
.4	.8	.382	.218	.237	.163	.600	.619	.725	.458
.6	.4	.361	.239	.278	.122	.600	.639	.714	.525
.6	.6	.364	.236	.272	.128	.600	.636	.712	.522
.7	.4	.355	.245	.290	.110	.600	.645	.707	.552
.7	.5	.357	.243	.286	.114	.600	.643	.705	.550
.7	.6	.359	.241	.283	.117	.600	.641	.703	.548
.7	.7	.360	.240	.279	.121	.600	.639	.702	.547
.7	.8	.362	.238	.276	.124	.600	.638	.700	.545
.8	.4	.351	.249	.298	.102	.600	.649	.698	.575
.8	.6	.355	.245	.290	.110	.600	.645	.693	.572
.8	.8	.359	.241	.282	.118	.600	.641	.689	.569
Group S									
.4	.8	.435	.219	.261	.085	.654	.696	.687	.713
.6	.4	.434	.223	.263	.080	.657	.697	.698	.696
.6	.6	.434	.224	.263	.079	.658	.697	.697	.698
.7	.4	.434	.227	.264	.075	.661	.698	.702	.691
.7	.5	.434	.227	.264	.075	.661	.698	.702	.692
.7	.6	.434	.228	.264	.074	.661	.698	.701	.693
.7	.7	.434	.228	.264	.074	.662	.698	.701	.694
.7	.8	.434	.228	.265	.073	.662	.699	.700	.695
.8	.4	.434	.231	.266	.069	.665	.700	.706	.686
.8	.6	.434	.231	.266	.069	.665	.700	.705	.689
.8	.8	.434	.231	.266	.069	.665	.700	.704	.691

TABLE 7.2

THEORETICAL PREDICTIONS FOR $\theta = .6$ AND OBSERVED MEAN PROPORTIONS OF A_iB_j RESPONSE PAIRS OVER THE LAST 60 TRIALS

Response Pair	Group M			Group S		
	Predicted	Observed	SD	Predicted	Observed	SD
$P(A_1B_1)$	.364	.369	.139	.434	.463	.104
$P(A_1B_2)$	.236	.215	.090	.224	.164	.117
$P(A_2B_1)$	.272	.254	.103	.263	.282	.121
$P(A_2B_2)$	.128	.186	.161	.079	.091	.081

$\theta_A = \theta_B = .6$ given by (7.2.3). Finally, Fig. 7.4 presents the mean proportion of B_1 responses given an A_i $(i = 1, 2)$ on the preceding trial. The asymptotic predictions for $\theta_A = \theta_B = .6$ given by (7.2.3) are also presented.

Table 7.1 presents predicted asymptotic quantities for various values of θ_A and θ_B. They are obtained by solving the 16 linear equations given by (1.3.17) and then making appropriate substitutions in (7.2.2) and (7.2.3).

Table 7.2 presents the predicted asymptotic values for $\theta_A = \theta_B = .6$ and the observed proportions and standard deviations over the last 60 trials for $P(A_1B_1)$, $P(A_1B_2)$, $P(A_2B_1)$, and $P(A_2B_2)$. If we take these observed proportions as estimates of asymptotic performance, the comparison with theoretical predictions is quite good. For Group M the largest discrepancy between predicted and observed values is .058 for the $P(A_2B_2)$ measure; relative to the observed standard deviation this difference is not large $(t = (.058)\sqrt{24}/.161 = 1.76)$. For Group S, the largest discrepancy is .060 for the $P(A_1B_2)$ measure. Dividing this difference by the standard error of the mean yields a value of 2.51; a formal statistical test of whether or not this difference is significant is a complex matter, and we do not attempt it. Further, an inspection of Fig. 7.1 suggests that the curve for $P(A_1B_2)$ in Group S is progressively moving toward the predicted values over the last three blocks and may not yet be at asymptote.

The asymptotic predictions of Table 7.1 for $\theta_A = \theta_B = .6$ are very close to those given in Chapter 3 for the corresponding reinforcement schedules. Comparison with (3.1.5) for Group M and with (3.1.13) for Group S is most appropriate. It is apparent that Player B's observation of Player A's responses after the trial does not much affect the asymptotic mean probabilities of response. Comparison of the last two columns of the table with the predictions $P(B_{1,n} \mid A_{1,n})$ and $P(B_{1,n} \mid A_{2,n})$ for Chapter 6 is also of interest. The differential effects of the discrimination are much sharper in Chapter 6, i.e., in Chapter 6 the quantity

$$|P(B_{1,n} \mid A_{1,n}) - P(B_{1,n} \mid A_{2,n})|$$

is greater than the quantity

$$|P(B_{1,n} \mid A_{1,n-1}) - P(B_{1,n} \mid A_{2,n-1})|$$

in the present chapter. This theoretical result, which is also confirmed by the data, accords well with intuitive and game-theory ideas. In order not to overemphasize the closeness of the asymptotic predictions of this chapter to the mean response predictions in Chapter 3, it should be mentioned that greater discrepancies exist when the predicted transition probabilities given in Table 7.3 are compared with those for Chapter 3 given by the general matrix (1.4.3). For example, the predicted matrices for this chapter each have two zeros, whereas the two derived from (1.4.3) have five.

7.5. Asymptotic Transition Data

We now compare the observed transition data with theoretical predictions for $\theta_A = \theta_B$. To begin with, we need the probabilities

$$P(A_{i,n+1}B_{j,n+1} \mid A_{i',n}B_{j',n}) \qquad (i, j = 1, 2).$$

As before, let $X_n = (A_{1,n}B_{1,n})$, $Y_n = (A_{1,n}B_{2,n})$, $Z_n = (A_{2,n}B_{1,n})$, and $W_n = (A_{2,n}B_{2,n})$. Further, let $C_{i,n}$ indicate the conditioning states on trial n, where i runs over the 16 states given in (7.2.1). Then

$$P(X_{n+1} \mid X_n) = \frac{1}{P(X_n)} \sum_{i,j} P(X_{n+1}C_{j,n+1}X_nC_{i,n})$$
$$= \frac{1}{P(X_n)} \sum_{i,j} P(X_{n+1} \mid C_{j,n+1})P(C_{j,n+1} \mid X_nC_{i,n})P(X_n \mid C_{i,n})P(C_{i,n}) .$$

Evaluating the components of the summation, we obtain

$$P(X_{n+1} \mid X_n) = \frac{(1-\theta)(u_{11^*1}^{(n)} + u_{11^*2}^{(n)} + u_{111^*}^{(n)}) + \theta a_1 u_{111^*}^{(n)}}{P(X_n)} ,$$

where $u_{ijk^*}^{(n)}$ is the probability of being in state ijk^* on trial n, given a fixed distribution on trial 1. As n becomes large, independent of the initial distribution,

$$\lim P(X_{n+1} \mid X_n) = \frac{(1-\theta)(u_{11^*1} + u_{11^*2} + u_{111^*}) + \theta a_1 u_{111^*}}{u_{11^*1} + u_{11^*2} + u_{111^*} + u_{121^*}} . \tag{7.5.1}$$

The analogous asymptotic transition probabilities can be derived by identical considerations. They are as follows:

$$\begin{aligned}
P_\infty(Y_{n+1} \mid X_n) &= [(1-\theta)u_{121^*} + \theta a_1(u_{11^*1} + u_{11^*2} + u_{121^*})]/P_\infty(X) ,\\
P_\infty(Z_{n+1} \mid X_n) &= \theta(1-a_1)(u_{11^*1} + u_{11^*2} + u_{111^*})/P_\infty(X) ,\\
P_\infty(W_{n+1} \mid X_n) &= \theta(1-a_1)u_{121^*}/P_\infty(X) ,\\
P_\infty(X_{n+1} \mid Y_n) &= [(1-\theta)u_{112^*} + \theta a_2(u_{12^*1} + u_{12^*2} + u_{112^*})]/P_\infty(Y) ,\\
P_\infty(Y_{n+1} \mid Y_n) &= [(1-\theta)(u_{12^*1} + u_{12^*2} + u_{122^*}) + \theta a_2 u_{122^*}]/P_\infty(Y) ,\\
P_\infty(Z_{n+1} \mid Y_n) &= \theta(1-a_2)u_{112^*}/P_\infty(Y) ,\\
P_\infty(W_{n+1} \mid Y_n) &= \theta(1-a_2)(u_{12^*1} + u_{12^*2} + u_{122^*})/P_\infty(Y) ,\\
P_\infty(X_{n+1} \mid Z_n) &= \theta(1-a_3)(u_{21^*1} + u_{211^*} + u_{221^*})/P_\infty(Z) ,\\
P_\infty(Y_{n+1} \mid Z_n) &= \theta(1-a_3)u_{21^*2}/P_\infty(Z) ,\\
P_\infty(Z_{n+1} \mid Z_n) &= [(1-\theta)(u_{21^*1} + u_{211^*} + u_{221^*}) + \theta a_3 u_{21^*1}]/P_\infty(Z) ,\\
P_\infty(W_{n+1} \mid Z_n) &= [(1-\theta)u_{21^*2} + \theta a_3(u_{21^*2} + u_{211^*} + u_{221^*})]/P_\infty(Z) ,\\
P_\infty(X_{n+1} \mid W_n) &= \theta(1-a_4)u_{22^*1}/P_\infty(W) ,\\
P_\infty(Y_{n+1} \mid W_n) &= \theta(1-a_4)(u_{22^*2} + u_{212^*} + u_{222^*})/P_\infty(W) ,\\
P_\infty(Z_{n+1} \mid W_n) &= [(1-\theta)u_{22^*1} + \theta a_4(u_{22^*1} + u_{212^*} + u_{222^*})]/P_\infty(W) ,\\
P_\infty(W_{n+1} \mid W_n) &= [(1-\theta)(u_{22^*2} + u_{212^*} + u_{222^*}) + \theta a_4 u_{22^*2}]/P_\infty(W) ,
\end{aligned} \tag{7.5.2}$$

where

$$\begin{aligned}
P_\infty(X) &= u_{11^*1} + u_{11^*2} + u_{111^*} + u_{121^*} ,\\
P_\infty(Y) &= u_{12^*1} + u_{12^*2} + u_{122^*} + u_{112^*} ,\\
P_\infty(Z) &= u_{21^*1} + u_{21^*2} + u_{211^*} + u_{221^*} ,\\
P_\infty(W) &= u_{22^*1} + u_{22^*2} + u_{212^*} + u_{222^*} .
\end{aligned} \tag{7.5.3}$$

For both experimental groups $a_2 = 1$; consequently, the transition probabilities $P_\infty(Z_{n+1} \mid Y_n)$ and $P_\infty(W_{n+1} \mid Y_n)$ are identically 0. All other transitions are possible.

In terms of (7.5.1–3) we may define a likelihood function similar to (6.5.7). Unfortunately, in order to find its maximum we need explicit expressions for the u's as functions of θ, which would be very difficult to obtain; and even if such expressions were available, it can be shown that the pseudomaximum-likelihood estimate would not have a simple form. We have investigated other estimates for which the transition probabilities are conditioned on several responses and reinforcing events, but they involved even more complicated expressions.

By way of compromise we shall compute the transition probabilities given by (7.5.1) and (7.5.2) for $\theta_A = \theta_B = .6$, and compare the predictions with observed data. Of course, this particular value of the conditioning parameter was selected because it is close to the estimates obtained in previous chapters.

In order to compute (7.5.1) and (7.5.2) we need the values of u_{ijk} for $\theta_A = \theta_B = .6$. For Group M

$$\begin{aligned} &u_{11*1} = .1818\,, && u_{21*1} = .0928\,, && u_{111*} = .0502\,, && u_{211*} = .0744\,, \\ &u_{11*2} = .1090\,, && u_{21*2} = .0436\,, && u_{112*} = .0462\,, && u_{212*} = .0663\,, \\ &u_{12*1} = .1021\,, && u_{22*1} = .0092\,, && u_{121*} = .0230\,, && u_{221*} = .0613\,, \\ &u_{12*2} = .0615\,, && u_{22*2} = 0\,, && u_{122*} = .0262\,, && u_{222*} = .0524\,. \end{aligned} \tag{7.5.4}$$

For Group S

$$\begin{aligned} &u_{11*1} = .1936\,, && u_{21*1} = .0826\,, && u_{111*} = .0816\,, && u_{211*} = .0988\,, \\ &u_{11*2} = .1402\,, && u_{21*2} = .0421\,, && u_{112*} = .0268\,, && u_{212*} = .0528\,, \\ &u_{12*1} = .1124\,, && u_{22*1} = .0056\,, && u_{121*} = .0187\,, && u_{221*} = .0397\,, \\ &u_{12*2} = .0815\,, && u_{22*2} = 0\,, && u_{122*} = .0031\,, && u_{222*} = .0205\,. \end{aligned} \tag{7.5.5}$$

Table 7.3 presents the predicted transition probabilities computed in terms of (7.5.1–5) and the observed transition proportions for the last block of 90 trials. The over-all agreement between predicted and observed quantities is not good. To provide a measure of the agreement we have calculated the χ^2 given by (2.4.4), replacing the theoretical quantity in the denominator by the observed quantity, since two of the theoretical transitions are 0. The obtained values of χ^2 were 405.2 for Group M and 319.5 for Group S. With 12 degrees of freedom these quantities are both extremely large.

These poor goodness-of-fit results for what is essentially the one-element model (on each trial B has exactly one element available for sampling, even though there are two elements in his set of stimuli) are not peculiar to the experimental situation of this chapter. Because of the many zeros in the transition matrices for the three groups of Chapter 3, the goodness-of-fit test was not even applied to the one-element model; and similarly for the two groups of Chapter 4. In these earlier chapters we could go on to consider multi-element models, as well as the generalized conditioning model, and obtain considerably better goodness-of-fit results. The extraordinary large values of χ^2 obtained above testify to the extreme sensitivity of the test, and thus add to the signifi-

TABLE 7.3

OBSERVED AND THEORETICAL TRANSITION VALUES

(Observed proportions are for the last 90 trials; theoretical probabilities, given in parentheses, are calculated for $\theta = .6$)

	Group M				Group S			
n \ $n+1$	X	Y	Z	W	X	Y	Z	W
X	.425 (.402)	.192 (.198)	.277 (.375)	.106 (.025)	.490 (.439)	.180 (.261)	.262 (.287)	.068 (.013)
Y	.479 (.612)	.219 (.388)	.199 (0)	.103 (0)	.502 (.640)	.235 (.360)	.155 (0)	.108 (0)
Z	.313 (.252)	.201 (.048)	.292 (.438)	.194 (.262)	.443 (.378)	.087 (.072)	.393 (.383)	.077 (.167)
W	.196 (.036)	.286 (.464)	.214 (.129)	.304 (.371)	.325 (.011)	.198 (.139)	.265 (.479)	.212 (.371)

cance of the good results obtained by using the more complicated models. Unfortunately, the computations involved in applying either the two-element model or the generalized conditioning model to the present experimental situation were too great to undertake. There is, however, no reason to think that these models would be any less satisfactory here than in earlier chapters.

7.6. Summary

In Chapter 6 we extended the degree of interaction between players in a two-person game situation by communicating Player A's response immediately to Player B, so that B could use this information in choosing his own response. In the present chapter, B is not shown A's response until after he has made his own response. Thus his knowledge of A's response can be used only on subsequent trials.

The theoretical analysis of this situation is based on the same discrimination learning assumptions that were used in Chapter 6, but interestingly enough the resulting Markov chain is completely different. The details of the theoretical analysis and predictions are given in Secs. 7.2 and 7.5. Also, in Sec. 7.2 we consider the application of the model to the theory of coalitions, as outlined by von Neumann and Morgenstern. In Sec. 7.3 we report details of the experimental method. In Secs. 7.4 and 7.5 theoretical predictions are compared with observed data. The observed mean proportions and some sequential effects are well described by the theory; however, certain of the predictions for transition data are not as satisfactory.

8

A Three-Person, Simple Majority Situation

8.1. General Description of the Situation

In the preceding five chapters we have reported different two-person experiments. Generally speaking, it is perfectly obvious how to extend the models to n-person experiments for $n > 2$. This is far from the case for other theories of multiperson situations. The difficulties of passing from two-person to three-person zero-sum games in the theory of games are well known. In classical economics the theory of monopolistic competition rather radically increases in complexity in passing from duopoly to oligopoly. Other theoretical examples are easily cited. For this reason it seemed of interest to execute at least one experiment involving the simultaneous interaction of three persons.

The study deals with a three-person game situation in which the outcome on a given trial is specified by a majority rule. In general terms, we will be investigating the changes that occur in the behavior of group members when the response made by the majority of the group is correct, and the changes that occur when the majority's response is incorrect.

We designate the three players as A, B, and C. At the start of a trial each of the three players independently chooses one of two responses: A_1 or A_2, B_1 or B_2, and C_1 or C_2. After the players have indicated their responses, the outcome of the trial is announced. At the start of an experimental session the subjects are informed of the game characteristics of the situation, and are instructed to maximize the number of trials on which their responses are correct. However, once the session is under way they are told only whether they were correct or not on a given trial; they are not told the other two players' responses. (The computational difficulties involved in permitting each player to see the responses of the other two are discussed in Sec. 7.2.)

The outcome of each trial is identical for the three players; i.e., either E_1 or E_2 occurs for all players. Further, the outcome on a given trial depends on whether the majority of responses are Type 1 (i.e., $A_1B_1C_1$, $A_1B_1C_2$, $A_1B_2C_1$,

$A_2B_1C_1$) or Type 2 (i.e., $A_2B_2C_2$, $A_2B_2C_1$, $A_2B_1C_2$, $A_1B_2C_2$). Specifically, the game can be characterized by the following matrix:

(8.1.1)

	E_1	E_2
Majority A_1	m_1	$1 - m_1$
Majority A_2	m_2	$1 - m_2$

If the majority of responses are of Type 1, the probability of an E_1 event for all players is m_1, and the probability of an E_2 event is $1 - m_1$; if the majority are of Type 2, the corresponding probabilities are m_2 and $1 - m_2$, respectively. In other words, the majority are correct with probability m_1 if they select Type 1 responses, and are correct with probability $1 - m_2$ if they select Type 2 responses.

In the study to be considered in this chapter, four groups were run. For all groups $m_1 = .6$. The groups differed with respect to m_2, which took the following values: .8 (Group I), .6 (Group II), .4 (Group III), and .2 (Group IV).

8.2. Model

As before, we assume one stimulus element for each player; hence, the stipulation of (8.1.1) determines the derivation of the Markov process from the axioms of Sec. 1.2. There are eight possible states, which will be represented by the ordered triples (1, 1, 1), (1, 1, 2), (1, 2, 1), (1, 2, 2), (2, 1, 1), (2, 1, 2), (2, 2, 1), and (2, 2, 2). The first member of each triple indicates the conditioning of Player A's stimulus element, the second Player B's, and the third Player C's.

To illustrate the application of our basic axioms to this problem we display in Fig. 8.1 one of the trees for the Markov process, the tree starting from state (1, 2, 2). The reason θ_B and θ_C are omitted from the lower part of the tree and θ_A from the upper part is that the effectiveness of conditioning is irrelevant when the response actually made is reinforced (a point discussed in Sec. 1.3). Construction of the analogous trees yields an 8×8 transition matrix, which represents an irreducible, aperiodic Markov chain.

We shall not pursue our investigation of this particular Markov process because a major simplification can be introduced if we assume that $\theta_A = \theta_B = \theta_C = \theta$. The simplification enables us to collapse states (1, 1, 2), (1, 2, 1), and (2, 1, 1) into a single state, and states (2, 2, 1), (2, 1, 2), and (1, 2, 2) into another single state, thus obtaining a four-state chain. That is, if we do not distinguish between the players in terms of their θ values, application of the axioms of Sec. 1.2 yields a Markov chain in which the states may be identified as 0, 1, 2, and 3, depending simply on the *number* of players who have their single stimulus element conditioned to a Type 1 response.

As a general rule, it would be unwise to assume the same θ for players A, B, and C, since it has been shown empirically that the value of the conditioning parameter may be related to the particular reinforcement schedule employed

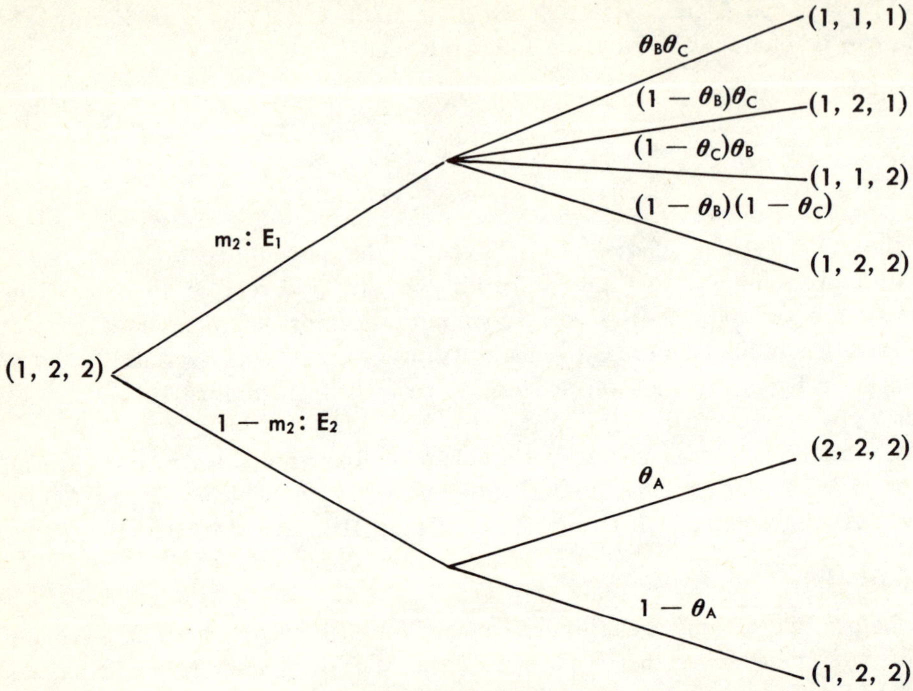

FIGURE 8.1

(e.g., θ_B was larger than θ_A for the two groups of Chapter 4). Nevertheless, despite the interactive character of the three-person setup, the symmetry in the payoff matrix is such that the three subjects are all on the same reinforcement schedule. Thus, although differences between subjects may exist in the values of θ, these differences will not be magnified by differential effects of reinforcement schedules.

Given the assumption of one stimulus element for each player and equal θ values, the derivation of the Markov chain is straightforward. As noted above, the states 0, 1, 2, and 3 designate the number of players who have their stimulus element conditioned to a Type 1 response. To indicate the structure of this particular application, two of the trees are given in Figs. 8.2 and 8.3. Construction of the analogous trees yields the following transition matrix:

(8.2.1)

	0	1	2	3
0	$(1 - m_2) + (1 - \theta)^3 m_2$	$3\theta(1 - \theta)^2 m_2$	$3\theta^2(1 - \theta)m_2$	$\theta^3 m_2$
1	$\theta(1 - m_2)$	$(1 - \theta)(1 - m_2) + (1 - \theta)^2 m_2$	$2\theta(1 - \theta)m_2$	$\theta^2 m_2$
2	$\theta^2(1 - m_1)$	$2\theta(1 - \theta)(1 - m_1)$	$(1 - \theta)m_1 + (1 - \theta)^2(1 - m_1)$	θm_1
3	$\theta^3(1 - m_1)$	$3\theta^2(1 - \theta)(1 - m_1)$	$3\theta(1 - \theta)^2(1 - m_1)$	$m_1 + (1 - \theta)^3(1 - m_1)$

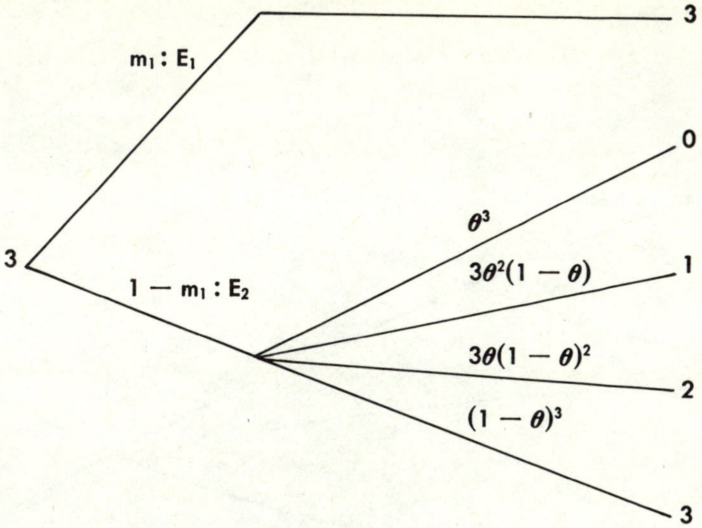

FIGURE 8.2

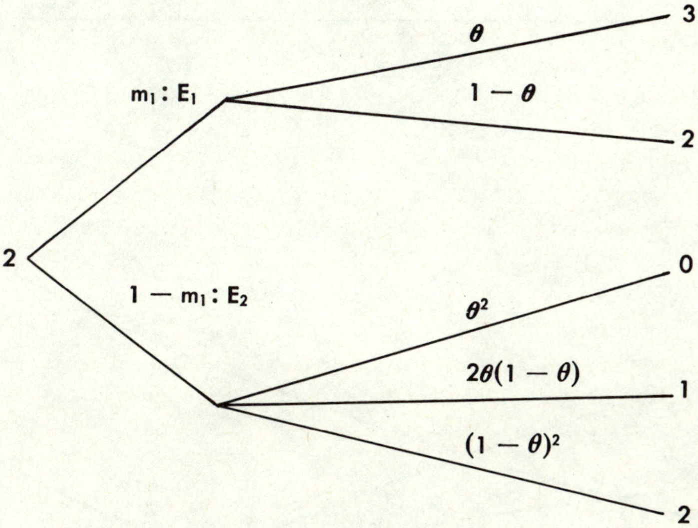

FIGURE 8.3

From the discussion in Sec. 1.3 it is clear that this matrix represents an irreducible, aperiodic case; thus the u_i's $(i = 0, 1, 2, 3)$ exist and are independent of initial conditions. Figure 8.4 presents u_0, u_1, u_2, and u_3 as a function of θ. An inspection of these curves indicates that the asymptotic quantities are highly sensitive to the value of θ, generally more so than in any of the earlier studies.

Another theoretical expression to be compared with data is the following linear combination of u_i's:

(8.2.2) $$C = u_0 + \tfrac{2}{3} u_1 + \tfrac{1}{3} u_2 .$$

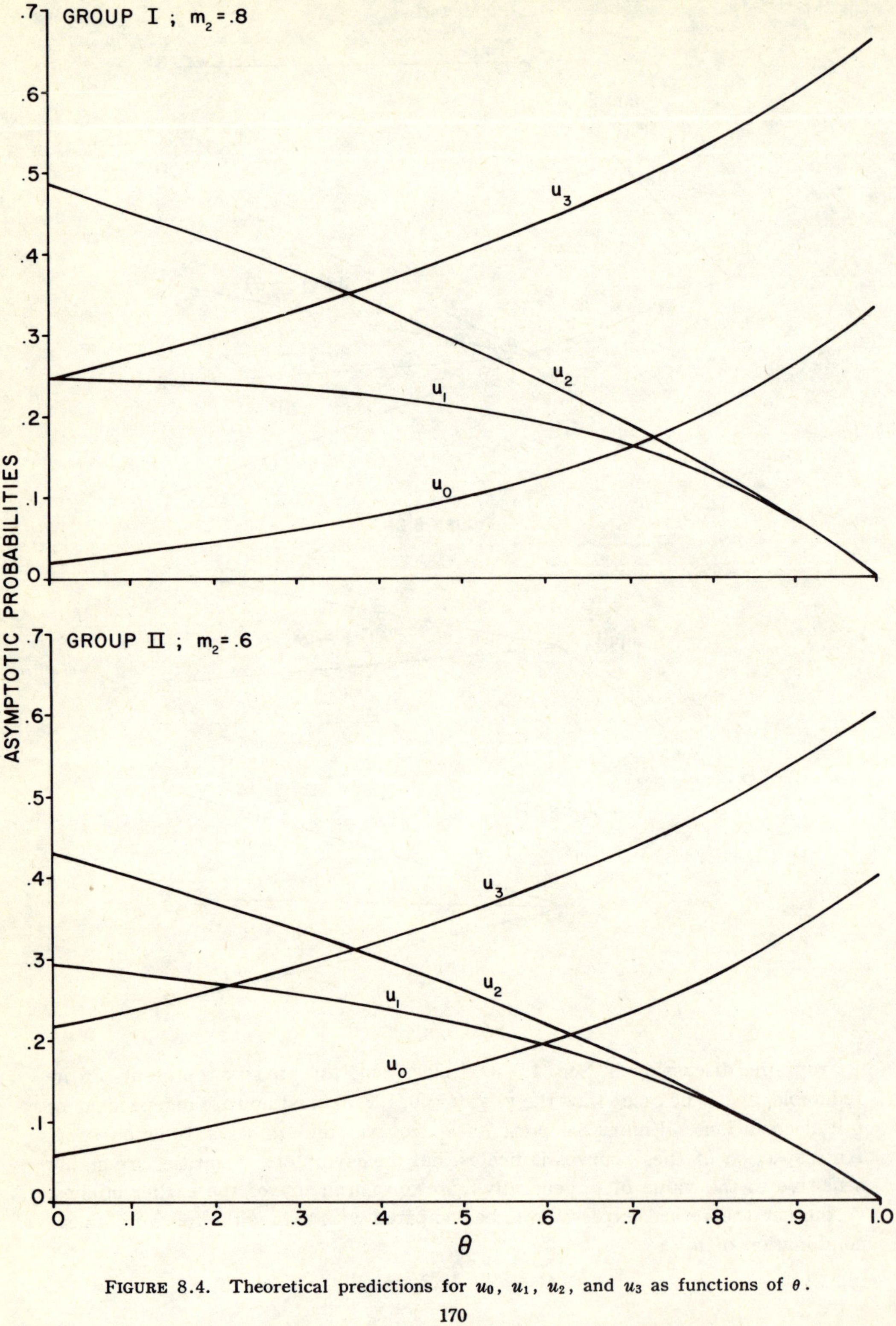

FIGURE 8.4. Theoretical predictions for u_0, u_1, u_2, and u_3 as functions of θ.

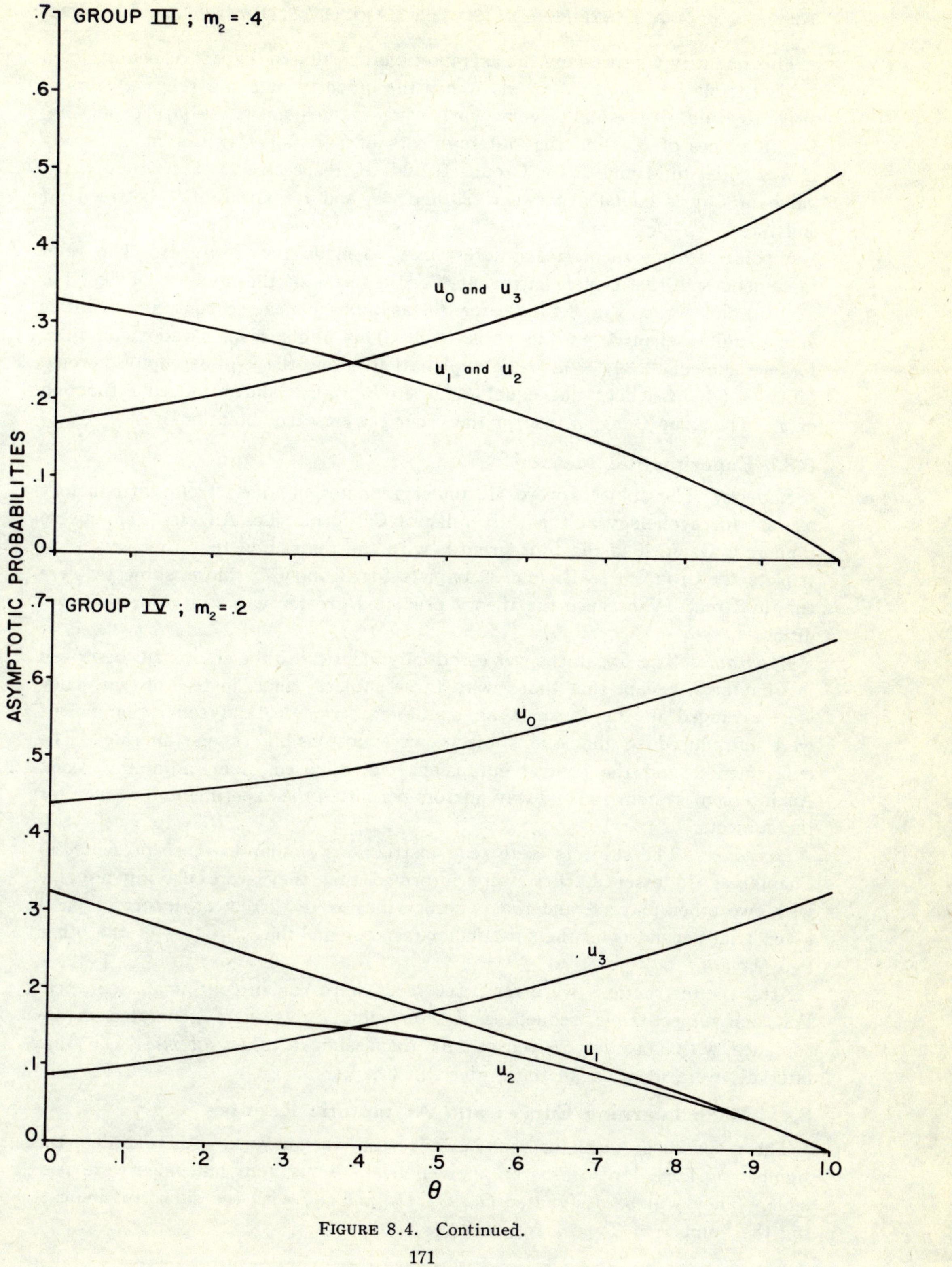

FIGURE 8.4. Continued.

This quantity C represents the asymptotic proportion of Type 2 responses for a subject-triple. In contrast to u_i, it has the property of being relatively insensitive to θ; i.e., it is equal or very nearly equal to the quantity $(1-m_1)/(1-m_1+m_2)$ for all values of θ. For the different sets of parameter values in our study C was equal to .4 and .5 for Groups II and III, respectively. For Group I, the value of C was bounded between .33 and .35; and for Group IV, between .67 and .71.

A point of some theoretical interest may be made for Group II. The reinforcement schedule is noncontingent (i.e., in terms of the notation of Sec. 1.3, $\pi = .6$), and yet, as Fig. 8.4 indicates, the asymptotic mean response probabilities u_i are quite sensitive to the value of θ. This might seem paradoxical, since for one subject $P_\infty(A_1) = \pi$. The explanation is that the joint asymptotic probability $P_\infty(A_1B_1C_1)$ does not equal $P_\infty(A_1)P_\infty(B_1)P_\infty(C_1)$, and is indeed a function of θ. The same thing is true of the other response combinations.

8.3. Experimental Method

Subjects. The subjects were 312 undergraduates obtained from introductory courses in psychology at the University of California, Los Angeles. They were randomly assigned to the four groups, with the restriction that there were 24 triplets for Groups I to III and 32 triplets for Group IV. More subjects were run in Group IV because the theory predicted greater variability for this condition.

Apparatus. The apparatus was essentially identical to the equipment described in Chapter 3, except that there were three subject panels instead of two. They were arranged in the form of an equilateral triangle, and separation panels were introduced so that the subjects were not visible to one another. The experimenter and the control equipment were housed in an adjoining room. An intercom system and one-way mirrors permitted the experimenter to monitor the subjects.

Procedure. The subjects were read instructions similar to those presented in Chapter 4. In essence, they were informed that they were playing a game with two other players, and that whether they were correct or incorrect on a given trial depended on their particular response and the responses of the other two players.

After the instructions were read, 200 trials were run in continuous sequence. For each subject-triple, sequences of reinforcing lights were generated in accordance with observed responses and the assigned values of m_i. The time intervals were identical to those given in Chapter 3.

8.4. Mean Learning Curves and Asymptotic Results

The response protocol for each subject-triple was analyzed with regard to the number of Type 1 responses on a given trial. Thus, four outcomes were possible. They will be designated O_0, O_1, O_2, and O_3, with the subscript indicating the number of Type 1 responses.

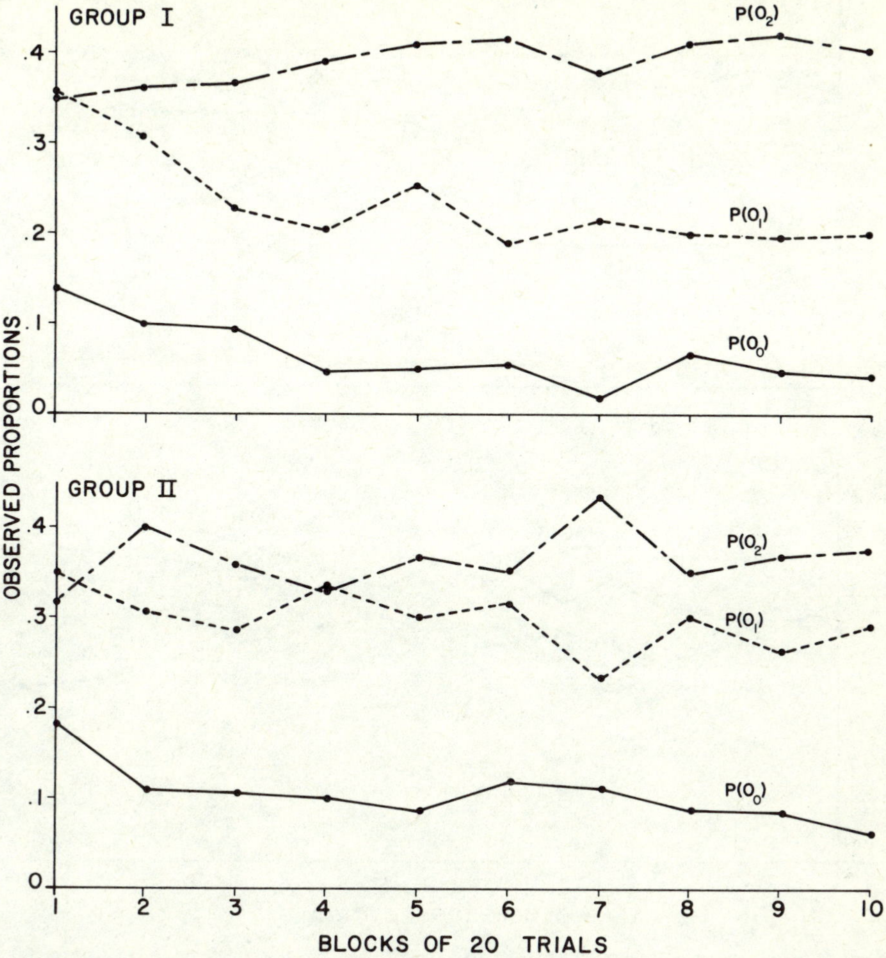

FIGURE 8.5. Observed mean proportions of O_i outcomes for Groups I and II in successive 20-trial blocks.

Figure 8.5 presents the average proportion of O_i outcomes [denoted $P(O_i)$] for Groups I and II. Figure 8.6 presents the proportion of O_i outcomes for Groups III and IV in successive blocks of 20 trials for the entire sequence of 200 trials. Figure 8.7 presents the values for the combination $\hat{C} = \hat{P}(O_0) + \frac{2}{3}\hat{P}(O_1) + \frac{1}{3}\hat{P}(O_2)$, again in successive 20-trial blocks and averaged over subject-triples. The theoretical predictions for the asymptotic value of C given by (8.2.2) are indicated at the far right.

In all figures the curves appear reasonably stable over the last three blocks of trials. The observations on $\hat{C}$ are the most reliable, since they are based on all the data; and for Groups I, II, and III there is every indication that the curves have reached a stable level. The plot of $\hat{C}$ for Group IV shows evidence

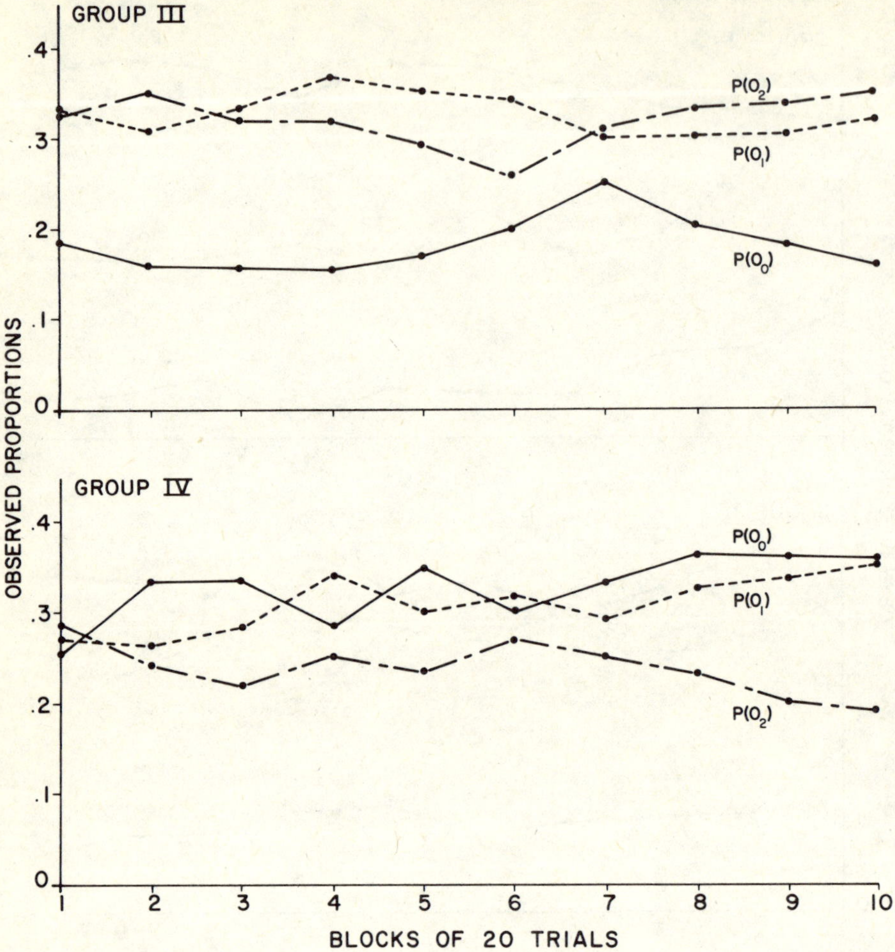

FIGURE 8.6. Observed mean proportions of O_i outcomes for Groups III and IV in successive 20-trial blocks.

of possibly increasing further, but seems close to a final level. In view of these observations we shall assume that a stable response level has been attained; consequently, the proportions computed over the last 60 trials will be used as estimates of asymptotic quantities. Table 8.1 presents the observed means and standard deviations for C and $P(O_i)$ in the last 60-trial block. Entries for Group IV are based on 32 triples; for the other groups, 24.

The predictions for C that were given in Sec. 8.2 are substantiated surprisingly well in Table 8.1. For Group I, $\hat{C} = .322$ compares well with the predicted interval (.33, .35); for Group II, $\hat{C} = .398$ is as close to the prediction of .4 as could possibly be expected; for Group III, $\hat{C} = .505$ is also remarkably close to the predicted .5; and for Group IV, $\hat{C} = .652$ is slightly below the predicted interval (.67, .71).

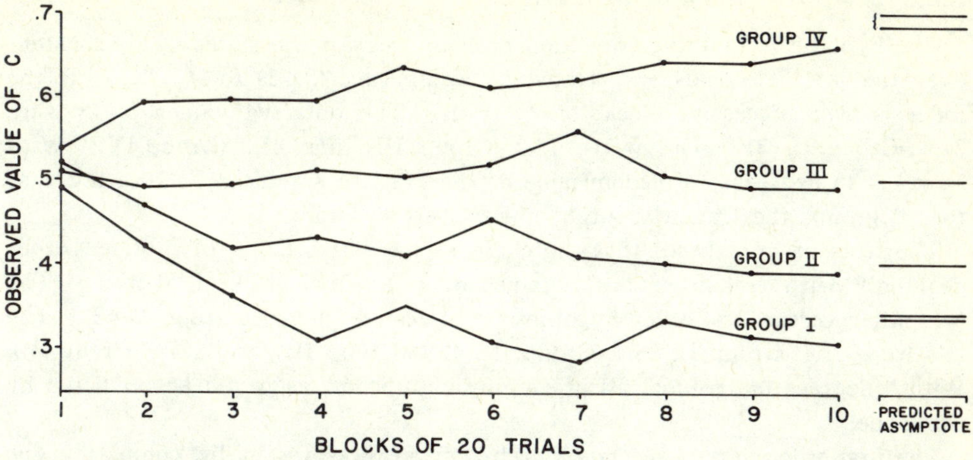

FIGURE 8.7. Observed mean values of C in successive 20-trial blocks, and predicted asymptotic values of C.

8.5. Estimates of θ in Terms of $\hat{P}(O_i)$

The observation that the u_i's in this particular experiment are highly sensitive to changes in θ suggests a procedure of estimating this parameter in terms of the observed mean asymptotic values of $\hat{P}(O_i)$ presented in Table 8.1. For each experimental group the following average absolute difference was computed as a function of θ:

$$D(\theta) = \frac{1}{4} \sum_{i=0}^{3} | u_i(\theta) - \hat{P}_\infty(O_i) | .$$

The computations were performed in the interval on θ from .01 to 1.0 in increments of .01. The average of $D(\theta)$ over the four groups is at a minimum when $\theta = .12$, the average absolute difference between observed and predicted values being .026. However, the individual group minimums show marked discrepancies from $\theta = .12$; specifically, $D(\theta)$ was minimal for $\theta = .27$ in Group I, $\theta = .19$ in Group II, $\theta = .06$ in Group III, and $\theta = .04$ in Group IV.

TABLE 8.1

OBSERVED MEAN VALUES FOR C AND $P(O_i)$ OVER THE LAST 60 TRIALS

Group	C	$P(0_0)$		$P(0_1)$		$P(0_2)$		$P(0_3)$	
	Mean	Mean	SD	Mean	SD	Mean	SD	Mean	SD
I	.322	.053	.045	.198	.069	.412	.092	.337	.114
II	.398	.082	.049	.289	.074	.371	.072	.258	.096
III	.505	.183	.066	.312	.073	.342	.074	.163	.064
IV	.652	.364	.186	.326	.117	.212	.098	.098	.068

8.6. Tests for Markov Properties

The hypothesis that the transition probabilities for the states O_i are constant over the last 100 trials was tested by computing (2.4.1) for $t = 1, \cdots, 5$; the index is over successive blocks of 20 trials. The obtained values of χ^2 were 37.6 (Group I), 31.5 (Group II), 18.2 (Group III), and 45.3 (Group IV). With $12 \times 4 = 48$ degrees of freedom none of the χ^2's are significant, indicating that for all groups the transition probabilities were constant.

The tests of hypotheses that the chain in the states O_i is of a given order deal only with transition data obtained over the final block of 20 trials. The zero-order versus first-order hypothesis was tested by calculating (2.4.2). The χ^2's were 32.4 (Group I), 36.4 (Group II), 42.1 (Group III), and 127.6 (Group IV). With 9 degrees of freedom all are highly significant, as would be predicted by the model.

The first-order versus second-order hypothesis was tested by computing the χ^2 of (2.4.3). The obtained values were 53.5 (Group I), 32.9 (Group II), 44.7 (Group III), and 57.2 (Group IV); for each there are 36 degrees of freedom. The only χ^2 significant at the .05 level is for Group IV. Thus, for Groups I–III there is no statistically significant improvement in prediction if one knows the response history of the subject-triple on the previous two trials as against only the immediately preceding trial. Of course, the one-element model presented in Sec. 8.2 is formulated as a one-stage process, and these findings support the model. However, as indicated earlier, this assumption is not necessary for our general approach; the results for Group IV simply argue for a multi-element analysis of the data.

8.7. Maximum-Likelihood Estimate of θ and Predicted Transition Probabilities

We have not obtained an explicit expression for the maximum-likelihood function, but instead have resorted to computing the function

$$L(\theta) = \sum_{i,j} n_{ij} \log p_{ij}(\theta)$$

for successive values of θ and then using the results to approximate its maximum. The n_{ij}'s, of course, refer to transition frequencies associated with the outcomes O_0, O_1, O_2, and O_3.

For each group two sets of computations were carried out. In all cases, $L(\theta)$ was computed for $\theta = .01k$ $(k = 1, \cdots 100)$. One set of computations dealt with aggregated n_{ij}'s obtained from the second block of 20 trials (i.e., trials 21–40), and the other dealt with n_{ij}'s obtained from the final block of 20 trials (i.e., trials 181–200). Table 8.2 presents the values of θ at which the maximum of $L(\theta)$ occurred. All these values are close to one another, and there is no reason to suspect that θ differs from group to group or between the two trial blocks. To simplify computation we shall take the estimated value of θ to be .5 for all groups.

TABLE 8.2
MAXIMUM-LIKELIHOOD ESTIMATES OF θ

Group	Trials 21–40	Trials 181–200
I	.51	.48
II	.52	.49
III	.51	.52
IV	.53	.53

Table 8.3 presents the theoretical matrices given by (8.2.1) for $\theta = .5$. The table also gives the observed transition probabilities for the last 20-trial block aggregated over subjects. To test the goodness of fit between observed and predicted values we compute the χ^2 of (2.4.4). The degrees of freedom are $12 - 1 = 11$, since θ has been estimated from these same data. The obtained values for χ^2 are 218.5 (Group I), 185.1 (Group II), 319.4 (Group III), and 632.5 (Group IV); all are highly significant.

Because the mean asymptotic probabilities $P_\infty(O_i)$ for $i = 0, 1, 2, 3$ are so sensitive to the value of θ (see Fig. 8.4), the predicted $P_\infty(O_i)$ for $\theta = .5$ do not

TABLE 8.3
OBSERVED AND THEORETICAL TRANSITION MATRICES
(The observed matrices are for the last block of 20 trials and the theoretical matrices are calculated for $\theta = .5$)

Group	$n \backslash n+1$	Observed 0	Observed 1	Observed 2	Observed 3	Theoretical 0	Theoretical 1	Theoretical 2	Theoretical 3
I	0	.073	.244	.341	.341	.300	.300	.300	.100
	1	.056	.266	.412	.266	.100	.300	.400	.200
	2	.032	.232	.426	.310	.100	.200	.400	.300
	3	.048	.121	.401	.430	.050	.150	.150	.650
II	0	.180	.328	.298	.194	.475	.225	.225	.075
	1	.086	.343	.376	.196	.200	.350	.300	.150
	2	.069	.271	.406	.254	.100	.200	.400	.300
	3	.053	.211	.366	.370	.050	.150	.150	.650
III	0	.223	.362	.292	.123	.650	.150	.150	.050
	1	.212	.350	.322	.116	.300	.400	.200	.100
	2	.116	.305	.404	.175	.100	.200	.400	.300
	3	.160	.253	.309	.278	.050	.150	.150	.650
IV	0	.540	.285	.132	.043	.825	.075	.075	.025
	1	.325	.397	.196	.082	.400	.450	.100	.050
	2	.196	.392	.273	.139	.100	.200	.400	.300
	3	.215	.280	.327.	.178	.050	.150	.150	.650

agree well with the observed proportions given in Table 8.1. Recall that in Sec. 8.5 we found that $\theta = .12$ is most suitable to predict the observed proportions of O_i in the last 60 trials. In previous chapters the mean asymptotic response probabilities have not shown this extreme sensitivity to the value of θ. In view of our other difficulties with the one-element stimulus-sampling model, it is not surprising that the estimate of θ from sequential data does not accurately predict mean asymptotes that are highly dependent on θ. Unfortunately, the computational problems were too great to extend the analysis of data to multi-element models or to the generalized conditioning model. It seems likely that these more complicated models would once again have yielded a much better fit if it had been possible to carry through the computations.

It is important to emphasize that the poor goodness-of-fit results for the transition probabilities and the relatively poor predictions of the mean probabilities $P_\infty(O_i)$ are not separate problems. Good results on the transition probabilities guarantee good results on the mean response probabilities, but not conversely, as we have seen in earlier chapters.

8.8. Game-Theory Comparisons

Because once the experiment started there was no communication between the three players, the game situation is most aptly described as non-cooperative; hence Nash's equilibrium-point theory is relevant. The game matrix for each of the four groups has at least one pure equilibrium point. It is evident from the results presented in earlier sections that no group came close to such a pure point; yet the qualitative results are generally in the right direction and are of some interest.

For Group I ($m_2 = .8$), the matrix for Player A may be written as follows:

	B_1C_1	B_1C_2	B_2C_1	B_2C_2
A_1	.6	.6	.6	.8
A_2	.4	.2	.2	.2

The unique equilibrium point is $\{A_1, B_1, C_1\}$, which also is the unique set of strategies selected by the even stronger sure-thing principle. From Table 8.1 we see that in the last 60 trials all three players selected a Type 1 response with probability .337, and two made this selection with probability .412; this means that the majority were using equilibrium-point strategies with a relative frequency of .749.

For Group II ($m_2 = .6$), the reinforcement schedule is actually noncontingent. The game matrix for Player A is

	B_1C_1	B_1C_2	B_2C_1	B_2C_2
A_1	.6	.6	.6	.6
A_2	.4	.4	.4	.4

Again the unique equilibrium point is $\{A_1, B_1, C_1\}$, which also satisfies the sure-thing principle. As Table 8.1 indicates, the majority used equilibrium-point strategies with a relative frequency of .629 in the last 60 trials.

For Group III ($m_2 = .4$), the game matrix for Player A is

	B_1C_1	B_1C_2	B_2C_1	B_2C_2
A_1	.6	.6	.6	.4
A_2	.4	.6	.6	.6

This situation is completely symmetric between Type 1 and Type 2 responses, and this symmetry is reflected in the predictions for learning theory and game theory, as well as in the empirical observations. The sets of strategies $\{A_1, B_1, C_1\}$ and $\{A_2, B_2, C_2\}$ are both equilibrium points. Moreover, from Table 8.1, $\hat{P}(O_0) + \hat{P}(O_1) = .495$ and $\hat{P}(O_2) + \hat{P}(O_3) = .505$. The learning-theory symmetry is accurately illustrated in the theoretical transition matrix for $\theta = .5$ given in Table 8.3.

For Group IV ($m_2 = .2$), the game matrix for Player A is

	B_1C_1	B_1C_2	B_2C_1	B_2C_2
A_1	.6	.6	.6	.2
A_2	.4	.8	.8	.8

Again there are two equilibrium points, $\{A_1, B_1, C_1\}$ and $\{A_2, B_2, C_2\}$, but here the latter is more favorable; this fact is also reflected in the observations. From Table 8.1, $\hat{P}(O_0) + \hat{P}(O_1) = .690$, which represents the relative frequency of choice by the majority of a strategy in the second equilibrium point.

Individual triples showed no tendency to settle on equilibrium-point strategies. For Groups I and II the observed maxima of $P(A_1B_1C_1) = P(O_3)$ for individual triples based on the last 60 trials were .55 and .48, respectively. For Groups III and IV the observed maxima of $P(A_2B_2C_2) = P(O_0)$ for individual triples based on the last 60 trials were .32 and .80, respectively.

The experiments reported in the chapter did not permit a reasonable analysis of coalition structure. As we noted in the last chapter, for the formation and dissolution of coalitions it is not necessary that the players be able to come to explicit "understandings" outside the framework of the game. There is the possibility that coalitions will develop without explicit agreements because of the prevailing schedule of reinforcements. However, it is difficult for this possibility to be realized when each player is not given information about the choices of the other players after each play of the game. The complete symmetry of the simple majority situation is also unfavorable to the formation of coalitions when there is no communication between the players.

The computational difficulties mentioned in the last chapter must be overcome before experimental situations that are conducive to the formation of coalitions can be studied in any detail from the standpoint of stimulus-sampling theory.

8.9. Summary

In this chapter the theory of behavior that was presented in Chapter 1 is extended to a three-person interaction situation. Again the experimental situation involves a series of 200 trials. At the start of a trial each of three players independently selects either his A_1 or A_2 response. After the players have indicated their responses, the outcome of the trial is announced. The outcome of a trial is specified by a simple majority rule; the majority of players are correct (and the minority incorrect) with probability m_1 if they select the A_1 response, or the majority are correct with probability $1 - m_2$ if they select the A_2 response. Four experiments are reported, each using different values of m_1 and m_2; the selection of these values is determined by game-theory considerations.

In Sec. 8.2 the theoretical analysis of this situation is carried out and various predictions are computed. A side result of this analysis is that several interesting and, at first glance, paradoxical predictions are obtained for one-person noncontingent learning situations. The experimental method is described in Sec. 8.3, and in Secs. 8.4 through 8.7 theoretical predictions are compared with the observed data; excellent agreement is obtained on various measures of asymptotic performance. Maximum-likelihood estimates of the conditioning parameter yield virtually identical values of θ (approximately .5) for the four different experimental groups; however, these estimates tend to be too high when compared with estimates of θ based on asymptotic response proportions alone. When the states are defined as the number of subjects making response A_1, it is found that the observed transition probabilities are constant over trials and that the Markov chain is a first-order chain.

In Sec. 8.8 the data are analyzed with regard to game-theory strategies. The game matrix for each of the four groups has at least one pure equilibrium point; however no group came close to such a strategy.

9

Display of Payoff Matrix

9.1. Theoretical Description

Of the three major restrictions on the interaction of subjects mentioned at the beginning of Chapter 3, one—not showing the payoff matrix to the subjects—has not been removed in any of the experiments reported thus far. This chapter investigates experimentally the effects of displaying the payoff matrix. Perhaps the best evidence for the absence of any effects would be close confirmation of the asymptotic mean probabilities of response in the one-element model, because these predictions were well confirmed in Chapters 3 and 4 even though they were independent of θ. On the other hand, it would be too much to expect an excellent fit of the one-element model to the sequential data, since no such fit was achieved by any of the experiments reported in the preceding chapters.

If the observed asymptotic response proportions do deviate from the predicted proportions, it would seem obvious to consider next the fit of the generalized conditioning model. However, following the structure of other chapters, we defer consideration of this model until the end of the chapter, and devote the remainder of this section to the predictions of the one-element model.

Two groups of subjects, designated E for experimental and C for control, were run with a zero-sum payoff matrix of the kind described in Chapter 3. The difference between the groups is that E was shown the numerical payoff matrix, and C was shown a matrix with only literal values p_i and q_i. The reinforcement probabilities were the same for both groups, and were given by the following matrix:

(9.1.1)

	B_1	B_2
A_1	.3	.8
A_2	.4	.1

Thus for response pair (A_1B_1) the probability was .3 that Player A would be reinforced and .7 that Player B would be reinforced.

It may be noted that this matrix has mixed minimax strategies for both players. This choice was deliberate in order to prevent subjects in Group E from being presented with a game that had a simple and obvious optimal solution.

Applying the theoretical results of Sec. 1.4, we obtain for the one-element model the following numerical quantities. When $\theta_A = \theta_B$,

(9.1.2)
$$u_1 = .348\,, \quad u_3 = .259\,, \quad \alpha = .589\,,$$
$$u_2 = .241\,, \quad u_4 = .152\,, \quad \beta = .607\,.$$

Moreover, as the ratio θ_A/θ_B varies from 0 to ∞, we obtain the the following bounds:

(9.1.3)
$$.302 < u_1 < .435\,, \quad .080 < u_3 < .353\,, \quad .584 < \alpha < .598\,,$$
$$.163 < u_2 < .282\,, \quad .063 < u_4 < .321\,, \quad .515 < \beta < .655\,.$$

Player A's minimax strategy is to respond A_1 with probability .375, and Player B's minimax strategy is to respond B_1 with probability .875. Matrix (9.1.1) was chosen so that these minimax strategies would fall outside the bounds given by (9.1.3). A comparison of these minimax strategies with experimental results is given in Sec. 9.5.

We emphasize that the experimental conditions used in this chapter are not the most unfavorable and therefore not the most stringent that we could have selected to test the application of stimulus-sampling theory to game situations. A payoff matrix that had optimal strategies satisfying the sure-thing principle could have been used. The instructions to the subjects could have contained an elementary discussion of optimal strategies from the standpoint of game theory. It seems obvious to us that experiments can be designed for which learning theory as applied in these pages will be wholly inadequate. This in no way means that the fundamental theory is incorrect, but rather that new and complicated patterns of stimulation that are highly dependent on the subject's past verbal experience have been introduced. At the present time quantitative analysis of these complex patterns is out of the question. The point of the present experiment is to explore cautiously the extent to which such patterns may be introduced without seriously affecting the results obtained by the methods of analysis used in earlier chapters.

9.2. Experimental Method

Subjects. In the first phase 80 undergraduates from introductory courses at Stanford University were used. They were randomly assigned to the Experimental Group (E') and the Control Group (C), with the restriction of 20 pairs in each group. After the first phase was completed, a replication of the Experimental Group was run using 56 subjects; this group will be denoted E''. This replication was to check the stability of the results for the experimental condition.

Apparatus. The apparatus described in Chapter 3 was used. In addition, in each subject's booth an 8 × 8 inch white card was attached to the vertical panel and centered between the reinforcing lights. Different cards were used in the E and C conditions. For Groups E' and E'' the cards were printed as follows:

	B_1	B_2
A_1	3/10	8/10
A_2	4/10	1/10

	B_1	B_2
A_1	7/10	2/10
A_2	6/10	9/10

The left array was used for Player A, the right for Player B. For Group C the cards were printed as follows:

	B_1	B_2
A_1	p_1	p_2
A_2	p_3	p_4

	B_1	B_2
A_1	q_1	q_2
A_2	q_3	q_4

Again, the left array was for Player A, and the right for Player B.

Also, labels were attached to each of the keys. Player A's labels read A_1 and A_2; Player B's read B_1 and B_2. The labels were interchangeable so that response-positions could be counterbalanced.

Procedure. For each of the three groups, four subgroups of equal size were formed by counterbalancing right and left positions of the A_i and B_j response keys.

The subjects in Groups E' and E'' were read the following instructions:

"This experiment is a game in which you will be playing against each other. It is similar to a real life situation in that what you gain or lose depends not only on what you do but also on what someone else—your opponent—does.

"To keep things straight I will call you 'Player A' and you 'Player B.' The experiment for each of you consists of a series of trials. The top center light on each of your panels will go on for two seconds to indicate the start of each trial. When this light goes on, you will each press one or the other of the two keys in front of you. That is, Player A will press either his A_1 or his A_2 key, and Player B will press his B_1 or B_2 key.

"On each trial one of you will be correct and one of you will be incorrect. Shortly after you have made your response, one of the two lower white lights on your panels will go on, indicating whether you were correct or incorrect. If the white light above the key you pressed goes on, then you were correct and the other person was incorrect. If the white light above the other key goes on, that is, above the key you didn't press, then you were incorrect and the other player was correct. Being correct on a given trial depends on the combination of your key press with the other player's key press.

"More specifically, the game is described by the chart in front of you. The numbers corresponding to various response pairs indicate the proportion of times that you can expect to be correct. For example, if Player A presses his A_1 key and Player B presses his B_1 key, then Player A will be correct 3 out of

10 times on the average and Player B will be correct 7 out of 10 times. Thus the entry corresponding to the (A_1B_1) response pair for Player A is 3/10 and the corresponding entry for Player B is 7/10. Similarly, if Player A presses his A_2 key and Player B presses his B_2 key, then Player A will be correct 1 out of 10 times and Player B will be correct 9 out of 10 times. Thus this chart tells you what your chances of being correct are. For example, if Player A knew that Player B was going to press B_1, then Player A would press A_2 to maximize his chance of being correct. Similarly, if Player B knew that Player A was going to press A_2, then Player B would press B_2 to maximize his chance of being correct. The catch is that you don't know what the other player is going to do. Consequently on each trial you must make a key choice that you think will maximize your chance of being correct.

"Your job, then, is to obtain as many correct responses as you possibly can. The trials move along rapidly, and you must make your key choice as soon as the signal light goes on. That is, as soon as the signal light goes on press one or the other key and release it before the signal light goes off. Then wait until one of the lower white lights goes on. If the light above the key you pressed goes on, you are correct and your opponent is incorrect. If, however, the white light above the key opposite from the one you pressed goes on, you are incorrect and the other player is correct.

"Try to do as well as you can, and above all don't give up. The success of the experiment depends on your doing the very best you can on all trials. It will take us about 45 minutes."

The instructions for Group C were identical except for the fourth paragraph, which read as follows:

"More specifically, the game is described by the chart in front of you. The p's and q's corresponding to various response pairs stand for the proportion of times that you can expect to be correct. For example, if Player A presses his A_1 key and Player B presses his B_1 key, then p_1 is the proportion of times that Player A will be correct, and q_1 is the proportion of times that Player B will be correct. Thus the entries corresponding to A_1B_1, that is, p_1 and q_1, together add up to 1. Similarly, if Player A presses his A_2 key and Player B presses his B_2 key, then the proportion of times Player A will be correct is p_4, and the proportion of times Player B will be correct is q_4. Thus if you had some knowledge about the values of these p's and q's, you would know your chances of being correct. For example, if one player knew what key his opponent was going to press, he could select his response to maximize his chance of being correct. The catch is, of course, that you don't know what the p and q values are, nor do you know what the other player is going to do. Consequently, depending on what you can find out, on each trial you must make a key choice that you think will maximize your chance of being correct."

After the instructions were read, 210 trials were run in continuous sequence. For each pair of subjects sequences of reinforcing lights were randomly generated in accordance with the assigned values of a_i and the observed response pairs.

9.3. Tests for Differences Among Experimental Conditions

Test of the hypothesis that Groups E' and E'' are from the same Markov chain of first order. As indicated in Sec. 9.2, the E' and E'' groups were run on the same experimental procedure; however, they were run about six months apart and by different experimenters. Since we should like to combine the data of these two groups, we want to determine whether they can be properly viewed as equivalent. We do this by testing the null hypothesis that the two samples are from the same first-order Markov chain.

The particular transition frequencies to be examined are described by the states of the matrix in (2.6.9). That is, we consider the observed probabilities

$$P(A_{i,n+1}B_{j,n+1} \mid A_{i',n}E_{k',n}B_{j',n}) \, , \tag{9.3.1}$$

and ask whether or not they are the same for Group E' and Group E''. In this notation, $E_{k'}$ indicates the reinforcing event for Player A; and, as remarked earlier, in the zero-sum case the responses of both subjects and the reinforcing event given one subject uniquely determine the reinforcing event for the other subject.

The χ^2 test of homogeneity given by (2.4.6) is appropriate here when modified for an 8×4 array. The reader is referred to the discussion related to (2.6.9) to justify the consideration of this reduced form compared with the 8×8 process when $E_{k,n+1}$ is not omitted in (9.3.1). The test is to be interpreted with $8 \times 3 = 24$ degrees of freedom, since eight states are specified on trial n and four on trial $n + 1$.

The $n_{ij}^{(E')}$'s and $n_{ij}^{(E'')}$'s $(i = 1, \cdots, 8;\ j = 1, \cdots, 4)$, for Groups E' and E'', respectively, were tabulated for the last block of 100 trials and aggregated over subject-pairs. The χ^2 given by (2.4.6) yielded a value of 35.7, which with 24 degrees of freedom is not significant at the .05 level of confidence. Further, the curves for A_1 and B_1 responses are virtually identical over trials for Groups E' and E''. In view of these results, we shall no longer distinguish between these two groups but will refer to them as Group E.

Test of the hypothesis that the E and C groups are from the same Markov chain of first order. In this section we test the hypothesis that the observed probabilities (9.3.1) are the same for Group E and Group C. Precisely the same χ^2 test of (2.4.6) was run. The $n_{ij}^{(C)}$'s for Group C were tabulated for the last 100 trials and aggregated over subject-pairs. The $n_{ij}^{(E)}$'s were obtained in terms of the $n_{ij}^{(E')}$'s and $n_{ij}^{(E'')}$'s specified above; i.e., $n_{ij}^{(E)} = n_{ij}^{(E')} + n_{ij}^{(E'')}$.

The observed probabilities (9.3.1) are given in Table 9.1 for Group E and Group C. The obtained value of χ^2 is 85.4; and with 24 degrees of freedom this value is highly significant, as would be anticipated by an inspection of the two arrays of numbers in Table 9.1. Thus, the different payoff displays used in this study have a statistically significant effect in terms of these transition

TABLE 9.1

OBSERVED TRANSITION PROBABILITIES OVER THE LAST 100 TRIALS FOR GROUPS E AND C

(The quantities in parentheses are for Group C)

n \ $n+1$	A_1B_1	A_1B_2	A_2B_1	A_2B_2
$A_1E_1B_1$	.532 (.447)	.319 (.367)	.099 (.116)	.050 (.070)
$A_1E_2B_1$	.473 (.486)	.143 (.101)	.288 (.352)	.096 (.061)
$A_1E_1B_2$	.535 (.405)	.312 (.370)	.107 (.153)	.046 (.072)
$A_1E_2B_2$	.313 (.242)	.331 (.418)	.163 (.154)	.193 (.186)
$A_2E_1B_1$	.531 (.535)	.124 (.104)	.268 (.292)	.077 (.069)
$A_2E_2B_1$	.318 (.258)	.161 (.187)	.364 (.327)	.157 (.228)
$A_2E_1B_2$	.312 (.213)	.353 (.444)	.140 (.106)	.195 (.237)
$A_2E_2B_2$	.462 (.305)	.205 (.217)	.231 (.348)	.102 (.130)

quantities. However, there seems to be no simple or obvious interpretation of these differences.

9.4. Mean Learning Curves and Asymptotic Results

Figure 9.1 presents the observed mean proportions of A_1 and B_1 responses in successive blocks of 30 trials. The curves for Group E are consistently above Group C on both the A_1 and B_1 responses, suggesting that the display of the actual payoff matrix gave rise to an increase in both A_1 and B_1 responses.

Figure 9.2 presents the observed mean proportions for the response pairs (A_1B_1), (A_1B_2), (A_2B_1), and (A_2B_2) in blocks of 30 trials. There is no marked differentiation between Groups E and C on response pairs (A_1B_2), (A_2B_1), and (A_2B_2), although in general the curves for Group C were above those for Group E. In contrast, on the (A_1B_1) pair the curve for Group E is consistently well above that for Group C.

Table 9.2 presents the observed means and standard deviations for the last block of 60 trials. The comments made with regard to Figs. 9.1 and 9.2 apply to these terminal proportions. Group E is above Group C on the response pair (A_1B_1) and responses A_1 and B_1; however, on all other response pairs Group C is above Group E.

A comparison of the asymptotic predictions generated by the one-element model with the observed terminal values for Group E indicates some marked

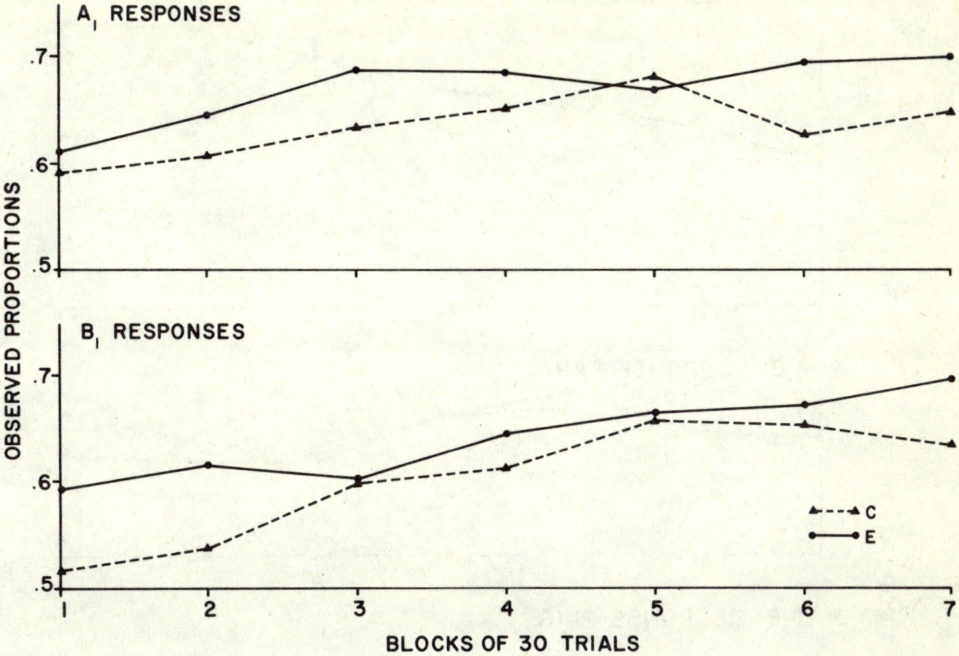

FIGURE 9.1. Observed mean proportions of A_1 and B_1 responses in successive 30-trial blocks for the Experimental and Control Groups.

discrepancies. On the pair (A_1B_1) and responses A_1 and B_1 the observed quantities are well above the maximum theoretical values given by (9.1.3). Further, an inspection of Figs. 9.1 and 9.2 suggests that all three of these quantities may still be increasing to a yet higher asymptotic level.

For Group C the comparison between the one-element predictions and the values in Table 9.2 is more favorable. All measures in the table fall within the appropriate bounds given by (9.1.3), except for $P(A_1)$. Further, for $P(A_1)$

TABLE 9.2
OBSERVED PROPORTIONS OVER THE LAST 60 TRIALS

Response Probability	Group E		Group C	
	Mean	SD	Mean	SD
$P(A_1B_1)$	.469	.097	.388	.092
$P(A_1B_2)$	.226	.118	.251	.142
$P(A_2B_1)$	.211	.117	.257	.095
$P(A_2B_2)$	.094	.077	.104	.059
$P(A_1)$	.691	.113	.638	.112
$P(B_1)$	.684	.149	.644	.145

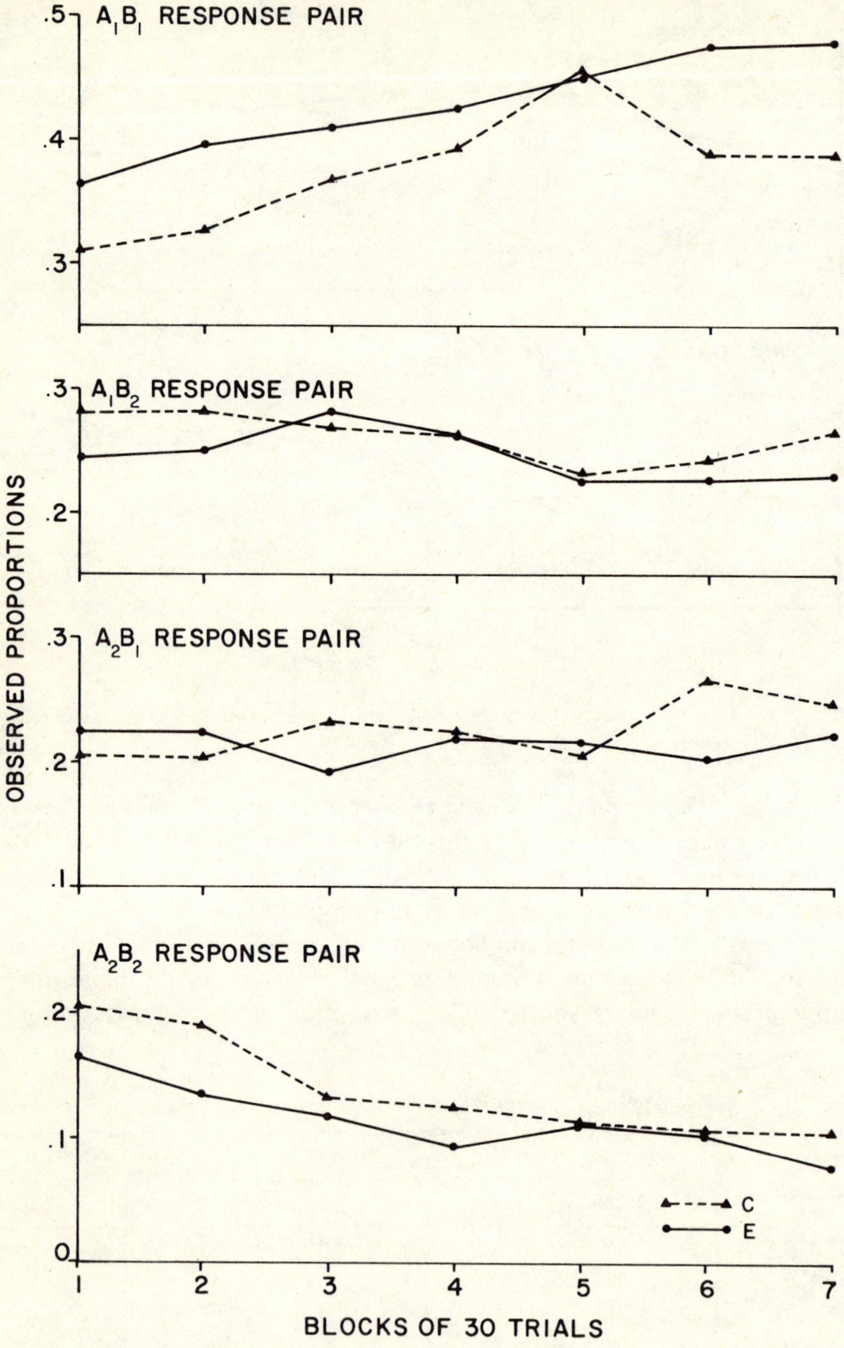

FIGURE 9.2. Observed mean proportions of the response pairs (A_1B_1), (A_1B_2), (A_2B_1), and (A_2B_2) in successive 30-trial blocks for the Experimental and Control Groups.

the discrepancy is not large when allowance is made for the observed variability; in particular, if $\theta_A = \theta_B$ the difference on the A_1 measure between theory and observation is .040, which yields $t = (.040\sqrt{20})/.112 = 1.59$. Thus, the overall correspondence between the data of Table 9.2 for Group C and the mean asymptotic predictions of the one-element model is fairly adequate.

9.5. Game-Theory Comparisons

Because the numerical payoff matrix is displayed to Group E, the game-theory comparisons are particularly pertinent in this chapter. The ways in which Group E deviated from the mean asymptotic predictions of the one-element model were indicated in the last section. Surprisingly enough, these deviations were not necessarily in the direction of the minimax strategies. Player A's minimax strategy of .375 is less than the one-element model's predicted α of .589. However the observed α of .691 is larger than the model's predicted value, rather than smaller. Thus there is no central tendency for the A players to move toward the minimax solution. The B players did tend to move toward minimax strategies, for the observed β of .684 is between the predicted β of .607 and the minimax strategy of .875, but considerably closer to the former.

A partial explanation of these results is provided by the numerical values given in Table 9.3. The game-theory and learning-theory values are close. The observed value for Group E (based on the last 60 trials), which is identical

TABLE 9.3

GAME-THEORY, LEARNING-THEORY, AND OBSERVED VALUES OF THE ZERO-SUM GAME

Type of Value	Value
Game Theory	$-.137$
Learning Theory	$-.084$
Observed: E	$-.084$
Observed: C	$-.069$

with the learning-theory value, indicates further that the expected outcome for the two players is not very sensitive to the exact mean strategies used. Player A subjects who are unversed in game theory might prefer response A_1 to A_2 on the ground that A_1 is only .1 worse than A_2 if B selects B_1, and .7 better than A_2 if B selects B_2. We do not actually know if considerations of this kind account for the rather large observed α for Group E; but it seems likely that they do. Similar considerations can be given to account for $\hat{\beta}$. In essence, what we are saying is that the particular mixed minimax strategies are unimportant; it is the intuitively obvious (even if misleading) numerical characteristics of the payoff matrix that account for deviations from the learning-theory predictions.

The $\hat{\alpha}$ and $\hat{\beta}$ for Group C are also larger than the α and β predicted by the one-element model, although the discrepancy is not as great as it was for Group E. This was already pointed out in the previous section. This systematic effect, together with the internal check on Group E provided by the two subgroups E' and E'', argues that display of the letter-valued payoff matrix had a slight effect on Group C, and that display of the numerical payoff matrix had a larger effect on Group E. Moreover, this effect seems to be manifested in a search for what appear to be more optimal strategies than those predicted by learning theory, even though these strategies are not the minimax ones of game theory.

9.6. Tests for Markov Properties

We begin by testing the hypothesis that the transition probabilities $P(A_{i,n+1}B_{j,n+1} \mid A_{i',n}B_{j',n})$ are constant over trials. For each group χ^2 was calculated by (2.4.1) for $t = 1, 2$; here $t = 1$ refers to the block of trials from 11 to 110, and $t = 2$ to the blocks of trials from 111 to 210. For both Group E and Group C the obtained values of χ^2 were significant at the .05 level. The same test was run for the last 50-trial block and the next-to-last 50-trial block. In this case neither χ^2 was significant at the .05 level, suggesting that these transition probabilities approached stationarity in the last 100 trials.

The test of the hypothesis that the chain is of a given order in the states (A_iB_j) deals only with transition frequencies for the final 100 trials. The zero-order versus first-order hypothesis was tested by calculating (2.4.2). The χ^2's were 150.3 for Group E and 272.0 for Group C. With 9 degrees of freedom the χ^2's for both groups are highly significant. The first-order versus second-order hypothesis was tested by computing the χ^2 of (2.4.3). The obtained values were 79.5 for Group E and 74.9 for Group C. With 36 degrees of freedom both χ^2's are significant. Thus, there is an improvement in prediction if one knows the responses of the subject-pair on the previous two trials as against knowing only the responses on the immediately preceding trial or only the over-all level of responding.

These results agree reasonably well with those obtained in earlier chapters, although in Chapters 3, 4, and 8 there was little support for rejecting the null hypothesis in the test of the first-order versus second-order chain.

*9.7. Generalized Conditioning Model

In this section we again consider the application of the model described in Sec. 2.6. The transition frequencies for the 8×4 matrix given in (2.6.9) were tabulated for the last 100 trials and aggregated over subjects. Using (2.6.10) we obtain maximum-likelihood estimates of c_{ij} and d_{ij} for both groups, which are presented in Table 9.4.

Given these estimates, the χ^2 goodness-of-fit test (2.4.4) can be applied to matrix (2.6.9), where the observed n_{ij}'s are those on which the estimates are made (see Sec. 2.6 for a discussion of this particular application of the goodness-of-fit

TABLE 9.4

MAXIMUM-LIKELIHOOD ESTIMATES OF c_{ij} AND d_{ij}

(Computed over the last 100 trials)

Parameter	Group E	Group C
c_{11}	.849	.790
c_{21}	.659	.647
c_{12}	.619	.598
c_{22}	.499	.454
d_{11}	.772	.835
d_{21}	.645	.564
d_{12}	.650	.573
d_{22}	.460	.346

test). There are $8 \times 3 = 24$ degrees of freedom for the matrix. However, eight parameters have been estimated from these data; hence, the χ^2's are to be interpreted with $24 - 8 = 16$ degrees of freedom. The obtained values of χ^2 are 16.7 for Group E and 15.0 for Group C. Of course, neither value approaches significance, indicating excellent agreement between the predicted and observed transition data.

Given these estimates of the conditioning parameters, other predictions can be made. We now consider $P(A_{i,n+1}B_{j,n+1} \mid A_{i',n}B_{j',n})$ for the generalized model. Theoretical expressions are given in (3.8.1). Table 9.5 presents the predicted probabilities based on the estimates in Table 9.4 and also the observed proportions computed over the last 100 trials. As would be expected from the χ^2 tests reported above, the correspondence between theoretical and observed values is quite good.

TABLE 9.5

PREDICTED AND OBSERVED TRANSITION MATRICES

(The observed quantities are given in parentheses and were computed over the last 100 trials)

	Group E				Group C			
n \ $n+1$	(1, 1)	(1, 2)	(2, 1)	(2, 2)	(1, 1)	(1, 2)	(2, 1)	(2, 2)
(1, 1)	.500 (.476)	.188 (.178)	.235 (.283)	.077 (.063)	.485 (.491)	.170 (.196)	.271 (.231)	.074 (.082)
(1, 2)	.495 (.371)	.308 (.380)	.113 (.153)	.084 (.096)	.398 (.494)	.354 (.316)	.123 (.117)	.125 (.073)
(2, 1)	.435 (.426)	.160 (.137)	.288 (.305)	.117 (.132)	.428 (.447)	.142 (.139)	.302 (.305)	.128 (.109)
(2, 2)	.305 (.224)	.338 (.417)	.173 (.135)	.184 (.224)	.227 (.327)	.400 (.338)	.141 (.149)	.232 (.186)

TABLE 9.6

PREDICTED AND OBSERVED ASYMPTOTIC QUANTITIES FOR GENERALIZED CONDITIONING MODEL

Quantity	Group E		Group C	
	Predicted	Observed	Predicted	Observed
u_1	.466	.469	.421	.388
u_2	.224	.226	.234	.251
u_3	.213	.211	.228	.257
u_4	.097	.094	.117	.104
α	.690	.691	.655	.638
β	.679	.684	.650	.644

In terms of the theoretical matrices given in Table 9.4, we may compute u_i, α, and β for each group by (1.3.17), (1.4.10), and (1.4.11). The results are presented in Table 9.6. Also, the observed quantities for the last 60 trials, originally given in Table 9.2, are presented for comparison. Again the agreement between observed and predicted values is good.

Qualitatively, the values $\hat{c}_{ij}$ and $\hat{d}_{ij}$ given in Table 9.4 agree with those reported in Chapters 3, 4, and 5 when $\hat{\alpha}$ and $\hat{\beta}$ are greater than $\frac{1}{2}$. Contrary to most of our results, $\hat{d}_{12} > \hat{d}_{21}$ for both Groups E and C, indicating that Player B's actual response was slightly more self-reinforcing than the experimenter-determined reinforcement. A similar effect was observed in the Sure Group of Chapter 4 and in the single experimental group of Chapter 5.

We now consider the case $c_{ij} = d_{ij}$ and use (2.6.11) to estimate the conditioning parameters. The $\hat{c}_{ij}$'s are given in Table 9.7. The same χ^2's reported above were run for the predictions based on these $\hat{c}_{ij}$'s. The obtained values were

TABLE 9.7

MAXIMUM-LIKELIHOOD ESTIMATES OF $c_{ij} = d_{ij}$

Parameter	Group E	Group C
c_{11}	.804	.816
c_{21}	.652	.608
c_{12}	.631	.588
c_{22}	.476	.392

48.9 for Group E and 31.2 for Group C. With 20 degrees of freedom the first value is highly significant, whereas the second just misses significance at the .05 level.

The values of $\hat{c}_{ij}$ given in Tables 9.4 and 9.7, as well as the values of $\hat{d}_{ij}$ given in Table 9.4, show conclusively that subjects do not tend to use a fixed strategy in the sense of game theory. The use of such a strategy would imply complete insensitivity to the responses and reinforcements on the preceding trial; the variation in $\hat{c}_{ij}$ and $\hat{d}_{ij}$ indicates that this is not the case. For example, in Table 9.7 $\hat{c}_{11} - \hat{c}_{22}$ is equal to .328 for Group E and .424 for Group C, both values being appreciably different from 0. Similar evidence against the use of a fixed strategy, whether optimal or not, is provided by the conditioning parameters obtained for the generalized model in Chapters 3, 4, and 5.

9.8. Summary

In Chapter 3 we discussed three restrictions that were imposed on our original experiments: (1) subjects were not told they were interacting; (2) subjects were not informed of each other's responses; and (3) the payoff matrix was not displayed. Restriction (1) was removed in Chapter 4 and thereafter; restriction (2) was removed in Chapters 6 and 7. In this chapter we eliminate restriction (3). Two groups of subjects were run in a zero-sum, two-person game situation. The difference between the groups is that one was shown the numerical payoff matrix (Group E) and the other was shown a matrix with only literal values p and q (Group C).

The payoff matrix had a mixed minimax strategy for both players. This choice was deliberate in order to prevent subjects in Group E from being presented with a game that has a simple and obvious solution.

The experimental method is presented in Sec. 9.2. In Secs. 9.3 and 9.4 observed differences between the two experimental groups are examined. In general, response proportions for subjects in Group C correspond to the predictions of stimulus-sampling theory as formulated in the preceding chapters. However, statistically significant differences were obtained between Groups E and C. In Sec. 9.5 these results are examined within the framework of game theory. The interesting observation is that although Group E deviated from learning-theory predictions, the deviation was not toward an optimal strategy. No detailed explanation is provided for these findings and it is noted that stimulus-sampling theory, at its present stage of development, is not able to incorporate an analysis of the quantitative effects of displaying the payoff matrix.

Finally, in Sec. 9.7 the data are analyzed in terms of the generalized conditioning model. The χ^2 goodness-of-fit tests are not significant for either group and, considering the large number of observations, this result is highly favorable to the model. Further, observed differences between Groups E and C are nicely reflected in the estimated values of the conditioning parameters.

10

Monetary Payoffs and Noncontingent Reinforcement Schedules

10.1. Introduction

At the beginning of Chapter 3 we discussed three restrictions that were imposed on the first experiment: (1) the payoff matrix was not displayed; (2) subjects were not informed of each other's responses; (3) subjects were not told that they were interacting in any way. The third restriction was removed in Chapter 4 and thereafter; the second was removed in Chapters 6 and 7, and the first in Chapter 9.

In addition to the restrictions specified above, all the experiments reported thus far have employed the same type of reinforcement; i.e., simply showing the subject the correct response. From the many studies using this kind of reinforcement, it may be concluded that there is no problem of the subject's being insufficiently motivated by reinforcing events of this nature. However, in terms of an experimental program designed to simulate real game situations it would be desirable to introduce a broader range of reinforcement. Consequently, in this chapter and the next we consider the effects of monetary rewards.

The empirical question of what effects, if any, monetary rewards will have is pertinent to the status and application of the fundamental theory of stimulus sampling. The basic axioms of Sec. 1.2 do not provide a natural method for dealing with reinforcements of different magnitudes. From a quantitative standpoint, techniques for handling the behavioral consequences of different levels of reinforcement are still at a rather primitive stage. We do not pretend to present a finished theory of these matters here.

For our investigation of the effects of different magnitudes of reinforcement, three groups of subjects were given identical treatment, with the exception that Group T won (if correct) or lost (if incorrect) ten cents on each trial, Group F won or lost five cents, and Group Z won or lost zero cents. As should be apparent, Group Z was run under the reinforcement conditions characteristic of the experiments reported in earlier chapters.

We shall analyze the experimental data for the three groups from a variety of standpoints in the later sections of this chapter. We restrict ourselves in this section to the one-element model. Subjects were run in pairs with game-type instructions, but in fact the reinforcement schedule for an individual subject was noncontingent, with $\pi = .6$. Thus the one-element matrix (1.3.4) is appropriate for the analysis of individual data. The many theoretical results for the noncontingent case given in Chapters 1 and 2 need not be repeated here.

Because the two subjects were not reinforced in a statistically independent fashion, both receiving an E_1 or both an E_2 on a given trial, the data may also be analyzed in terms of a two-person situation with features not possessed by a noncontingent experiment in which subjects are run individually. For example, the mean asymptotic probability of a pair of A_1 responses is a function of θ, and is not simply π^2. (A similar observation was made at the end of Sec. 8.2 in connection with Group II of the three-person study.) As a two-person study the model is simply that of Chapter 4 for non-zero-sum games, with the parameters a_i, b_i, c_i, and d_i assuming the following values:

$$
\begin{array}{llll}
a_1 = \pi\,, & a_2 = 0\,, & a_3 = 0\,, & a_4 = 1 - \pi\,, \\
b_1 = 0\,, & b_2 = \pi\,, & b_3 = 1 - \pi\,, & b_4 = 0\,, \\
c_1 = 0\,, & c_2 = 1 - \pi\,, & c_3 = \pi\,, & c_4 = 0\,, \\
d_1 = 1 - \pi\,, & d_2 = 0\,, & d_3 = 0\,, & d_4 = \pi\,.
\end{array}
\tag{10.1.1}
$$

And the two-person transition matrix for the one-element model is as follows:

(10.1.2)

	(1, 1)	(1, 2)	(2, 1)	(2, 2)
(1, 1)	$\pi + (1 - \pi)(1 - \theta)^2$	$(1 - \pi)\theta(1 - \theta)$	$(1 - \pi)\theta(1 - \theta)$	$(1 - \pi)\theta^2$
(1, 2)	$\pi\theta$	$(1 - \theta)$	0	$(1 - \pi)\theta$
(2, 1)	$\pi\theta$	0	$(1 - \theta)$	$(1 - \pi)\theta$
(2, 2)	$\pi\theta^2$	$\pi\theta(1 - \theta)$	$\pi\theta(1 - \theta)$	$(1 - \pi) + \pi(1 - \theta)^2$

In the later sections of this chapter, as noted above, the data will be analyzed from the standpoint of a number of different models. Section 10.2 describes the experimental method. Section 10.3 presents the mean learning curves and asymptotic results. Here we are particularly concerned with determining whether there are significant differences between the three reinforcement conditions. Section 10.4 is devoted to the Markov characteristics of the data from a two-person standpoint. Section 10.5 considers the generalized conditioning model, first in terms of pairs of subjects and then in terms of individual subjects. In the latter case, changes in the conditioning parameters over trials are also analyzed. Section 10.6 applies the multi-element model to Group Z, with particular emphasis on the adequacy of various sequential predictions. Section

10.7 covers the same ground in terms of the linear model defined in Sec. 1.5, and Sec. 10.8 in terms of the hypothesis models discussed in Sec. 1.8. Section 10.9 introduces a "memory" model, which is related to both the generalized conditioning model and the hypothesis model. The chapter contains no separate section on game-theory comparisons, but various remarks on this subject are interspersed among the sections.

10.2. Experimental Method

Subjects. The subjects were 90 undergraduates from introductory courses at Stanford University. They were randomly assigned to the three experimental groups with the restriction of 15 pairs per group.

Apparatus. The apparatus described in Chapter 9 for Group C was used.

Procedure. Subjects in Groups F and T were read the following instructions:

"This experiment is a game in which you will be playing against each other. Each of you will win or lose x cents[1] on each play of the game. The game is similar to a real-life situation in that what you gain or lose depends not only on what you do, but also on what someone else—your opponent—does.

"To keep things straight I will call you 'Player A' and you 'Player B.' The experiment for each of you consists of a series of trials. The top center light on each of your panels will go on for two seconds to indicate the start of each trial. When this light goes on, you will each press one or the other of the two keys in front of you. That is, Player A will press either his A_1 or his A_2 key; Player B will press his B_1 or B_2 key. Then wait until one of the lower lights goes on. If the light above the key you pressed goes on, your prediction was correct; if the light above the key opposite the one you pressed goes on, you were incorrect. The player who is correct wins x cents, and the player who is incorrect loses x cents. Please tabulate your wins and losses on the sheet of paper in front of you. I'll compare your tabulations with mine at the end of the experiment.

"Being correct or incorrect on a given trial depends on the key you press and also on the key the other player presses. With some combinations of your key choice with the other player's key choice, you may both be correct; with other combinations, one player will be correct and the other incorrect; for still other combinations, you may both be incorrect. More specifically, the game is described by the chart in front of you. The p's and q's, corresponding to various response pairs, stand for the proportion of times that you can expect to be correct. For example, if Player A presses his A_1 key and Player B presses his B_1 key, then p_1 is the proportion of times that Player A will be correct and q_1 the proportion of times that Player B will be correct. Thus, if you had some knowledge about the values of these p's and q's, you would know your chances of being correct. For example, if one player knew what key his opponent was going to press, he could select his response to maximize his chance of

[1] For x read "five" for Group F and "ten" for Group T.

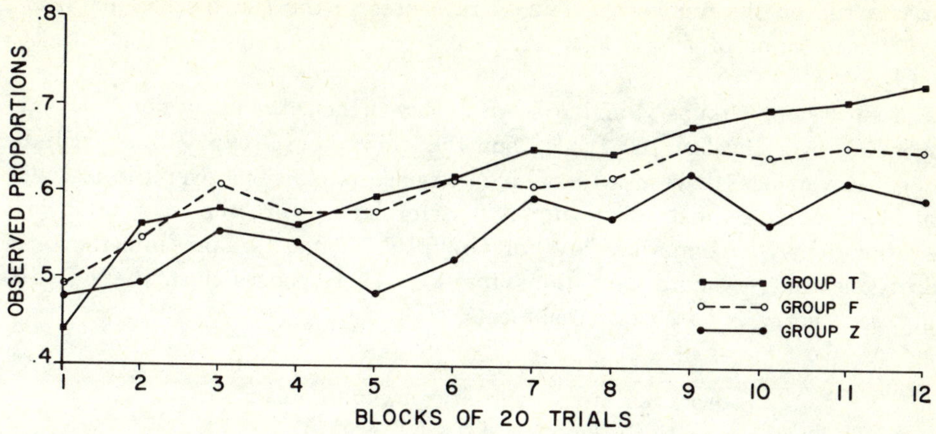

FIGURE 10.1. Proportion of Type 1 responses in successive 20-trial blocks for Groups Z, F, and T.

being correct. The catch is, of course, that you don't know what the p and q values are, nor do you know what the other player is going to do.

"Your job, then, is to obtain as many correct responses as you possibly can. If you play intelligently, you may win two or three dollars. The exact amount will be paid to you at the end of the experiment. The trials move along rapidly, and you must make your key choice as soon as the signal light goes on. That is, when the signal light goes on, press one or the other key and release it before the signal light goes off. Then wait until one of the lower white lights goes on. If the light above the key you pressed goes on, you are correct; and if the light above the key opposite from the one you pressed goes on, you are incorrect."

The instructions for Group Z were similar, except that no mention was made of money.

After the instructions were read, 240 trials were run in continuous sequence. Fifteen reinforcement schedules were used. Each was constructed randomly, with the restriction of exactly 144 E_1's and 96 E_2's. Both members of a subject-pair received the same schedule of E_1's and E_2's. Further, once a given schedule had been used for a subject-pair, it was not used again in that particular experimental group. Thus, within each group two subjects received identical reinforcement schedules; and across groups each schedule was represented exactly once.

10.3. Mean Learning Curves and Asymptotic Results

Figure 10.1 presents the mean proportion of Type 1 responses in successive 20-trial blocks for the three experimental groups; each point is based on observations from 30 subjects. An inspection of the figure indicates that the three groups are approaching separate asymptotes. To check this observation, an F-

test was run on the number of Type 1 responses in the last block of 100 trials; the obtained value of F was 3.33 (d.f. = 2,87), which is significant at the .05 level.

The curves for Groups Z and F seem to have reached a fairly stable level of responding over the last 100 trials, but the curve for Group T may still be rising. Nevertheless, we shall examine response proportions over the last 100-trial block as an estimate of asymptotic performance, with the reservation that the estimate may be somewhat low for Group T. Table 10.1 presents the mean proportions and standard deviations for Type 1 responses over the last 100 trials; each entry is based on 30 subjects.

TABLE 10.1
OBSERVED MEAN PROPORTIONS AND STANDARD DEVIATIONS FOR TYPE 1 RESPONSES

Group	Mean	SD
Z	.593	.110
F	.644	.121
T	.690	.128

In view of these results, it seems quite clear that the amount of monetary reinforcement significantly affected behavior in the experimental situation. The stimulus-sampling model described by the axioms of Sec. 1.2 predicted that the probability of a Type 1 response approaches π as the number of trials becomes large, independent of θ and the number N of stimulus elements. For Group Z the asymptotic prediction of $\pi = .6$ fits the observed terminal values for the proportion of A_1 responses almost perfectly. However, for Groups F and T it is clear that asymptotic performance is markedly above π. Thus the stimulus-sampling model developed in Chapters 1 and 2 does not fit the experimental data at even the grossest level of analysis when monetary payoffs are involved. In later sections we consider other formulations, among them the generalized conditioning model, that offer more promise.

It would be interesting to investigate the effect of monetary reinforcement when the reinforcement schedule is more complex than the one employed in this experiment. For example, when the payoff matrix is the mixed minimax type considered in Chapter 9, will the subject move away from the stimulus-sampling model prediction and toward the optimal game-theory strategy as the monetary reward increases? Without further experimental evidence the question cannot be answered. In the present study the subjects tended to approach an optimal game strategy as the monetary reward increased, and it may well be that this result can be directly generalized to more complex reinforcement schedules. However, results of the type described in Chapter 9 (where subjects were shown the payoff matrix and deviated from learning-theory predictions, but not in the direction predicted by game theory) leave the issue open to further analysis.

10.4. Tests for Markov Properties

Table 10.2 presents the observed transition matrix $[\hat{p}_{ij}]$ computed over the last 100 trials corresponding to (10.1.2). As may be seen from (10.1.2), the transition probabilities $P(A_{2,n+1}B_{1,n+1} \mid A_{1,n}B_{2,n})$ and $P(A_{1,n+1}B_{2,n+1} \mid A_{2,n}B_{1,n})$ are identically 0 in the one-element model for the particular reinforcement procedure used in this experiment. An inspection of Table 10.2 indicates that the observed values are substantially different from 0.

TABLE 10.2
OBSERVED TRANSITION MATRIX $[\hat{p}_{ij}]$ COMPUTED OVER THE LAST 100 TRIALS

	Group Z				Group F				Group T			
n \ $n+1$	(1, 1)	(1, 2)	(2, 1)	(2, 2)	(1, 1)	(1, 2)	(2, 1)	(2, 2)	(1, 1)	(1, 2)	(2, 1)	(2, 2)
(1, 1)	.453	.214	.194	.139	.514	.178	.207	.101	.590	.163	.182	.065
(1, 2)	.330	.284	.193	.193	.402	.236	.173	.189	.428	.348	.088	.136
(2, 1)	.365	.185	.249	.201	.373	.098	.389	.140	.445	.075	.345	.135
(2, 2)	.280	.204	.254	.262	.333	.142	.283	.242	.295	.160	.267	.278

The hypothesis that the transition probabilities are constant over trials was tested by computing (2.4.1) for $t = 1, 2$; $t = 1$ refers to the block of trials from 1 to 140, and $t = 2$ refers to the block of trials from 141 to 240. The obtained values of χ^2 were 34.1, 47.4, and 70.36 for Groups Z, F, and T, respectively. With 12 degrees of freedom all three values are significant, indicating that the probabilities $P(A_{i,n+1}B_{j,n+1} \mid A_{i',n}B_{j'n})$ are not constant over trials. The same test was run for the last 50-trial block and the next-to-last 50-trial block. For this analysis none of the groups yielded significant χ^2's, indicating that the transition probabilities approached stationarity in the last 100 trials.

Tests of hypotheses concerning the order of the chain deal only with the transition data over the final block of 100 trials. The zero-order versus first-order hypothesis was tested by calculating (2.4.2). The χ^2's were 48.3, 105.5, and 206.4 for Groups Z, F, and T, respectively; with 9 degrees of freedom all are highly significant, indicating that the outcomes of trials are not statistically independent.

The first-order versus second-order hypothesis was tested by computing the χ^2 of (2.4.3). The obtained χ^2's were 40.7, 97.1, and 92.6 for Groups Z, F, and T, respectively. With 36 degrees of freedom the results for Groups F and T are significant; for Group Z the result is not significant. Thus for Groups F and T better predictions of responses can be made by observing responses on the two immediately preceding trials rather than simply on the single immediately preceding trial. For Group Z this is not the case, and the finding for this group is in accord with the non-significant results on the same test for Group II in Chapter 8. As noted in Chapter 8, Group II had a noncontingent reinforcement schedule with $\pi = .6$.

*10.5. Generalized Conditioning Model

We now apply the model described in Sec. 2.6 to the data of our three groups. The transition numbers $n(A_iB_j \mid A_{i'}E_kB_{j'}E_{k'})$ for the 16×4 matrix (4.8.1) were tabulated over the last 100 trials and aggregated over subjects. Using (4.8.2), we obtained maximum-likelihood estimates of c_{ij} and d_{ij}, which are presented in Table 10.3. Given these estimates, the χ^2 goodness-of-fit test (2.4.4) can be applied to matrix (4.8.1), where the observed n_{ij}'s are those on which the estimates of c_{ij} and d_{ij} are based (see Sec. 2.6 for a discussion of this application of the χ^2 test). The reinforcement schedule was such that both Player *A* and Player *B* received the same reinforcing event on a given trial; hence, transitions are not possible in rows 2, 3, 6, 7, 10, 11, 14, and 15 of (4.8.1), and

TABLE 10.3

MAXIMUM-LIKELIHOOD ESTIMATES OF c_{ij} AND d_{ij}

Parameter	Group *Z*	Group *F*	Group *T*
c_{11}	.714	.779	.849
c_{21}	.582	.569	.625
c_{12}	.549	.539	.622
c_{22}	.421	.306	.312
d_{11}	.716	.800	.838
d_{21}	.603	.670	.654
d_{12}	.523	.647	.688
d_{22}	.402	.466	.345

TABLE 10.4

PREDICTED TRANSITION PROBABILITIES FOR THE GENERALIZED CONDITIONING MODEL

(The observed values computed over the last 100 trials are given in parentheses)

	Group *Z*				Group *F*				Group *T*			
n \\ $n+1$	(1, 1)	(1, 2)	(2, 1)	(2, 2)	(1, 1)	(1, 2)	(2, 1)	(2, 2)	(1, 1)	(1, 2)	(2, 1)	(2, 2)
(1, 1)	.421 (.453)	.227 (.214)	.217 (.194)	.135 (.139)	.513 (.514)	.170 (.178)	.225 (.207)	.092 (.101)	.598 (.590)	.160 (.163)	.180 (.182)	.062 (.065)
(1, 2)	.347 (.330)	.301 (.284)	.176 (.193)	.176 (.193)	.413 (.402)	.269 (.236)	.175 (.173)	.143 (.189)	.420 (.428)	.339 (.348)	.111 (.088)	.130 (.136)
(2, 1)	.338 (.365)	.180 (.185)	.300 (.249)	.182 (.201)	.352 (.373)	.112 (.098)	.386 (.389)	.150 (.140)	.400 (.445)	.100 (.075)	.378 (.345)	.122 (.135)
(2, 2)	.279 (.280)	.239 (.204)	.244 (.254)	.238 (.262)	.286 (.333)	.178 (.142)	.302 (.283)	.234 (.242)	.289 (.295)	.211 (.160)	.242 (.267)	.258 (.278)

consequently there are $8 \times 3 = 24$ degrees of freedom associated with the matrix. However, since eight parameters have been estimated from these data for each group, χ^2 is to be interpreted with $24 - 8 = 16$ degrees of freedom. The obtained values of χ^2 were 28.37 for Group Z, 14.98 for Group F, and 23.24 for Group T. The value for Group Z is significant at the .05 level; the other two values are not significant.

Table 10.4 presents the predicted values of $P(A_{i,n+1}B_{j,n+1} \mid A_{i',n}B_{j',n})$ based on the estimated c_{ij}'s and d_{ij}'s; the observed quantities are also presented and are based on the last 100 trials. Table 10.5 presents the asymptotic values of

TABLE 10.5

ASYMPTOTIC PREDICTIONS FOR THE GENERALIZED CONDITIONING MODEL AND OBSERVED VALUES BASED ON THE LAST 100 TRIALS

Asymptotic Value	Group Z		Group F		Group T	
	Predicted	Observed	Predicted	Observed	Predicted	Observed
$P_\infty(A_1B_1)$	.360	.374	.422	.433	.488	.496
$P_\infty(A_1B_2)$	.236	.221	.172	.162	.186	.177
$P_\infty(A_2B_1)$	.231	.217	.271	.259	.217	.210
$P_\infty(A_2B_2)$	.173	.188	.135	.146	.109	.117

$P(A_{i,n}B_{j,n})$ that have been computed in terms of the estimates of c_{ij} and d_{ij} by (1.3.17); the observed values for the last 100 trials are also given.

The empirical adequacy of the generalized conditioning model for the three groups is roughly comparable to the results obtained in earlier chapters. It is interesting that the fit for Group Z is actually slightly worse than the fit for Groups F and T; the introduction of monetary reward has not impaired the adequacy of the model. On the other hand, the effects of such reinforcement are strikingly exhibited in the conditioning-parameter values given in Table 10.3. Taking Z and T as the extreme groups, $c_{ij}(Z) < c_{ij}(T)$ and $d_{ij}(Z) < d_{ij}(T)$, except for c_{22} and d_{22}. This exception is of considerable interest. If all the inequalities were in the same direction, a simple explanation would be that the more often occurring A_1 and E_1 events have a secular effect on conditioning. Such secular effects are no doubt present, but they are discouraging to the theorist because they suggest that a time-dependent mechanism must be incorporated into an adequate theory. What is surprising in the present data is the tendency in the other direction when a correct prediction is made, i.e., when $i = j$. As reward increases, the probability of making an A_2 response increases when it is preceded by an A_2 and an E_2. We return to this point later in discussing individual rather than pair data. It may also be noted from Table 10.3 that the monotonicity relations discussed in Sec. 3.8 hold up very well. For all three groups $c_{11} > c_{21} > c_{12} > c_{22}$ and $d_{11} > d_{21} > d_{12} > d_{22}$, with only one exception: $d_{21}(T) < d_{12}(T)$.

Since the experimental setup is perfectly symmetric for both players, we should not expect systematic differences between subjects run as Player A and those run as Player B. Consequently, as in previous chapters, we now estimate $c_{ij} = d_{ij}$ for the generalized conditioning model. The estimate is given by (4.8.3) for the appropriate two-person interaction situation. However, in theory the two subjects are not affecting each other in the present experimental procedure. We shall accordingly estimate c_{ij} as though the protocol of responses and reinforcements for each subject were collected in an individual learning situation of the noncontingent type; that is, (2.6.5) will be used to estimate the conditioning parameters. Further, since the reinforcement schedule used in this study is relatively simple, it will be of interest to estimate c_{ij} at several points in the learning process and to note the character of change in c_{ij} over trials.

For each subject the transition matrix corresponding to (2.6.6) was tabulated in successive blocks of 20 trials. These transition matrices were then summed over the 30 subjects in each group, and estimates of c_{ij} were obtained in successive 20-trial blocks by (2.6.5). Aggregating the transition data in this fashion and estimating c_{ij} from the aggregated transition frequencies, of course, assumes that all subjects in a given group have the same values for the conditioning parameters. The merits and defects of this procedure have been commented on earlier and we shall not repeat the discussion here; however, later in this section we compare some aggregate estimates with individual estimates.

Figure 10.2 presents the estimate of c_{ij} in successive 20-trial blocks for Groups Z, F, and T. For Groups Z and F the curves appear to have approached a stable level over the last four or five blocks of trials. However, for Group T, curves c_{11}, c_{12}, and c_{21} appear to be still rising; of course, we would expect that some of the c_{ij}'s are still changing for Group T, since the proportion of A_1 responses also has not reached a stable level for this group.

There are a number of comments to be made about these results. First of all, over the last four or five blocks of trials, in general

$$c_{11}(T) > c_{11}(F) > c_{11}(Z)\,, \quad c_{12}(T) > c_{12}(F) > c_{12}(Z)\,, \quad c_{21}(T) > c_{21}(F) > c_{21}(Z)\,.$$

However, in marked contrast, $c_{22}(T) < c_{22}(F) < c_{22}(Z)$. These results are consistent with the results obtained in the first part of this section. Increasing the monetary reward leads to an increase in the probability of an A_1 response given A_1E_1, A_1E_2, and A_2E_1. However, for A_2E_2 the reverse effect occurs; the greater the monetary payoff, the less likely an A_1 given A_2E_2.

Over the last few blocks of trials $c_{11} > c_{21} > c_{12} > c_{22}$ for Group Z, and $c_{11} > c_{21} \approx c_{12} > c_{22}$ for Groups F and T. The surprising observation is that the order $c_{21} > c_{12}$ for Group Z does not hold for Groups F and T; indeed over the terminal trials, it appears that with increasing monetary reward the effects of A_1E_2 and A_2E_1 on the probability of an A_1 response become progressively more equivalent. It is more surprising still that for the first 120 trials $c_{12} > c_{21}$ for both Group F and Group T; whereas for Group Z the reverse is true. For Group T the

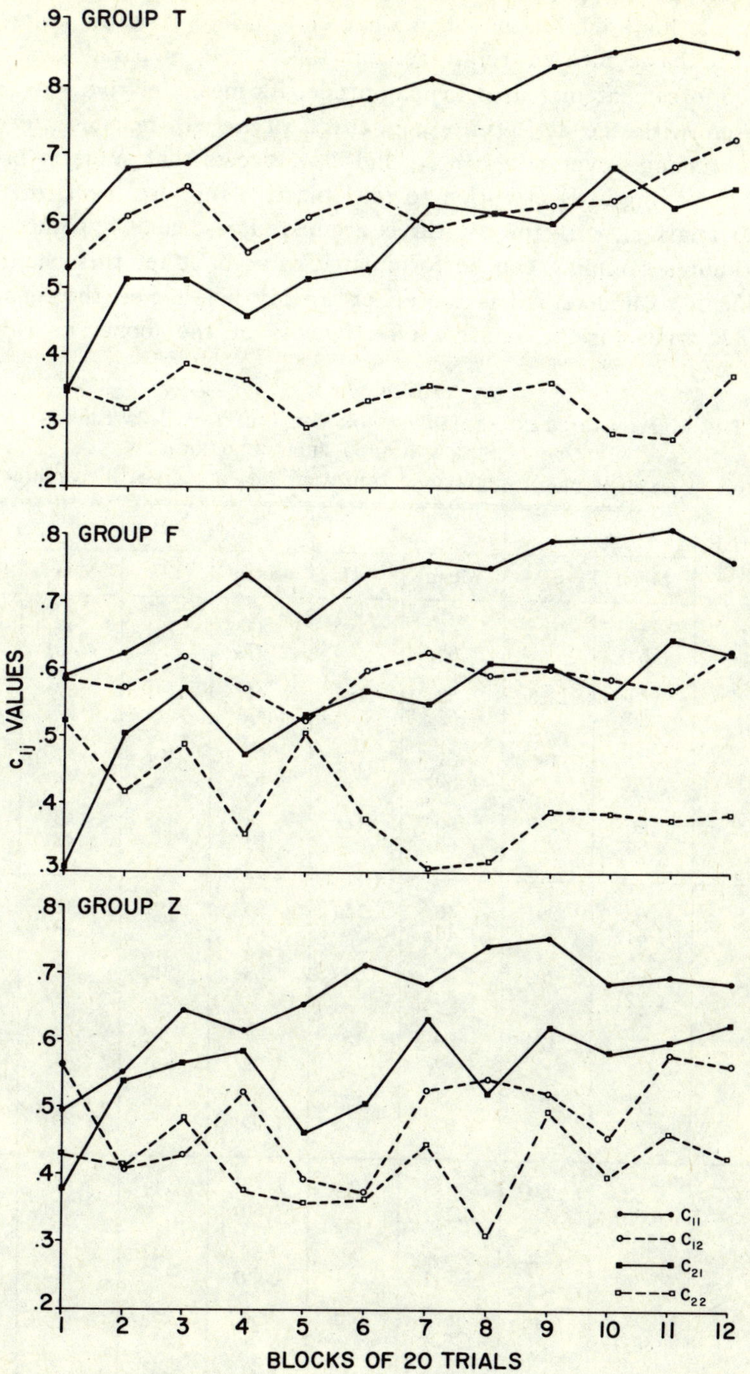

FIGURE 10.2. Estimates of c_{ij} in successive 20-trial blocks for Groups Z, F, and T.

differences are particularly large, as is easily seen from Fig. 10.2. A very slight effect of this kind was noted in Chapter 9. Here it is so large that it challenges the basic idea of reinforcement theory. It is paradoxical that as objective reinforcement increases in magnitude, its mean relative effectiveness, in comparison with the self-reinforcing effect of an A_1 response, decreases. The relative change over trials in c_{21} and c_{12} suggests that subjects begin the experiment by paying little attention to the objective reinforcement; in the first block of 20 trials $c_{21} < .4$ for all three groups. Reasonable hypotheses about selective stimulus sampling can be formulated to account for this phenomenon. What is difficult to understand is the effect already noted, i.e., the tendency in the initial 120 trials for the relative effectiveness of the monetary reward to

TABLE 10.6

THE MEAN AND STANDARD DEVIATION FOR INDIVIDUAL ESTIMATES OF c_{ij} IN SUCCESSIVE 60-TRIAL BLOCKS

(Estimates of c_{ij} in terms of aggregated transition data are given in parentheses)

Group	Trial Block	c_{11}		c_{21}		c_{12}		c_{22}	
		Mean	SD	Mean	SD	Mean	SD	Mean	SD
Z	1	.57 (.57)	.179	.51 (.50)	.182	.47 (.47)	.164	.44 (.44)	.190
	2	.64 (.66)	.223	.53 (.52)	.188	.41 (.43)	.191	.37 (.37)	.243
	3	.72 (.73)	.199	.61 (.59)	.174	.51 (.53)	.177	.42 (.42)	.262
	4	.66 (.69)	.248	.60 (.60)	.171	.54 (.54)	.185	.43 (.43)	.243
F	1	.63 (.64)	.201	.46 (.46)	.143	.60 (.60)	.191	.47 (.48)	.189
	2	.69 (.72)	.202	.55 (.53)	.089	.55 (.57)	.246	.43 (.42)	.274
	3	.75 (.77)	.124	.60 (.60)	.151	.57 (.60)	.252	.34 (.34)	.272
	4	.76 (.79)	.213	.64 (.62)	.177	.57 (.60)	.228	.40 (.39)	.272
T	1	.63 (.65)	.231	.48 (.45)	.179	.58 (.60)	.219	.35 (.35)	.164
	2	.71 (.76)	.253	.55 (.51)	.202	.58 (.60)	.202	.38 (.33)	.276
	3	.77 (.81)	.249	.67 (.62)	.221	.58 (.61)	.255	.41 (.36)	.261
	4	.84 (.86)	.217	.68 (.65)	.239	.65 (.68)	.224	.35 (.32)	.268

vary inversely with the amount. Further experimental and theoretical investigation of this paradox seems called for.

In order to compare the estimates of c_{ij} from aggregated transition frequencies with estimates from individual protocols we have also computed c_{ij}'s for each subject in successive blocks of 60 trials. Sixty-trial blocks were used rather than 20-trial blocks to ensure that the estimates would be fairly reliable. The means and standard deviations for these individual estimates are given in Table 10.6; shown in parentheses are the estimates of c_{ij} in successive 60-trial blocks based on the aggregated transition data. In general, the mean of the individual estimates and the estimate from the aggregate data are very close, as with similar comparisons given elsewhere in this book. Another feature of this table is that the observed standard deviations do not appear to be changing in any systematic way over the four blocks of trials. If, in fact, there are significant differences initially between subjects on c_{ij}, one might expect such differences to become less pronounced over trials; however, this does not appear to be the case.

From Fig. 10.2 and Table 10.6 it is evident that the c_{ij}'s are changing systematically over trials; the phenomena, however, are complex. For all three groups the graph of c_{11} has the form of an exponential learning curve, but it is not clear whether this is the case for the other conditioning parameters. Since it would take us far afield to describe our present ideas for developing a theory to account for these changes, we confine ourselves to a few general comments. In postulating a theory whose mechanism of learning is expressed directly in terms of the conditioning parameters c_{ij}, we would, in a sense, be halfway between the Markov stimulus-sampling model and the linear model. The linear model directly postulates changes in the probability of a response *given* the sequence of preceding responses and reinforcements received by the individual subject. The Markov model postulates changes in the states of stimulus conditioning. For both models one of the most important theorems asserts a recursion for the *mean* probability of an A_1 response. A theory postulated in terms of changing c_{ij}'s would take as axioms recursions in the conditional mean probabilities c_{ij}. Independence-of-path assumptions would assert that the past history of the organism before the last occurrence of an A_iE_j combination does not affect the c_{ij}. A theory developed in this manner would be at a different level of abstraction from that of the stimulus-sampling and linear models.

10.6. Application of the Multi-Element Model to Group *Z*

In the remaining sections of this chapter we no longer distinguish between Player *A* and Player *B*. That is, hereafter we view each subject's protocol of responses and reinforcements as being collected in an individual learning situation of the noncontingent type discussed in Sec. 1.3. For each subject two responses are available, and the probability of E_1 on any trial is constant and independent of the subject's responses; i.e., $P(E_{1,n}) = \pi$. In order that the notation in the following sections may correspond to that in Sec. 1.3, we label the two responses A_1 and A_2, independent of whether the subject was run as Player A or Player B.

In this section and the next two sections the data of Group Z will be analyzed extensively. First we consider the multi-element stimulus-sampling model described by the axioms of Sec. 1.2. Then we apply the linear model of Sec. 1.5 to the data, and in Sec. 10.8 we examine a simple hypothesis model of the sort discussed in Sec. 1.7. Finally, in Sec. 10.9 we consider a new model for the data of Groups F and T.

We now turn to an analysis of the data for Group Z in terms of the axioms of Sec. 1.2. Specifically, we consider the general case in which there are N elements and the sampling of elements is equiprobable. In Sec. 2.3 a procedure for making a joint estimate of N and θ was described, and the details of the technique were presented for the noncontingent case. This method is applied here.

As in (2.3.26), let $x = (1 - \theta)/N$ and $y = \theta/N$. The pseudo-likelihood function for x is given in terms of (2.3.27):

$$L(x) = n_{11} \log [\pi + x(1 - \pi)] + n_{12} \log (1 - x)(1 - \pi) + n_{21} \log (1 - x)\pi + n_{22} \log [1 - \pi(1 - x)] . \tag{10.6.1}$$

Hence, the following equation yields the pseudomaximum-likelihood estimate of x:

$$\frac{dL(x)}{dx} = \frac{n_{11}(1 - \pi)}{\pi + x(1 - \pi)} + \frac{n_{22}\pi}{1 - \pi(1 - x)} - \frac{n_{12} + n_{21}}{1 - x} = 0 . \tag{10.6.2}$$

Using the n_{ij}'s for the last 100 trials aggregated over subjects, we obtain $x = (-641.2 \pm 812.0)/1440$. We are interested in the positive root; hence $x^* = .1186$. [Note that solving (10.6.2) for x is equivalent to solving (2.2.8) for $(1 - \theta)$.]

Next consider the estimate of y defined with reference to (2.3.28). Matrices (2.3.28) and (2.3.27) are identical, except that x is replaced by y in (2.3.28). Consequently, the estimate of y is given by (10.6.2), where x is replaced by y and where the n_{ij}'s now refer to the number of $A_{j,n+1}$'s given an $E_{i,n}$ at asymptote. Using the observed number of $A_{i,n}$'s given $E_{j,n-1}$ for the last 100 trials aggregated over subjects we obtain $y = (-599.8 \pm 849.8)/1440$. The estimate is the positive root, and hence $y^* = .174$. By definition

$$N^* = \frac{1}{x^* + y^*} = 3.42 , \qquad \theta^* = \frac{y^*}{x^* + y^*} = .594 . \tag{10.6.3}$$

Predictions for $P_\infty(A_{i,n+1} \mid E_{j,n} A_{k,n})$. Once θ and N have been estimated, predictions can be made concerning almost all observable quantities.[2] We begin by examining $P(A_{j,n+1} \mid E_{i,n} A_{k,n})$ for large n.

In order to derive a theoretical expression for these quantities we need again

[2] Of course, if the initial conditions are also known, then predictions can be made concerning all observable quantities.

the notation of Sec. 2.3. Let $C_{i,n}$ $(i = 0, 1, \cdots, N)$ denote the conditioning state on trial n, where i elements are connected to A_1 and $N - i$ elements are connected to A_2. Then, using methods similar to those employed to establish (2.3.17) and (2.3.18), we have

(10.6.4)

$$\begin{aligned} P(A_{1,n+1} \mid E_{1,n} A_{1,n}) &= \frac{1}{P(E_{1,n}A_{1,n})} \sum_{i,j} P(A_{1,n+1}C_{i,n+1}E_{1,n}A_{1,n}C_{j,n}) \\ &= \frac{1}{P(E_{1,n}A_{1,n})} \sum_{i,j} P(A_{1,n+1} \mid C_{i,n+1})P(C_{i,n+1} \mid E_{1,n}A_{1,n}C_{j,n}) \\ &\qquad \cdot P(E_{1,n} \mid A_{1,n}C_{j,n})P(A_{1,n} \mid C_{j,n})P(C_{j,n}) . \end{aligned}$$

Further,

$$P(C_{j,n+1} \mid E_{1,n}A_{1,n}C_{i,n}) = \begin{cases} 1 & (i = j) , \\ 0 & (i \neq j) . \end{cases}$$

That is, the element sampled is conditioned to A_1 (since an A_1 occurred on trial n) and, given an E_1 reinforcement, the conditioning of the sampled element cannot change. In addition, $P(A_{1,n} \mid C_{i,n}) = i/N$, and $P(E_{1,n} \mid A_{j,n}) = P(E_{1,n} \mid A_{j,n}C_{i,n}) = \pi$. Hence, (10.6.4) can be rewritten as follows:

$$P(A_{1,n+1} \mid E_{1,n}A_{1,n}) = \frac{1}{P(A_{1,n})} \sum_i \left(\frac{i}{N}\right)^2 P(C_{i,n}) .$$

It can easily be shown (see Sec. 2.3) that

(10.6.5)

$$\begin{aligned} \lim_{n\to\infty} P(C_{i,n}) &= \binom{N}{i} \pi^i (1 - \pi)^{N-i} , \\ \lim_{n\to\infty} P(A_{1,n}) &= \pi . \end{aligned}$$

Therefore, as $n \to \infty$

$$\begin{aligned} P_\infty(A_{1,n+1} \mid E_{1,n}A_{1,n}) &= \frac{1}{\pi} \sum_i \left(\frac{i}{N}\right)^2 \binom{N}{i} \pi^i (1 - \pi)^{N-i} \\ &= \frac{1}{N^2\pi} \sum_i i^2 \binom{N}{i} \pi^i (1 - \pi)^{N-i} . \end{aligned}$$

The summation is the second raw moment of the binomial distribution with parameter π and sample size N. Therefore

(10.6.6)

$$P_\infty(A_{1,n+1} \mid E_{1,n}A_{1,n}) = \frac{1}{N^2\pi}[N^2\pi^2 + N\pi(1 - \pi)] = \pi + \frac{1 - \pi}{N} .$$

By identical methods (cf. Estes [1959]) one can establish that

(10.6.7)

$$\begin{aligned} P_\infty(A_{1,n+1} \mid E_{1,n}A_{2,n}) &= \pi\left(1 - \frac{1}{N}\right) + \frac{\theta}{N} , \\ P_\infty(A_{1,n+1} \mid E_{2,n}A_{1,n}) &= \pi\left(1 - \frac{1}{N}\right) + \frac{1 - \theta}{N} , \\ P_\infty(A_{1,n+1} \mid E_{2,n}A_{2,n}) &= \pi\left(1 - \frac{1}{N}\right) . \end{aligned}$$

The observed transition frequencies $n(A_{i,n+1} \mid E_{j,n}A_{k,n})$ for the last 100 trials aggregated over subjects are as follows:

(10.6.8)

	A_1	A_2
A_1E_1	748	298
A_1E_2	394	342
A_2E_1	462	306
A_2E_2	186	264

From the estimated values of θ and N, the predicted frequencies can be obtained in terms of (10.6.6), (10.6.7), and the row sums of (10.6.8). They are as follows:

(10.6.9)

	A_1	A_2
A_1E_1	749.9	296.1
A_1E_2	399.8	336.2
A_2E_1	459.5	308.5
A_2E_2	191.1	258.9

The χ^2 test given by (2.4.4) for the goodness of fit between (10.6.8) and (10.6.9) yielded a value of .475. There are four degrees of freedom associated with the array of numbers, but since two parameters have been estimated from these data, χ^2 is to be interpreted with only two degrees of freedom. As might be expected in view of the excellent agreement between (10.6.8) and (10.6.9), the χ^2 is not significant.

Predicted learning curves. We now generate theoretical learning curves corresponding to observed results presented in Fig. 10.1 for Group Z. In order to do this, we need some additional theoretical quantities. In the noncontingent case

$$P(C_{j,n+1} \mid C_{i,n}) = \begin{cases} \dfrac{(N-i)\theta\pi}{N} & (j = i+1)\,, \\ \dfrac{i\theta(1-\pi)}{N} & (j = i-1)\,, \\ (1-\theta) + \dfrac{\theta[i\pi + (N-i)(1-\pi)]}{N} & (i = j)\,, \\ 0 & (\text{otherwise})\,. \end{cases} \tag{10.6.10}$$

Further,

$$\begin{aligned} P(A_{1,n}) &= \sum_{i=0}^{N} \frac{i}{N} P(C_{i,n}) \\ &= \sum_{i=0}^{N} \frac{i}{N} \sum_{j=0}^{N} P(C_{i,n} \mid C_{j,n-1}) P(C_{j,n-1})\,. \end{aligned}$$

Substituting in (10.6.10) and summing, we obtain

$$P(A_{1,n}) = \left(1 - \frac{\theta}{N}\right) P(A_{1,n-1}) + \frac{\theta}{N}\pi\,,$$

which has the solution

$$P(A_{1,n}) = \pi - [\pi - P(A_{1,1})]\left(1 - \frac{\theta}{N}\right)^{n-1}. \tag{10.6.11}$$

In Fig. 10.1 the data are plotted in terms of the proportion of A_1 responses in 20-trial blocks. The theoretical expression that describes this empirical curve is readily obtained from (10.6.11). Let m be the ordinal number of a 20-trial block running from trial $n+1$ through trial $n+20$, and $P(m)$ the expected proportion of A_1's in the block. Then

$$\begin{aligned} P(m) &= \frac{1}{20}\left[\sum_{n=1}^{20m} P(A_{1,n}) - \sum_{n=1}^{20(m-1)} P(A_{1,n})\right] \\ &= \pi - \frac{1}{20\theta}[\pi - P(A_{1,1})]\left[1 - \left(1 - \frac{\theta}{N}\right)^{20}\right] N\left(1 - \frac{\theta}{N}\right)^{20(m-1)}. \end{aligned} \tag{10.6.12}$$

The value of $P(A_{1,1})$ should be in the neighborhood of .5 if response bias does not exist. However, to avoid sampling deviations we eliminate $P(A_{1,1})$ in favor of the observed value of $P(1)$. This can be done in the following way. Note that

$$P(1) = \pi - \frac{N}{20\theta}[\pi - P(A_{1,1})]\left[1 - \left(1 - \frac{\theta}{N}\right)^{20}\right].$$

Solving for $[\pi - P(A_{1,1})]$ and substituting the result in (10.6.12), we obtain

$$P(m) = \pi - [\pi - P(1)]\left(1 - \frac{\theta}{N}\right)^{20(m-1)}. \tag{10.6.13}$$

The observed value of $P(1)$ is .478. Using the estimated values of θ and N, we obtain $P(2) = .597$, $P(3) = .600, \cdots$. A comparison of these values with the observed values given in Fig. 10.1 for Group Z shows that the predicted curve approaches asymptote much more rapidly than the observed curve. This result agrees with those reported in earlier chapters and with the analysis of the c_{ij}'s as a function of trials in the preceding section. In earlier chapters it was possible to maintain that some of the discrepancies between the observed and theoretical mean learning curves were mainly due to the fact that we restricted ourselves to the one- or two-element model. No such justification is possible here; hence, we must conclude that the experimental evidence suggests a more complicated conceptualization of the processes of stimulus sampling and conditioning over trials.

It is not appropriate at this point to examine alternative schemes in any detail, but it is worth mentioning that an adaptation of Restle's notion [1955] of relevant stimuli seems *a priori* promising. The essential idea would be that the number N of relevant stimuli is large at the beginning of the experiment. It decreases as the experiment progresses because the subject "adapts out" the stimuli that are not relevant to the experiment. Thus initially, because N is large, the learning rate θ/N will be small, and will better account for the slope of the learning curves reported in this book. The Restle model is difficult to work with mathematically and does not have exactly the desired properties just mentioned. A simpler, more empirical approach, which we hope to try in some future studies, is to estimate N and θ for every block of ten or twenty trials, as was done for the c_{ij}'s in the preceding section. However, there are still theoretical and computational difficulties in such a program. The procedures given in Sec. 2.3 were dependent on the assumption that the subjects were already at asymptote. Also the discussions in later parts of this section indicate the desirability of estimating N and θ in terms of more sequential information, i.e., in terms of more complicated conditional probabilities.

Predicted asymptotic variance. We next consider the variance of the response proportions over a block of K trials once the subject is at asymptote. To obtain the appropriate expression several additional theoretical results are required. Define σ_n^2 as the variance of the response distribution on trial n, i.e.,

$$\sigma_n^2 = \sum_{i=0}^{N} \left(\frac{i}{N}\right)^2 P(C_{i,n}) - [P(A_{1,n})]^2 .$$

For the noncontingent case, it can be shown (cf. Estes [1959]) that as $n \to \infty$

$$\sigma_\infty^2 = \frac{\pi(1-\pi)}{N} . \tag{10.6.14}$$

We also need the following expression, which can be easily obtained and is stated without proof:

$$P(A_{1,n+k}A_{1,n}) = \pi P(A_{1,n}) - [\pi P(A_{1,n}) - P(A_{1,n+1}A_{1,n})]\left(1 - \frac{\theta}{N}\right)^{k-1} .$$

Using (2.3.17) and the fact that $P(A_{1,n}) \xrightarrow{n} \pi$, we obtain for $n \to \infty$

$$P_\infty(A_{1,n+k}A_{1,n}) = \pi^2 + \frac{\pi(1-\pi)(1-\theta)}{N}\left(1 - \frac{\theta}{N}\right)^{k-1} . \tag{10.6.15}$$

Using the notation introduced in Sec. 1.3, let $\bar{\boldsymbol{A}}_K$ be the random variable for the number of A_1 responses in K trials at asymptote. Following the development used to establish (1.3.31),

$$\operatorname{var}(\bar{\boldsymbol{A}}_K) = K\sigma_\infty^2 + 2 \sum_{1 \leqq j,k \leqq K} [P_\infty(A_{1,j}A_{1,k}) - \pi^2] .$$

Substituting in (10.6.14) and (10.6.15) and summing, we obtain

$$\operatorname{var}(\bar{\boldsymbol{A}}_K) = \frac{K\pi(1-\pi)}{N} + \frac{2\pi(1-\pi)(1-\theta)}{\theta}\left\{K - \frac{N}{\theta}\left[1 - \left(1 - \frac{\theta}{N}\right)^K\right]\right\} . \tag{10.6.16}$$

Dividing (10.6.16) by K^2 yields the theoretical expression that corresponds to the square of the standard deviation reported in Table 10.1 for Group Z. Using the estimated values of θ and N given by (10.8.2) and $K = 100$, we obtain

$$\frac{\operatorname{var}(\bar{\boldsymbol{A}}_K)}{K^2} \approx .0038 .$$

The observed variance (see Table 10.1) is .0121, about three times larger than the predicted value. Of course, this prediction is based on the assumption that θ and N are the same for all subjects; if this is not the case, the predicted variance would be larger. However, the magnitude of the difference between observed and predicted variance probably is too large to be accounted for in terms of individual differences. Further, remember that the prediction of the variance is a prediction about individual subjects and not simply about group properties. If some individuals are not satisfying the theory and are deviating widely from the mean asymptote of π, the variance will naturally increase.

Other sequential effects. We now consider some additional sequential statistics for the N-element model. The following result for $n \to \infty$ can be easily obtained, and is given without proof:

$$P_\infty(A_{1,n+1} \mid E_{1,n}A_{1,n} \cdots E_{1,n-k}A_{1,n-k}) = \frac{\mu_{k+2}}{N\mu_{k+1}},$$

where μ_i is the ith raw moment of the binomial distribution with parameter π and sample size N. The moment-generating function for this binomial distribution is $m(t) = (1 - \pi + \pi e^t)^N$, and the desired moment is given by $\mu_i = d^i m(t)/dt\,|_{t=0}$. For the present, only the first three moments will be required. They are $\mu_1 = N\pi$; $\mu_2 = \mu_1[(N-1)\pi + 1]$; and $\mu_3 = \mu_2[(N-2)\pi + 1] + \mu_1^2$. The particular sequential probability to be examined is

$$P_\infty(A_{1,n+1} \mid E_{1,n}A_{1,n}E_{1,n-1}A_{1,n-1}) = \frac{\mu_3}{N\mu_2} = \frac{1}{N}(1 - 2\pi) + \pi\frac{(N-1)\pi + 2}{(N-1)\pi + 1}. \tag{10.6.17}$$

Also, by considerations of symmetry,

$$P_\infty(A_{2,n+1} \mid E_{2,n}A_{2,n}E_{2,n-1}A_{2,n-1}) = \frac{1}{N}(2\pi - 1) + (1 - \pi)\frac{(N-1)(1-\pi) + 2}{(N-1)(1-\pi) + 1}. \tag{10.6.18}$$

Using the values of θ and N given in (10.6.3), we obtain

$$\begin{aligned} P_\infty(A_{1,n+1} \mid E_{1,n}A_{1,n}E_{1,n-1}A_{1,n-1}) &\approx .786\,, \\ P_\infty(A_{2,n+1} \mid E_{2,n}A_{2,n}E_{2,n-1}A_{2,n-1}) &\approx .662\,. \end{aligned} \tag{10.6.19}$$

Figure 10.3 presents the observed proportions for $P(A_1 \mid E_1A_1E_1A_1)$ and $P(A_2 \mid E_2A_2E_2A_2)$ in successive blocks of 30 trials.[3] An inspection of the figure indicates that the observed curve for $P(A_2 \mid E_2A_2E_2A_2)$ may be approaching a terminal level that is very close to or slightly below the predicted value of .662. However, the observed curve for $P(A_1 \mid E_1A_1E_1A_1)$ is far below the predicted value of .786, and in general appears to be below the observed curve for $P(A_2 \mid E_2A_2E_2A_2)$. We return to a consideration of these results later.

[3] When there is no possibility for ambiguity, we shall often simplify our notation by deleting the trial subscript.

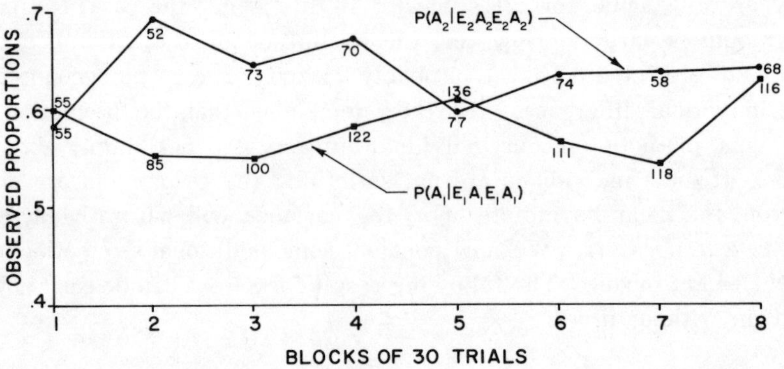

FIGURE 10.3. Observed proportions for $P(A_1 \mid E_1A_1E_1A_1)$ and $P(A_2 \mid E_2A_2E_2A_2)$ in successive 30-trial blocks. (The numbers appended to each point indicate the number of observations involved in making the computation.)

We next examine $P(A_{i,n} \mid E_{i,n-1}E_{i,n-2} \cdots E_{i,n-m})$. For $n \to \infty$, this expression will be denoted as $P_\infty(A_i \mid E_i^{(m)})$. By the methods used above, it can be shown that

$$P_\infty(A_1 \mid E_1^{(m)}) = 1 - (1-\pi)\left(1 - \frac{\theta}{N}\right)^m ,$$

(10.6.20)

$$P_\infty(A_2 \mid E_2^{(m)}) = 1 - \pi\left(1 - \frac{\theta}{N}\right)^m .$$

Using the pseudomaximum-likelihood estimates given by (10.6.3), we obtain for $m = 4$

(10.6.21) $$P_\infty(A_1 \mid E_1^{(4)}) \approx .813 , \qquad P_\infty(A_2 \mid E_2^{(4)}) \approx .720 .$$

Figure 10.4 presents the observed proportions for $P(A_1 \mid E_1^{(4)})$ and $P(A_2 \mid E_2^{(4)})$ in successive 60-trial blocks. As the number of trials increases, these curves should approach the respective asymptotic values given by (10.6.21). However, it appears that the observed value for $P(A_1 \mid E_1^{(4)})$ has leveled off at a value well below .813, whereas the value for $P(A_2 \mid E_2^{(4)})$ is approaching an asymptote far above .720. In this case we have not only poor correspondence between the predicted and observed quantities but also an inversion in the order of prediction; i.e., $P(A_1 \mid E_1^{(4)})$ is predicted to be greater than $P(A_2 \mid E_2^{(4)})$, but in fact the opposite ordering is observed.

The marked discrepancy between the predicted and observed proportions of A_1 responses following a run of four E_1 reinforcements is not entirely unexpected. To restate the finding, it appears that the A_1 response following four E_1's occurs with less frequency than is expected when the theoretical computations are based on θ and N estimated from data involving only one-stage transitions. The finding is related to the results of other investigators, who

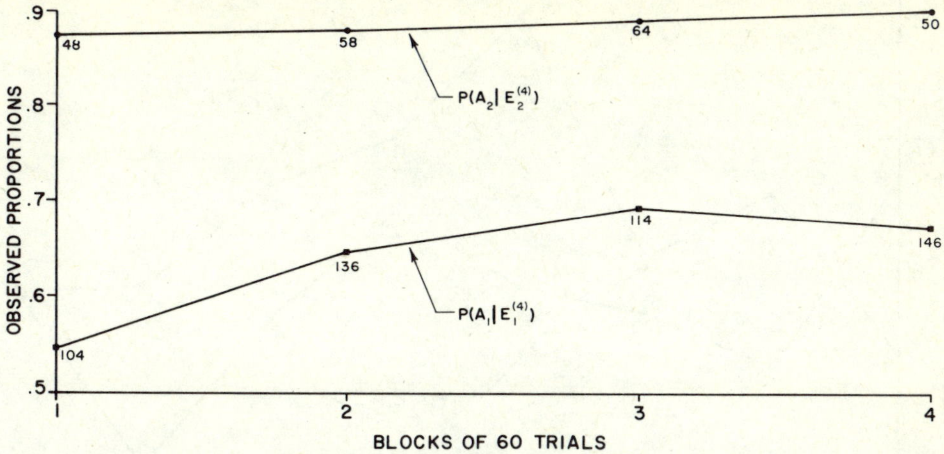

FIGURE 10.4. Observed proportions for $P(A_1 \mid E_1^{(4)})$ and $P(A_2 \mid E_2^{(4)})$ in successive 60-trial blocks. (The numbers appended to each point indicate the number of observations involved in making the computation.)

have found, in a two-choice experiment of the Humphreys type, that subjects may select A_1 less often after a run of $k+1$ reinforcing events E_1 than after a run of only k such events. For example, Goodnow, Rubinstein, and Lubin (1957) have obtained this result for $k=3$. Similar findings are reported by Jarvik (1951), Ross and Levy (1958), Anderson (1960), Atkinson and Sommer (1960), and others.

This phenomenon has been referred to as the negative recency effect, or the gambler's fallacy. Unfortunately, we do not have enough observations to perform an adequate statistical analysis regarding the extent to which such an effect is reflected in our data. However, the picture probably is not as simple as the negative-recency hypothesis might suggest. For this set of data it appears that subjects tend to undershoot on A_1's following a run of E_1's (where the E_1's are dominant in the sequence) and overshoot on A_2's following a run of E_2's. These results are particularly discouraging for any reasonably simple reinforcement-type theory; for if the reinforcements are predominantly E_1's, it is difficult to obtain $P_\infty(A_{2,n} \mid E_{2,n}^{(m)}) > P_\infty(A_{1,n} \mid E_{1,n}^{(m)})$ for $m = 3, 4, 5$, say. This difficulty is indeed severe in conjunction with the requirement that $P_\infty(A_{1,n}) = \pi$, as was observed for Group Z.

We next examine $P(A_1 \mid E_1A_1E_2A_1)$ and $P(A_1 \mid E_1A_1E_1A_2)$. As n becomes large, the theoretical expressions for these quantities become

$$(10.6.22)\quad P_\infty(A_{1,n+1} \mid E_{1,n}A_{1,n}E_{2,n-1}A_{1,n-1}) = \frac{(1-\theta)\mu_3 + \theta(\mu_3 - 2\mu_2 + \mu_1)}{N^3 P_\infty(A_{1,n} \mid E_{2,n-1}A_{1,n-1})P_\infty(A_{1,n-1})},$$

and

$$(10.6.23)\quad P_\infty(A_{1,n+1} \mid E_{1,n}A_{1,n}E_{1,n-1}A_{2,n-1}) = \frac{(1-\theta)(N\mu_2 - \mu_3) - \theta[N - \mu_3 + \mu_2(N-2) + \mu_1(2N-1)]}{N^3 P_\infty(A_{1,n} \mid E_{1,n-1}A_{2,n-1})P_\infty(A_{2,n-1})},$$

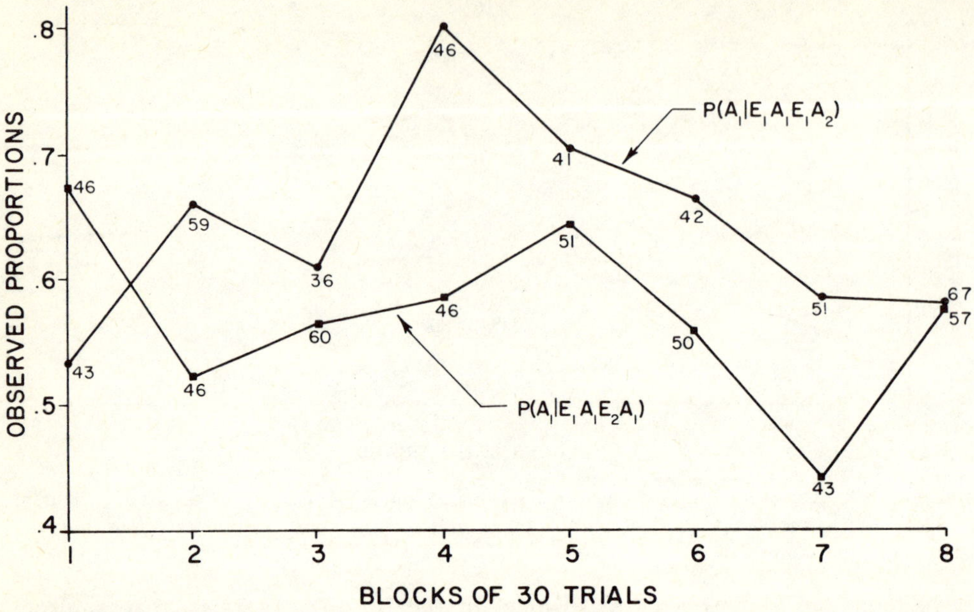

FIGURE 10.5. Observed proportions for $P(A_1 \mid E_1A_1E_2A_1)$ and $P(A_1 \mid E_1A_1E_1A_2)$ in successive 30-trial blocks. (The numbers appended to each point indicate the number of observations involved in making the computation.)

where μ_i is the ith raw moment of the binomial distribution with parameter π and sample size N. From (10.6.7) and the estimates of θ and N given in (10.6.3), the values for (10.6.22) and (10.6.23) are approximately equal to .672 and .714, respectively. The observed proportions for these quantities are presented in Fig. 10.5 in successive 30-trial blocks. For both measures the asymptotic predictions tend to be well above the observed terminal values. However, the appropriate ordering holds; i.e., $P(A_1 \mid E_1A_1E_1A_2)$ is above $P(A_1 \mid E_1A_1E_2A_1)$ for all but the first block of trials.

To conclude the analysis of sequential data we consider $P(A_{1,n+1} \mid A_{1,n})$. The expression for this quantity, as n becomes large, is given by (2.3.17). Using the estimated values of θ and N, we obtain

$$\lim_{n\to\infty} P(A_{1,n+1} \mid A_{1,n}) \approx .647.$$

Figure 10.6 presents the observed proportions in successive blocks of 30 trials. Over the terminal trial blocks the fit between the observed values and .647 is very close.

A comparison of the fit of this last result with the decidedly poorer fits for the longer sequential effects warns us against assuming the complete adequacy of models (like our generalized conditioning model) that are Markov chains in the observable events. This point is worth amplifying. Suppose we have a generalized conditioning model that fits the observed transition data very well; and suppose further that on the statistical test of first-order versus second-order

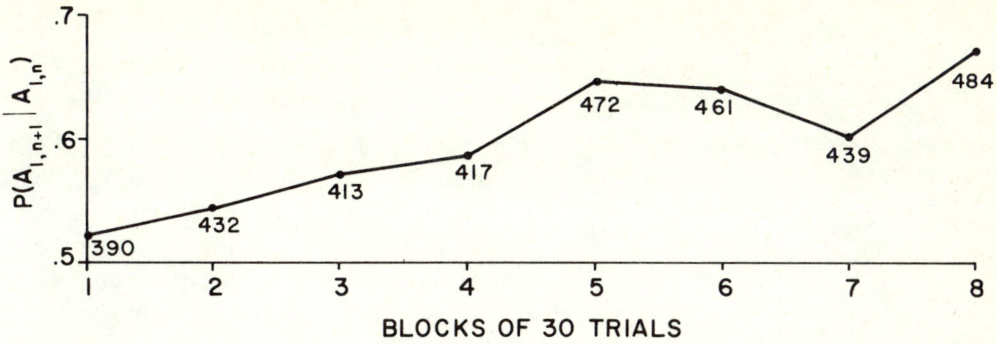

FIGURE 10.6. Observed proportions for $P(A_{1,n+1} \mid A_{1,n})$ in successive 30-trial blocks. (The numbers appended to each point indicate the number of observations involved in making the computation.)

Markov chains the data do not lead us to reject the null hypothesis. Then, generally speaking, we can claim to have a model that fits the full set of data. Predictions from this model of long sequential effects do not play a serious role in statistical tests of the model's adequacy, since such events rarely occur. Yet it is unsatisfactory that such a model does not provide for their consideration. A model like the multi-element model, which is a chain of infinite order in the observables, is both conceptually more satisfying and probably more useful in making further theoretical constructions.

What these remarks suggest is that in view of the relatively good results obtained for the generalized conditioning model, it may be fruitful to analyze in detail the multi-element model with conditioning parameters c_{ij}. Methods of estimation will not be optimal, but where workable recursions can be obtained pseudomaximum-likelihood methods can be used. For exact fit of data it will probably be necessary, even in this model, to let either the conditioning parameters or the number of relevant stimuli change in time.

**Empirical analysis of the two-element and one-element models.* In Sec. 2.5 procedures were presented for estimating the transition numbers associated with the unobservable states of conditioning in the multi-element model. The analysis was carried out in detail for the two-element model and in Sec. 3.7 the method was applied to the data of Chapter 3. We now apply those methods to the data of Group Z.

The transition matrix for the two-element, noncontingent case is given by (2.3.2). The states are designated 0, 1, and 2 to indicate the number of elements conditioned to A_1. As in Sec. 2.5 we use m_{ij} $(i, j = 0, 1, 2)$ to denote the transition frequencies associated with the unobservable states of conditioning, and n_{kl} $(k, l = 1, 2)$ to denote the transition frequencies in observed responses. The least-squares estimate of m_{ij} is given in terms of (2.5.18). In solving for m_{ij} we use the n_{kl}'s of (10.6.8); that is, $n_{11} = 1142$, $n_{12} = 640$, $n_{21} = 648$, and $n_{22} = 570$. The obtained values of m_{ij} are as follows:

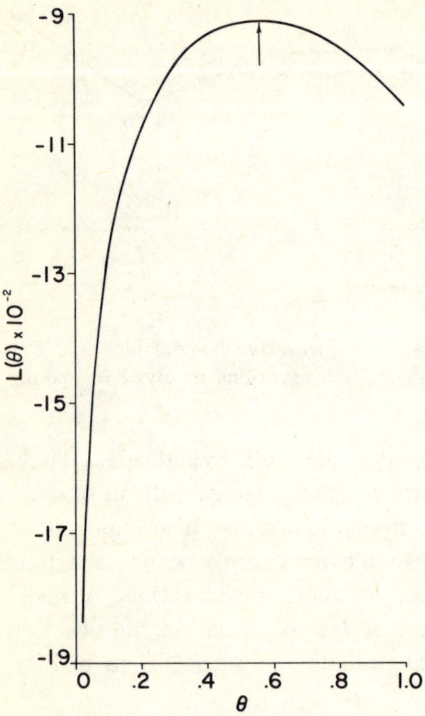

FIGURE 10.7. $L(\theta)$ for the noncontingent two-element model.

$$m_{00} = 407.7\,, \quad m_{01} = 169.1\,,$$
$$m_{10} = 169.1\,, \quad m_{11} = 1307.6\,, \quad m_{12} = 253.7\,,$$
$$m_{21} = 253.7\,, \quad m_{22} = 439.1\,.$$

The results of applying the method developed in Sec. 2.5 for estimating the transition numbers are better here than in Chapter 3. Above all, none of the transition numbers are negative, and all of the estimated numbers are intuitively reasonable.

Given these estimated transition numbers m_{ij}, we may now make a maximum-likelihood estimate of θ by (2.5.19). The function $L(\theta)$ is given in Fig. 10.7, and the maximum occurs at $\theta = .561$. Table 10.7 presents asymptotic predictions for the two-element model for this value of θ. For purposes of comparison the table also presents asymptotic predictions for the multi-element and one-element models. For the one-element model θ was estimated by (2.2.8) for the same n_{kl} given above; the maximum-likelihood estimate of θ was .891. When this value of θ was used, the

TABLE 10.7

COMPARISONS OF ASYMPTOTIC SEQUENTIAL PREDICTIONS FOR THE N-ELEMENT ($N = 3.42$), TWO-ELEMENT, AND ONE-ELEMENT MODELS

(Observed values are based on the last 100 trials)

Asymptotic Value	Observed	Predicted		
		N-Element	Two-Element	One-Element
$P(A_1 \mid A_1)$	.641	.647	.688	.643
$P(A_2 \mid A_2)$	.468	.471	.532	.465
$P(A_1 \mid E_1A_1)$	.715	.717	.800	1
$P(A_1 \mid E_1A_2)$	.602	.598	.580	.891
$P(A_1 \mid E_2A_1)$	.535	.543	.520	.109
$P(A_1 \mid E_2A_2)$	.413	.425	.300	0
$P(A_1 \mid E_1A_1E_1A_1)$	.580	.786	.875	1
$P(A_2 \mid E_2A_2E_2A_2)$	.637	.662	.786	1
$P(A_1 \mid E_1E_1E_1E_1)$	.675	.813	.892	.999
$P(A_2 \mid E_2E_2E_2E_2)$	.895	.720	.839	.999
$P(A_1 \mid E_1A_1E_2A_1)$	.533	.672	.700	1
$P(A_1 \mid E_1A_1E_1A_2)$	.602	.714	.790	1

goodness-of-fit test (2.4.4) applied to (1.3.4) yielded $\chi^2 = .09$. The parameter θ has been estimated from the same transition data, and therefore the χ^2 for the one-element model is to be interpreted with $2 - 1 = 1$ degrees of freedom. The significance level is .76, which is remarkably good. In fact, this result is by far the best for the one-element model in any of the experiments. As the table shows, however, the sequential predictions of the one-element model are very poor.

Table 10.7 offers a comparison of the three models with different numbers of stimuli. Twelve sequential predictions were made. The N-element model is superior to the two-element model in 11 cases, and to the one-element model in 9 cases. The two-element model is, in turn, superior to the one-element model in 10 of the 12 cases.

*10.7. Application of the Linear Model

In this section we apply Axioms L1 and L2 to the data of Group Z. The description of the linear model in Sec. 1.5 is adequate for our purposes, but for a detailed discussion of the foundations of the model the reader is referred to Estes and Suppes [1959*a*]. In order to derive predictions for this model some additional notation is required. We define the sample space X for a one-person experiment as the set of sequences of ordered pairs $\langle 0, 1\rangle$, $\langle 1, 1\rangle$, $\langle 0, 2\rangle$, $\cdots$, where the first member of each pair indicates the response, and the second, the reinforcement [see (1.2.1) and (1.2.2)]. Further, x_n is defined as the set of all sequences having the same responses and reinforcing events as a given sequence x on the first n trials. In addition, we require notation for a partition χ_n of the sample space X into sets of sequences having the same responses and reinforcing events on the first n trials; i.e., $\chi_n = \{x_n : x \in X\}$. Then, following the notation of Estes and Suppes [1959*a*], the raw moments of the conditional response probability distribution are

(10.7.1) $$V_{i,n+1} = \sum_{\chi_n} P(A_{1,n+1} \mid x_n)^i P(x_n) ,$$

and

(10.7.2) $$V_i = \lim_{n\to\infty} V_{i,n} .$$

For the noncontingent case, it can easily be shown that

(10.7.3) $$\begin{aligned} V_1 &= \pi , \\ V_2 &= \pi^2 + \frac{\theta\pi(1-\pi)}{2-\theta} , \\ V_3 &= \pi^3 + \frac{\pi(1-\pi)\theta^3(1-2\pi)}{1-(1-\theta)^3} . \end{aligned}$$

As in Sec. 10.6 we use a pseudomaximum-likelihood procedure to estimate based on the quantities

$$\lim_{n\to\infty} P(A_{i,n+1} \mid E_{j,n} A_{k,n}) = P_\infty(A_{i,n+1} \mid E_{j,n} A_{k,n}) .$$

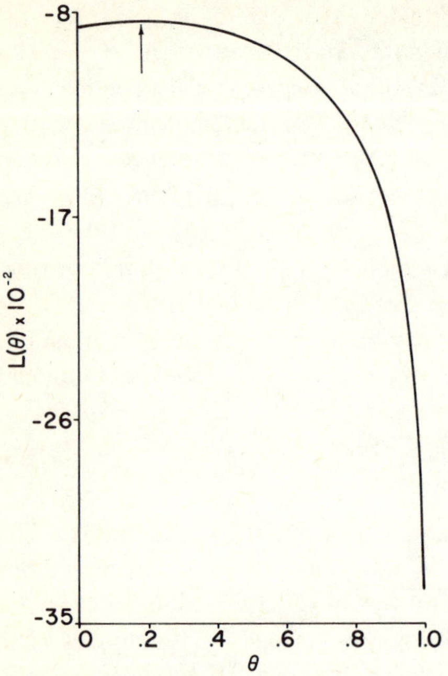

FIGURE 10.8. The pseudo-likelihood function, $L^*(\theta)$, for the linear model.

Expressions for these quantities are given in Estes and Suppes [1959*a*]; we restate them here for the noncontingent case:

(10.7.4)

$$P_\infty(A_{1,n+1} \mid E_{1,n}A_{1,n}) = (1-\theta)a + \theta\,,$$
$$P_\infty(A_{1,n+1} \mid E_{2,n}A_{1,n}) = (1-\theta)a\,,$$
$$P_\infty(A_{2,n+1} \mid E_{1,n}A_{2,n}) = (1-\theta)b\,,$$
$$P_\infty(A_{2,n+1} \mid E_{2,n}A_{2,n}) = (1-\theta)b + \theta\,,$$

where $a = [2\pi(1-\theta)+\theta]/(2-\theta)$ and $b = [2(1-\pi)(1-\theta)+\theta]/(2-\theta)$. Then the function to be maximized is defined in terms of (10.7.4) and the observed asymptotic transition frequencies given by (10.6.8). Specifically,

(10.7.5)

$$\begin{aligned} L^*(\theta) = {} & 748 \log[(1-\theta)a+\theta] \\ & + 298 \log[1-(1-\theta)a-\theta] \\ & + 394 \log[(1-\theta)a] \\ & + 342 \log[1-(1-\theta)a] \\ & + 462 \log[1-(1-\theta)b] + \cdots. \end{aligned}$$

Rather than differentiate this function and solve for θ we have resorted to computing $L^*(\theta)$ for $\theta = .01i$ $(i = 1, \cdots, 100)$; then we used these values to approximate the maximum. The results are presented in Fig. 10.8, and the maximum is at $\theta^* = .19$. For the remainder of this section this value will be taken as the pseudomaximum-likelihood estimate of θ. It should be noted that $L^*(\theta)$ is quite flat between .02 and .30, indicating that the model is not highly sensitive to the exact value of θ^* in this range.

From the estimated value of θ, the predicted frequencies given by (10.7.4) and the row sums of (10.6.8) are as follows:

(10.7.6)

	A_1	A_2
A_1E_1	742.7	303.3
A_1E_2	382.7	353.3
A_2E_1	480.0	288.0
A_2E_2	195.8	254.2

The χ^2 test for goodness of fit between (10.6.8) and (10.7.6) yields a value of 3.49. There are four degrees of freedom, but one parameter has been estimated from these data; consequently, the χ^2 is to be interpreted with d.f. = 3. Of course, the χ^2 is not significant.

Predicted learning curve. The mean learning curve for the linear model is

(10.7.7) $$P(A_{1,n}) = \pi - [\pi - P(A_{1,1})](1-\theta)^{n-1} .$$

A comparison of (10.7.7) with (10.6.11) indicates that the predicted curves for the N-element sampling model and the linear model are identical except that θ/N in the sampling model is replaced by θ in the linear model. For the sampling model the estimated value of θ/N is .174, and for the linear model the estimated value of θ is .19. Thus, the predicted learning curves for the two models are almost identical, with the linear model approaching asymptote somewhat more rapidly. As noted in the previous section, the observed mean learning curves approach asymptote at a much slower rate than either model predicted.

Predicted asymptotic variance. The quantity corresponding to (10.6.16) for the linear model is given by Estes and Suppes [1959*a*]. It is as follows:

$$\operatorname{var}(\bar{A}_K) = \frac{\pi(1-\pi)}{(2-\theta)\theta}\{K\theta(4-3\theta) - 2(1-\theta)[1-(1-\theta)^K]\} .$$

Dividing by K^2, we find that the variance of the proportion of A_1 responses, for large K, is approximately

(10.7.8) $$\frac{\pi(1-\pi)(4-3\theta)}{K(2-\theta)} .$$

Using $\theta = .19$ and $K = 100$, we obtain .0045 for the predicted variance, which is markedly smaller than the observed value of .0121 (see Table 10.1) but approximately the same as predicted by the N-element sampling model in the preceding section.

Other sequential effects. We now consider the same sequential statistics we examined in Sec. 10.6. First, we derive the expression for $P_\infty(A_1 \mid E_1A_1E_1A_1)$. Namely,

$$\begin{aligned} P(A_{1,n+1} \mid E_{1,n}A_{1,n}E_{1,n-1}A_{1,n-1})&P(E_{1,n}A_{1,n}E_{1,n-1}A_{1,n-1}) \\ &= \sum_{x_{n-2}} P(A_{1,n+1}E_{1,n}A_{1,n}E_{1,n-1}A_{1,n-1}x_{n-2}) \\ &= \pi^2 \sum_{x_{n-2}} P(A_{1,n+1} \mid E_{1,n}A_{1,n}E_{1,n-1}A_{1,n-1}x_{n-2})P(A_{1,n} \mid E_{1,n-1}A_{1,n-1}x_{n-2}) \\ &\qquad \cdot P(A_{1,n-1} \mid x_{n-2})P(x_{n-2}) \\ &= \pi^2 \sum_{x_{n-2}} [P(A_{1,n-1} \mid x_{n-2})^3(1-\theta)^3 + P(A_{1,n-1} \mid x_{n-2})^2\theta(1-\theta)(3-2\theta) \\ &\qquad + P(A_{1,n-1} \mid x_{n-2})\theta^2(2-\theta)]P(x_{n-2}) . \end{aligned}$$

Hence, as $n \to \infty$,

(10.7.9) $$P_\infty(A_{1,n+1} \mid E_{1,n}A_{1,n}E_{1,n-1}A_{1,n-1}) = \frac{(1-\theta)^3V_3 + \theta(1-\theta)(3-2\theta)V_2 + \theta^2(2-\theta)\pi}{(1-\theta)V_2 + \theta\pi} .$$

The expression for $P_\infty(A_2 \mid E_2A_2E_2A_2)$ is identical to (10.7.9) except that π is replaced everywhere by $1-\pi$.

Using the estimated value of $\theta = .19$, we obtain

(10.7.10)
$$P_\infty(A_{1,n+1} \mid E_{1,n}A_{1,n}E_{1,n-1}A_{1,n-1}) \approx .725 ,$$
$$P_\infty(A_{2,n+1} \mid E_{2,n}A_{2,n}E_{2,n-1}A_{2,n-1}) \approx .599 .$$

These predicted values are to be compared with the observed results in Fig. 10.3. The same remarks apply here as in the discussion of the stimulus-sampling model. That is, the two models are not far off on the prediction of $P(A_2 \mid E_2A_2E_2A_2)$, but both models markedly overpredict the observed terminal values of $P(A_1 \mid E_1A_1E_1A_1)$.

We now turn to an evaluation of $P_\infty(A_1 \mid E_1^{(m)})$. In terms of the linear model

$$\begin{aligned} P(A_{1,n} \mid E_1^{(m)})P(E_1^{(m)}) &= \sum_{x_{n-m-1}} P(A_{1,n}E_{1,n-1} \cdots E_{1,n-m}x_{n-m-1}) \\ &= \sum_{x_{n-m-1}} P(A_{1,n} \mid E_{1,n-1} \cdots E_{1,n-m}x_{n-m-1})\pi^m P(x_{n-m-1}) \\ &= \sum_{x_{n-m-1}} [(1-\theta)^m P(A_{1,n-m} \mid x_{n-m-1}) + 1 - (1-\theta)^m]\pi^m P(x_{n-m-1}) . \end{aligned}$$

Hence, as $n \to \infty$,

(10.7.11)
$$P_\infty(A_1 \mid E_1^{(m)}) = 1 - (1-\pi)(1-\theta)^m .$$

By identical considerations

(10.7.12)
$$P_\infty(A_2 \mid E_2^{(m)}) = 1 - \pi(1-\theta)^m .$$

Using the estimated value of θ, we obtain for $m = 4$

$$P_\infty(A_1 \mid E_1^{(4)}) \approx .828 , \qquad P_\infty(A_2 \mid E_2^{(4)}) \approx .742 .$$

These predictions are to be compared with the observed results in Fig. 10.4. As was the case for the stimulus-sampling model, the prediction for $P_\infty(A_1 \mid E_1^{(4)})$ overestimates, whereas the prediction for $P_\infty(A_2 \mid E_2^{(4)})$ markedly underestimates the observed terminal data.

Next we consider $P_\infty(A_1 \mid E_1A_1E_2A_1)$ and $P_\infty(A_1 \mid E_1A_1E_1A_2)$. For the linear model the derivations are as follows:

$$\begin{aligned} &P(A_{1,n+1} \mid E_{1,n}A_{1,n}E_{2,n-1}A_{1,n-1})P(E_{1,n}A_{1,n}E_{2,n-1}A_{1,n-1}) \\ &\quad = \pi(1-\pi)\sum_{x_{n-2}} P(A_{1,n+1} \mid E_{1,n}A_{1,n}E_{2,n-1}A_{1,n-1}x_{n-2})P(A_{1,n} \mid E_{2,n-1}A_{1,n-1}x_{n-2}) \\ &\qquad\qquad \cdot P(A_{1,n-1} \mid x_{n-2})P(x_{n-2}) \\ &\quad = \pi(1-\pi)\sum_{x_{n-2}} [(1-\theta)^2 P(A_{1,n-1} \mid x_{n-2}) + \theta](1-\theta)P(A_{1,n-1} \mid x_{n-2})^2 P(x_{n-2}) \\ &\quad = \pi(1-\pi)\sum_{x_{n-2}} [(1-\theta)^3 P(A_{1,n-1} \mid x_{n-2})^3 + \theta(1-\theta)P(A_{1,n-1} \mid x_{n-2})^2]P(x_{n-2}) . \end{aligned}$$

Hence, as $n \to \infty$,

$$\begin{aligned} P_\infty(A_{1,n+1} \mid E_{1,n}A_{1,n}E_{2,n-1}A_{1,n-1}) &= \frac{(1-\theta)^3 V_3 + \theta(1-\theta)V_2}{(1-\theta)V_2} \\ &= \frac{(1-\theta)^2 V_3}{V_2} + \theta . \end{aligned}$$

Similarly,

$$P(A_{1,n+1} \mid E_{1,n}A_{1,n}E_{1,n-1}A_{2,n-1})P(E_{1,n}A_{1,n}E_{1,n-1}A_{2,n-1})$$
$$= \pi^2 \sum_{x_{n-2}} P(A_{1,n+1} \mid E_{1,n}A_{1,n}E_{1,n-1}A_{2,n-1}x_{n-2})P(A_{1,n} \mid E_{1,n-1}A_{2,n-1}x_{n-2}) \cdot P(A_{2,n-1} \mid x_{n-2})P(x_{n-2})$$
$$= \pi^2 \sum_{x_{n-2}} [(1-\theta)^2 P(A_{1,n-1} \mid x_{n-2}) + \theta(1-\theta) + \theta][(1-\theta)P(A_{1,n-1} \mid x_{n-2}) + \theta] \cdot P(A_{2,n-1} \mid x_{n-2})P(x_{n-2}) .$$

Hence, as $n \to \infty$,

$$P_\infty(A_{1,n+1} \mid E_{1,n}A_{1,n}E_{1,n-1}A_{2,n-1}) = \frac{\theta^2(2-\theta) + \pi[3\theta(1-\theta)^2 - \theta^2] + (1-\theta)[(1-\theta)(1-3\theta) - \theta]V_2 - (1-\theta)^3 V_3}{(1-\pi)\left[\dfrac{2(1-\theta)(\theta+\pi-\theta\pi)+\theta^2}{2-\theta}\right]} .$$

Using the estimated value of θ, we obtain

$$P_\infty(A_{1,n+1} \mid E_{1,n}A_{1,n}E_{2,n-1}A_{1,n-1}) \approx .557 ,$$
$$P_\infty(A_{1,n+1} \mid E_{1,n}A_{1,n}E_{1,n-1}A_{2,n-1}) \approx .813 .$$

These theoretical predictions are to be compared with the observed results in Fig. 10.5. It is somewhat surprising that the prediction for $P_\infty(A_1 \mid E_1A_1E_2A_1)$ is less than π, which was not true for the multi-element model. On the other hand, the observed value is also below π when averaged over the last 90 or 120 trials.

We conclude the analysis of the linear model by considering $P_\infty(A_{1,n+1} \mid A_{1,n})$. The theoretical expression for this quantity, as given by Estes and Suppes [1959*a*], is

$$P_\infty(A_{1,n+1} \mid A_{1,n}) = \frac{(1-\theta)V_2 + \theta\pi^2}{\pi} .$$

Using the estimated value of $\theta = .19$, we obtain $P_\infty(A_{1,n+1} \mid A_{1,n}) \approx .634$. The relevant observed data are given in Fig. 10.6. Over the terminal trial blocks there is good agreement between the observed values and the predicted value of .634.

A comparison of the sequential predictions of the linear and multi-element models with each other and with the observed data averaged over the last 100 trials is given in Table 10.8. The multi-element sampling model comes out slightly better in the comparison; for 8 of the 11 quantities it is closer than the linear model. However, it is evident that both models are predicting rather poorly for the longer-range sequential effects. The last column of the table gives the predictions of the hypothesis model discussed in the next section.

At the end of Sec. 10.5 we made certain remarks about possible directions to move in order to construct a better theory. We should like now to add our belief that Markov models offer better opportunity for future theorizing than linear models because of their vastly greater mathematical flexibility. We have seen in Sec. 1.5 that the linear model is exceedingly difficult to work with, even in the simplest two-person situations.

Above all, however, it is important to emphasize that there is no really

TABLE 10.8
COMPARISON OF ASYMPTOTIC SEQUENTIAL PREDICTIONS FOR THE MULTI-ELEMENT, LINEAR, AND HYPOTHESIS MODELS WITH OBSERVED DATA OVER THE LAST 100 TRIALS IN GROUP Z

Asymptotic Value	Observed	Multi-Element	Linear	Hypothesis
$P(A_1 \mid A_1)$	.641	.647	.634	.635
$P(A_1 \mid E_1A_1)$	.715	.717	.710	.648
$P(A_1 \mid E_1A_2)$	.602	.598	.625	.528
$P(A_1 \mid E_2A_1)$	.535	.543	.520	.632
$P(A_1 \mid E_2A_2)$	.413	.425	.435	.552
$P(A_1 \mid E_1A_1E_1A_1)$	.580	.786	.725	1.000
$P(A_2 \mid E_2A_2E_2A_2)$	.637	.662	.599	1.000
$P(A_1 \mid E_1E_1E_1E_1)$	.675	.813	.828	.997
$P(A_2 \mid E_2E_2E_2E_2)$	.895	.720	.742	.995
$P(A_1 \mid E_1A_1E_2A_1)$	.533	.672	.557	.190
$P(A_1 \mid E_1A_1E_1A_2)$	.602	.714	.813	1.000

fundamental difference between the conceptual bases of the two models. It may be shown (see Estes and Suppes [1959*b*]) that the linear model is the limiting case of stimulus-sampling models as $N \to \infty$ under rather broad sampling assumptions. But these assumptions are not broad enough to include the "pattern" model of sampling exactly one element on each trial, for as N increases, it is obvious that the learning effects of this sampling scheme would diminish to zero. Unfortunately, limitations of space and time prohibited us from deriving estimates of N and θ, and the resulting predictions, for stimulus-sampling models that use different sampling schemes. It would be interesting, for example, to compare the results given in Table 10.8 with those that may be obtained under the assumption of independent sampling of elements or the assumption that a fixed-size sample of two, three, or four elements is drawn on each trial. Both of these assumptions lead in the limit (as $N \to \infty$) to the linear model.

*10.8. Application of a Hypothesis Model

In this section we analyze the data of Group Z in terms of one of the simpler hypothesis models described in Sec. 1.7. As indicated earlier, a hypothesis model may be interpreted as a stimulus-sampling model derivable from the basic postulates of Sec. 1.2. In analyzing the hypothesis model, we shall use stimulus-sampling terminology, but the reader may, if he chooses, translate to hypothesis language in the manner indicated in Sec. 1.7.

Our model assumes that there are two stimulus elements, s_1 and s_2, available for sampling. Further, the restrictions on sampling are such that s_1 is selected with probability 1 on trial n if E_1 occurred on trial $n - 1$, and s_2 is sampled with probability 1 on trial n if E_2 occurred on trial $n - 1$. A transition matrix for this model for the noncontingent case is presented in (1.7.3). However, the

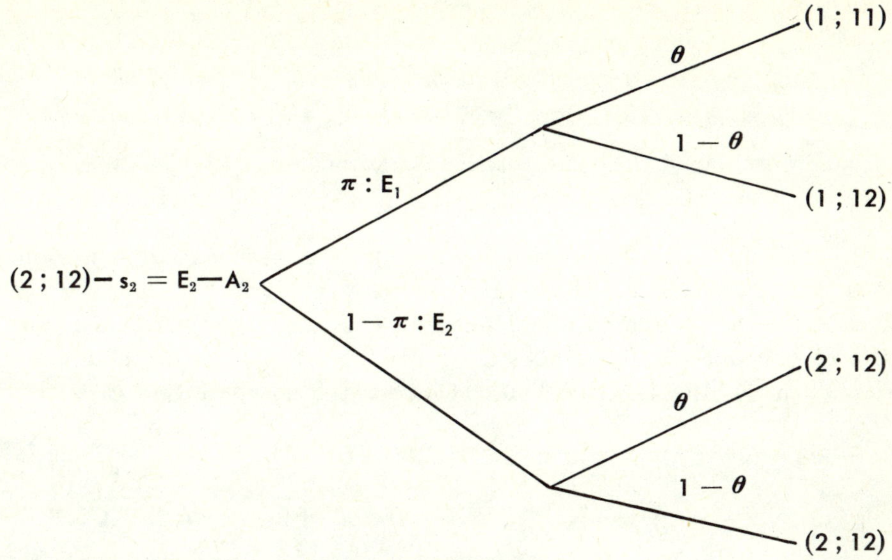

FIGURE 10.9

analysis of sequential data requires a more detailed representation. Specifically, we shall consider an eight-state Markov chain and use the notation $(i;jk)$ to represent the states. The numbers $i, j,$ and k are either 1 or 2; i indicates which reinforcing event occurred on the previous trial, and thus which stimulus is sampled; j and k indicate the conditioning of s_1 and s_2, respectively. Thus $(1;21)$ indicates that E_1 occurred on the previous trial, and hence that s_1 is sampled; further, that s_1 is conditioned to A_2, so that when the process is in this state, an A_2 will occur with probability 1.

Following the analysis laid down for other cases, the appropriate trees can be constructed. As an example, the tree for the state $(2;12)$ is given in Fig. 10.9. The transition matrix for the eight possible states has the following form:

(10.8.1)

	(1 ; 11)	(1 ; 12)	(1 ; 21)	(1 ; 22)	(2 ; 11)	(2 ; 21)	(2 ; 12)	(2 ; 22)
(1 ; 11)	π				$(1-\theta)(1-\pi)$	$\theta(1-\pi)$		
(1 ; 12)		π					$(1-\theta)(1-\pi)$	$\theta(1-\pi)$
(1 ; 21)	$\theta\pi$		$(1-\theta)\pi$			$1-\pi$		
(1 ; 22)		$\theta\pi$		$(1-\theta)\pi$				$1-\pi$
(2 ; 11)	π				$(1-\theta)(1-\pi)$		$\theta(1-\pi)$	
(2 ; 21)			π			$(1-\theta)(1-\pi)$		$\theta(1-\pi)$
(2 ; 12)	$\theta\pi$	$(1-\theta)\pi$					$1-\pi$	
(2 ; 22)			$\theta\pi$	$(1-\theta)\pi$				$1-\pi$

For convenience, we number the states from 1 to 8 in the order given by the matrix. The expressions for the asymptotic probabilities u_i are easily obtained. They are as follows:

$$
(10.8.2) \qquad
\begin{aligned}
u_1 &= \pi^2[1-(1-\pi)(1-\theta)]\,, & u_5 &= \pi^2(1-\pi)(1-\theta)\,,\\
u_2 &= \pi^2(1-\pi)(1-\theta)\,, & u_6 &= \pi(1-\pi)[1-\pi(1-\theta)]\,,\\
u_3 &= \pi(1-\pi)[1-(1-\pi)(1-\theta)]\,, & u_7 &= \pi(1-\pi)^2(1-\theta)\,,\\
u_4 &= \pi(1-\pi)^2(1-\theta)\,, & u_8 &= (1-\pi)^2[1-\pi(1-\theta)]\,.
\end{aligned}
$$

From an inspection of the states of the Markov chain, it is clear that

$$(10.8.3) \qquad \lim_{n\to\infty} P(A_{1,n}) = u_1 + u_2 + u_5 + u_6 = \pi\,.$$

As in the previous two sections, we employ a pseudomaximum-likelihood method for estimating θ, which involves the quantities $P(A_{i,n+1} \mid E_{k,n}A_{j,n})$ for large n. The derivation of these quantities is straightforward, and will be illustrated only for one term. Let $C_{i,n}$ $(i = 1, \cdots, 8)$ denote the states of the Markov chain in (10.8.1) on trial n. Then, for the noncontingent case

$$
\begin{aligned}
P(A_{1,n+1} \mid E_{1,n}A_{1,n}) &= \frac{1}{P(E_{1,n}A_{1,n})} \sum_{i,j} P(A_{1,n+1}C_{i,n+1}E_{1,n}A_{1,n}C_{j,n})\\
&= \frac{1}{P(E_{1,n}A_{1,n})} \sum_{i,j} P(A_{1,n+1} \mid C_{i,n+1})P(C_{i,n+1} \mid E_{1,n}A_{1,n}C_{j,n})\\
&\qquad\qquad \cdot P(E_{1,n} \mid A_{1,n}C_{j,n})P(A_{1,n} \mid C_{j,n})P(C_{j,n})\\
&= \frac{1}{P(A_{1,n})} \sum_{i,j} P(A_{1,n+1} \mid C_{i,n+1})P(C_{i,n+1} \mid E_{1,n}A_{1,n}C_{j,n})\\
&\qquad\qquad P(A_{1,n} \mid C_{j,n})P(C_{j,n})\,.
\end{aligned}
$$

Noting that all of the conditional probabilities are 0 or 1, and carrying out the summation, we obtain

$$P(A_{1,n+1} \mid E_{1,n}A_{1,n}) = \frac{P(C_{1,n}) + P(C_{2,n}) + P(C_{5,n})}{P(A_{1,n})}\,.$$

And as $n \to \infty$,

$$(10.8.4) \qquad P_\infty(A_{1,n+1} \mid E_{1,n}A_{1,n}) = \frac{u_1 + u_2 + u_5}{u_1 + u_2 + u_5 + u_6} = \pi + \pi(1-\pi)(1-\theta)\,.$$

By similar arguments we obtain

$$
(10.8.5) \qquad
\begin{aligned}
P_\infty(A_{1,n+1} \mid E_{2,n}A_{1,n}) &= \frac{u_1 + (1-\theta)(u_5+u_6)}{u_1+u_2+u_5+u_6} = \pi + (1-\pi)^2(1-\theta)\,,\\
P_\infty(A_{1,n+1} \mid E_{1,n}A_{2,n}) &= \frac{\theta(u_3+u_4) + u_7}{u_3+u_4+u_7+u_8} = \pi - \pi^2(1-\theta)\,,\\
P_\infty(A_{1,n+1} \mid E_{2,n}A_{2,n}) &= \frac{u_3}{u_3+u_4+u_7+u_8} = \pi - \pi(1-\pi)(1-\theta)\,.
\end{aligned}
$$

From the observed transition frequencies over the last 100 trials given by (10.6.8), the pseudo-likelihood function to be maximized is as follows:

$$
\begin{aligned}
L^*(\theta) = {} & 748 \log[\pi + \pi(1-\pi)(1-\theta)] + 298 \log[1-\pi-\pi(1-\pi)(1-\theta)]\\
& + 394 \log[\pi + (1-\pi)^2(1-\theta)] + 342 \log[1-\pi-(1-\pi)^2(1-\theta)]\\
& + 462 \log[\pi - \pi^2(1-\theta)] + \cdots.
\end{aligned}
$$

As in Sec. 10.7, rather than differentiate this function and solve for θ, we computed $L^*(\theta)$ for $\theta = .01i$ $(i = 1, \cdots, 100)$ and used these values to approximate the maximum. The maximum occurred at $\theta^* = .80$; for the remainder of this section this value will be taken as the estimate of θ on which predictions are based.

The predicted frequencies, for $\theta = .80$, given by (10.8.4), (10.8.5), and the row sums of (10.6.8), are as follows:

(10.8.6)

	A_1	A_2
A_1E_1	677.8	368.2
A_1E_2	465.2	270.8
A_2E_1	405.5	362.5
A_2E_2	248.4	201.6

The correspondence between these values and the observed frequencies in (10.6.8) is extremely poor. The χ^2 goodness-of-fit test given by (2.4.4) yields a value of 101.9. It is clear from Table 10.8 that the hypothesis model is not predicting the conditional probabilities $P_\infty(A_{i'} \mid E_j A_i)$ nearly as well as the multi-element stimulus-sampling and linear models are. In making this comparison it is important to realize that we are considering here only one particular hypothesis model. One that took account of more complex hypotheses (i.e., considered more reinforcements and perhaps also prior responses) would undoubtedly fit the data better.

Other sequential statistics. We next consider the sequential statistics examined in Secs. 10.6 and 10.7. Here the order of analysis, however, will be somewhat different from the order in those two sections, since the model yields several completely deterministic predictions *contradicted by the data*. Specifically, for the hypothesis model presented in this section

$$\begin{aligned} &P(A_{1,n+1} \mid E_{1,n}A_{1,n}E_{1,n-1}A_{1,n-1}) = 1 \,, \\ &P(A_{2,n+1} \mid E_{2,n}A_{2,n}E_{2,n-1}A_{2,n-1}) = 1 \,, \\ &P(A_{1,n+1} \mid E_{1,n}A_{1,n}E_{1,n-1}A_{2,n-1}) = 1 \,. \end{aligned} \tag{10.8.7}$$

The explanation of these predictions is straightforward. Consider, for example, the first equation of (10.8.7). The occurrence of $E_{1,n-1}$ specifies that s_1 is sampled on trial n; $A_{1,n}$ indicates that s_1 is conditioned to A_1; further, $E_{1,n}$ ensures that the conditioning of s_1 will not change and also that s_1 again will be sampled on trial $n+1$. Hence, $A_{1,n+1}$ will occur with probability 1. The observed data presented in Figs. 10.3 and 10.5 clearly contradict the predictions of (10.8.7).

These theoretical results represent precisely the same type of overdeterminism generated by some of the one-element stimulus-sampling models; for example, the prediction in the one-element model that $P(A_{1,n+1} \mid E_{1,n}A_{1,n}) = 1$ [see (1.3.5)]. As noted in Chapter 1, several resolutions of the difficulty can be offered. In this particular case one could either postulate more than one element associated with each reinforcing event, or consider hypothesis models that encompassed more of the past sequence of responses and reinforcements. Note, however that this latter suggestion will not eliminate deterministic prediction; it will only reduce the number of such predictions, as should be evident from the present model.

We now turn to the remaining sequential statistics analyzed in Secs. 10.6 and 10.7. First consider

$$\lim_{n\to\infty} P(A_{1,n} \mid E_{1,n-1}E_{1,n-2}\cdots E_{1,n-m})\,,$$

which we have written as $P_\infty(A_1 \mid E_1^{(m)})$. Let $C_{i,n}$ $(i = 1, \cdots, 8)$ denote the states of the Markov chain in (10.8.1) on trial n. Then, we have

$$\begin{aligned} P(A_{1,n} \mid E_1^{(m)}) &= \frac{1}{\pi^m}\sum_{i,j} P(A_{1,n}C_{i,n}E_{1,n-1}\cdots E_{1,n-m}C_{j,n-m-1}) \\ &= \frac{1}{\pi^m}\sum_{i,j} P(A_{1,n} \mid C_{i,n})P(C_{i,n} \mid E_{1,n-1}\cdots E_{1,n-m}C_{j,n-m-1})\pi^m P(C_{j,n-m-1})\,. \end{aligned}$$

Carrying out the summation, we obtain

$$\begin{aligned} P(A_{1,n} \mid E_1^{(m)}) = P(C_{1,n-m-1}) &+ P(C_{2,n-m-1}) \\ &+ [1-(1-\theta)^m][P(C_{3,n-m-1}) + P(C_{4,n-m-1})] \\ &\quad + P(C_{5,n-m-1}) + P(C_{7,n-m-1}) \\ &\qquad + [1-(1-\theta)^{m-1}][P(C_{6,n-m-1}) + P(C_{7,n-m-1})]\,. \end{aligned}$$

And as $n \to \infty$, using (10.8.2), we obtain

$$P_\infty(A_1 \mid E_1^{(m)}) = 1 - (1-\pi)(1-\theta)^{m-1}\,.$$

And by identical considerations

$$P_\infty(A_2 \mid E_2^{(m)}) = 1 - \pi(1-\theta)^{m-1}\,.$$

Note that in these equations the exponent is $m-1$ and not m as in (10.6.20) and (10.7.11). Using the estimated value of θ, we obtain for $m = 4$

$$P_\infty(A_1 \mid E_1^{(4)}) \approx .997\,, \qquad P_\infty(A_2 \mid E_2^{(4)}) \approx .995\,.$$

These results are to be compared with the observed data in Fig. 10.4 and Table 10.8.

We next evaluate $P(A_1 \mid E_1A_1E_2A_1)$. Employing the methods used above, we can show, for the noncontingent case, that

$$P(A_{1,n+1} \mid E_{1,n}A_{1,n}E_{2,n-1}A_{1,n-1}) = \frac{(1-\theta)[P(C_{1,n-1}) + P(C_{5,n-1})]}{P(A_{1,n} \mid E_{2,n-1}A_{1,n-1})P(A_{1,n-1})}\,.$$

Hence, as $n \to \infty$,

$$P_\infty(A_{1,n+1} \mid E_{1,n}A_{1,n}E_{2,n-1}A_{1,n-1}) = \frac{\pi(1-\theta)}{\pi + (1-\pi)^2(1-\theta)}\,.$$

Using the estimated value of θ, we obtain $P_\infty(A_1 \mid E_1A_1E_2A_1) \approx .190$.

We conclude the analysis by considering $P(A_{1,n+1} \mid A_{1,n})$:

$$\begin{aligned} P(A_{1,n+1} \mid A_{1,n}) &= \frac{1}{P(A_{1,n})}\sum_{i,j} P(A_{1,n+1}C_{i,n+1}A_{1,n}C_{j,n}) \\ &= \frac{1}{P(A_{1,n})}\sum_{i,j} P(A_{1,n+1} \mid C_{i,n+1})P(C_{i,n+1} \mid A_{1,n}C_{j,n})P(A_{1,n} \mid C_{j,n})P(C_{j,n}) \\ &= \frac{P(C_{1,n}) + \pi P(C_{2,n}) + [1-\theta(1-\pi)]P(C_{5,n})}{P(A_{1,n})}\,. \end{aligned}$$

And as $n \to \infty$,

$$P_\infty(A_{1,n+1} \mid A_{1,n}) = \pi + (1-\theta)(1-\pi)[\pi^2 + (1-\theta)(1-\pi)] .$$

Using the estimated value of θ, we obtain $P_\infty(A_{1,n+1} \mid A_{1,n}) \approx .635$.

It is evident from Table 10.8 that the predictions of the hypothesis model are decidedly worse than those of the stimulus-sampling and linear models. It seems clear that for probabilistic reinforcement schedules the simpler hypothesis models are of little value.

*10.9. A Memory Model

In this section we discuss a model that specifically incorporates a device to account for the observed differences in the three experimental groups. We have already seen in Sec. 10.5 how the generalized conditioning model handles these differences with considerable success. Now we consider the notion that as the amount of reinforcement increases, the subject will recall and use the pattern of reinforcements over an ever-increasing number of preceding trials. To this end, we reformulate Axiom C2 as follows:

AXIOM C2$''$. *If a stimulus element is sampled on trial n, conditioning can occur with probability* θ. *The sampled element will be conditioned with probability* ω_i *(where* $\sum \omega_i = 1$*) to the response that was most frequently reinforced on the last i trials* ($n - i + 1$ *to* n)*; if no one reinforcement occurred more frequently than any other in the last i trials, no change in conditioning will take place.*

To illustrate the application of C2$''$ consider the sequence of reinforcements $E_{2,n}E_{1,n-1}E_{1,n-2}\cdots$. If conditioning occurs on trial n, then (1) with probability ω_1 the element is conditioned to A_2, (2) with probability ω_2 no change in conditioning occurs, (3) with probability ω_3 the element is conditioned to A_1, and so on.

It should be evident that in several respects the memory model is a special kind of hypothesis model. Simply counting the number of E_1's in the last i trials markedly reduces the number of effectively different strategies or states of conditioning. The probabilities ω_i provide a means of treating the memory span of the subject as a random variable. In this respect the present model is a generalization of the hypothesis models considered earlier, for they required a memory of fixed length throughout the experiment.

In this section we consider only a one-element model and assume that $\omega_i = 0$ for $i > 3$. For a noncontingent situation the states of the Markov chain will indicate the reinforcements on trials $n-2$ and $n-1$ and also the response on trial n (i.e., the response to which the stimulus element is conditioned). Specifically, let $(ij\,;k)$ denote the states of the chain, where i and j indicate the reinforcing events on trials $n-2$ and $n-1$, respectively, and k the response on trial n. A typical tree for the chain is given in Fig. 10.10. On the upper branch of the tree no change can occur. On the lower branch a change in conditioning is possible with probability θ; further, with probability ω_1 the element is conditioned to A_2, since $E_{2,n}$ is postulated; with probability ω_2 it

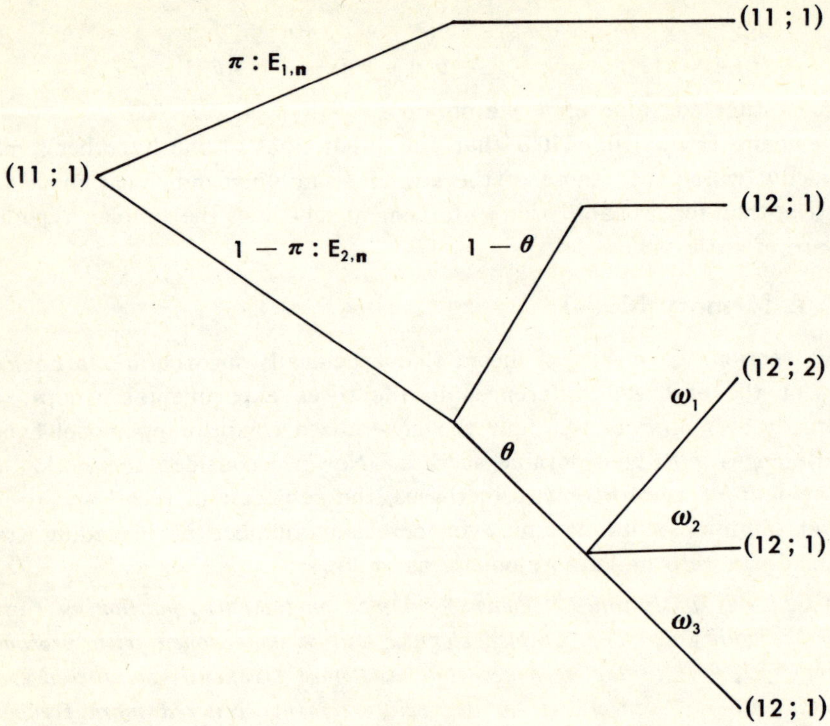

FIGURE 10.10

remains conditioned to A_1, since we have $E_{2,n}E_{1,n-1}$; and with probability ω_3 it remains conditioned to A_1, since we have $E_{2,n}E_{1,n-1}E_{1,n-2}$.

Construction of the analogous trees yields the transition matrix (10.9.1). From (10.9.1)

	11 ; 1	12 ; 1	21 ; 1	22 ; 1	11 ; 2	12 ; 2	21 ; 2	22 ; 2
11 ; 1	π	$(1-\pi)(1-\omega_1\theta)$				$(1-\pi)\omega_1\theta$		
12 ; 1			π	$(1-\pi)(1-\theta)$				$(1-\pi)\theta$
21 ; 1	π	$(1-\pi)[1-(1-\omega_2)\theta]$				$(1-\pi)(1-\omega_2)\theta$		
22 ; 1			$\pi(1-\omega_3\theta)$	$(1-\pi)(1-\theta)$			$\pi\omega_3\theta$	$(1-\pi)\theta$
11 ; 2	$\pi\theta$	$(1-\pi)\omega_3\theta$			$\pi(1-\theta)$	$(1-\pi)(1-\omega_3\theta)$		
12 ; 2			$\pi(1-\omega_2)\theta$				$\pi[1-(1-\omega_2)\theta]$	$1-\pi$
21 ; 2	$\pi\theta$				$\pi(1-\theta)$	$1-\pi$		
22 ; 2			$\pi\omega_1\theta$				$\pi(1-\omega_1\theta)$	$1-\pi$

the discussion in Sec. 1.3 it is clear that the limiting quantities u_{ijk} exist and are independent of the initial probability distribution. In terms of these quantities the probability of an A_1 response at asymptote is simply

$$P_\infty(A_1) = u_{111} + u_{121} + u_{211} + u_{221} .$$

In this model, as in other one-element models, there is a set of completely deterministic predictions of the type described by (1.3.5). These deterministic predictions, as usual, are contrary to intuition and experimental data. They are as follows:

$$P(A_{2,n+1} \mid E_{1,n}A_{1,n}E_{1,n-1}E_{1,n-2}) = 0\,, \qquad P(A_{1,n+1} \mid E_{2,n}A_{2,n}E_{2,n-1}E_{1,n-2}) = 0\,,$$
$$P(A_{2,n+1} \mid E_{1,n}A_{1,n}E_{2,n-1}E_{1,n-2}) = 0\,, \qquad P(A_{1,n+1} \mid E_{2,n}A_{2,n}E_{1,n-1}E_{2,n-2}) = 0\,,$$
$$P(A_{2,n+1} \mid E_{1,n}A_{1,n}E_{1,n-1}E_{2,n-2}) = 0\,, \qquad P(A_{1,n+1} \mid E_{2,n}A_{2,n}E_{2,n-1}E_{2,n-2}) = 0\,.$$

Of course, as in other cases of this sort, the difficulty can be overcome either by positing more than one stimulus element or by allowing the subject to scan more than just three prior reinforcements. Extensions of this sort quickly become unwieldy, and we do not pursue them here.

However, to get some hold on formulations of this type we have applied the model developed in this section to the data of Group F. A transition frequency matrix corresponding to (10.9.1) was tabulated for the first 140 trials and the last 100 trials. In terms of these transition frequencies and (10.9.1) the following likelihood function is defined:

$$L(\theta, \omega_1, \omega_2, \omega_3) = n_{11} \log \pi + n_{12} \log (1 - \pi)(1 - \omega_1\theta) + \cdots + n_{87} \log \pi(1 - \omega_1\theta) + n_{88} \log (1 - \pi)\,.$$

The maximum of this function was approximated by computing the value of the function for an array of 4,000 values of θ, ω_1, ω_2, and ω_3. Specifically, $\theta = .1i$ $(i = 1, \cdots, 10)$, $\omega_1 = .05i$ $(i = 1, \cdots, 20)$, and $\omega_2 = .05i$ $(i = 1, \cdots, 20)$. All combinations of these values were computed for the aggregated n_{ij}'s tabulated over the first 140 trials and over the last 100 trials. Over the first 140 trials, the maximum-likelihood estimates were as follows:

$$\text{(10.9.2)} \qquad \hat{\theta} = .6\,, \qquad \hat{\omega}_1 = .6\,, \qquad \hat{\omega}_2 = .1\,, \qquad \hat{\omega}_3 = .3\,.$$

Unexpectedly, the same estimates, to this degree of accuracy, were also obtained for the last 100 trials. Without further analysis it is not clear if the nonmonotonic character of the ω_i's is an artifact attributable to the particular rule for ties, or if it genuinely reflects the fact that subjects tend to look back three trials rather than two.

On the basis of (10.9.2) the predicted value of $P_\infty(A_1)$ is .62; this is somewhat lower than the observed value of .64 for Group F, but is, as would be expected, greater than $\pi = .6$.

We present only these very sketchy results for the memory model. However, in concluding, there are two related models about which we want to make some brief comments. One version of the memory model involves the introduction of the ω_i, not as probabilities of looking back i trials, but as weights. In this model, for $i = 1, 2, 3$, all three reinforcements play a role in conditioning but with different magnitude of effect as given by the weights ω_i. The sum of the weights is then interpreted in terms of probabilities. For example, if $\omega_1 = .6$, $\omega_2 = .3$, $\omega_3 = .1$, and if the sequence $E_{1,n}E_{2,n-1}E_{1,n-2}$ occurs, then an E_1-type reinforcement occurs on trial n with probability .7, and an E_2 occurs with probability .3. Reconstruction of the tree in Fig. 10.10 and the transition matrix (10.9.1) is a straightforward matter.

In Estes and Suppes [1959*a*, pp. 152–53] there is introduced the notion of a *recall span* for the linear model. This span is treated there not as a random variable but as fixed in length. Reinforcement is defined as follows: if the majority of reinforcements in the recall span are E_1's, the subject interprets the sequence as an E_1; if the majority are E_2's, the sequence is interpreted as an E_2; if the E_1's and E_2's are equally divided, the probability is .5 that A_1 will be reinforced. A span of $k = 1$ is the ordinary linear model, with $P_\infty(A_{1,n}) = \pi$.[4] If $k = 2$, also $P_\infty(A_{1,n}) = \pi$. If $k = 3$,

$$P_\infty(A_{1,n}) = \pi^3 + 3\pi^2(1 - \pi) .$$

Thus for $k = 3$ and $\pi = .6$, we have $P_\infty(A_{1,n}) = .648$, which is very close to the observed asymptote for Group F. If $k = 4$,

$$P_\infty(A_{1,n}) = \pi^4 + 4\pi^3(1 - \pi) + 3\pi^2(1 - \pi)^2 .$$

Thus for $\pi = .6$, we have $P_\infty(A_{1,n}) = .741$, which is slightly higher than the terminal response proportion for Group T; but, as is evident from Fig. 10.1, this group has not yet reached asymptote. Of course, how well the linear model with a recall span of 3 or 4 would fit the sequential data is an entirely different question.

Still another approach to the problem of choice behavior in relation to reinforcement schedules involving monetary payoffs has been proposed by Atkinson [1960*b*]. The essential idea is to introduce a strength-of-conditioning variable for each stimulus element in place of the all-or-none conditioning process. However, the details of this approach are too lengthy to examine in this book.

In this section we have sought only to sketch a class of memory models that may be useful in analyzing the effects of differential monetary reward; it has not been possible to evaluate their merits and weaknesses in detail.

10.10 Summary

In all the experiments reported in earlier chapters, the reinforcing event has consisted simply of indicating to the subject whether his response is correct or incorrect for the trial. Although there is no problem of the subject's being insufficiently motivated by reinforcing events of this type, in terms of an experimental program to simulate real game situations it is desirable to consider a wider class of reinforcements. Hence, this chapter analyzes the effect of monetary reward.

Three groups of subjects were run in a two-person, non-zero-sum interaction situation of the sort described in Chapter 4. Subjects in Group Z were simply informed on each trial whether their response was correct or incorrect; in Groups F and T the subjects, in addition to being so informed, won five and ten cents, respectively, when correct, and lost the same amount when incorrect. Subjects were run in pairs with game-type instructions; however, the reinforce-

[4] Here $k - 1 = k'$ in Estes and Suppes.

ment schedule for an individual subject could be considered noncontingent with $\pi = .6$.

The experimental method is presented in Sec. 10.2. In Secs. 10.3 and 10.4 certain comparisons are made, and significant differences are found between the groups, notably in the asymptotic proportion of Type 1 responses: for Group Z the observed proportion over the last 100 trials is .59; for Group F, .64; and for Group T, .69. The stimulus-sampling theory presented in Chapter 1 predicts .6 as an asymptote, and it is clear for this experiment that as the magnitude of reward increases, the difference between the predicted and observed values also increases.

Because of the absence of a well-defined theory for different magnitudes of reinforcement these data are analyzed from a variety of standpoints. In Sec. 10.5 the generalized conditioning model is applied to the data of the three groups. The conditioning parameters are estimated at several points during the course of learning, and some highly provocative results are obtained on the relation of these parameters to reinforcement magnitude. In Secs. 10.6–8 the data of Group Z are analyzed in detail for the multi-element stimulus-sampling model, the linear model, and a hypothesis model. The sequential predictions of the hypothesis model are much worse than those of the multi-element stimulus-sampling model ($N^* = 3.42$, $\theta^* = .59$) and the linear model ($\theta^* = .19$). As between the latter two, the stimulus-sampling model fares slightly better. Finally, Sec. 10.9 discusses a memory-span model that specifically incorporates a device for accounting for the observed differences in the three experimental groups.

11

Utility Functions

11.1. Introduction

In the second edition of *Theory of Games and Economic Behavior* von Neumann and Morgenstern added an appendix containing an axiomatic derivation of numerical utility. For obvious reasons game theory and utility theory have been closely related from the beginning. In order to study the problem of selecting an optimal game strategy, it is necessary to have clearly in mind which quantity is to be optimized. It is generally agreed that the appropriate quantity is expected utility; i. e., the average outcome of a strategy in terms of utility, where the "averaging" is over random moves of the game and the strategies to be adopted by the other players.

The sections in earlier chapters on game-theory comparisons have been presented in a manner that avoids the role of utility. The reason is not hard to find. An adequate utility theory for game-theory purposes has yet to be constructed. The axioms given by von Neumann and Morgenstern establish only the existence of a numerical utility for each individual, which is unique up to a linear transformation (an interval scale). The difficult problem of comparing utilities among individuals is not solved, although its solution must be assumed for the theory of most games (for comments on this problem see Luce and Raiffa [1957]). In practice one is forced to equate monetary value with utility, or in theoretical developments simply to assume a common numerical utility function without attempting to justify its basis.

There are two major reasons why the derivation of utility functions from learning-theory postulates is of conceptual interest. The first has already been implied. If an empirically successful theory can be given for deriving utility functions in terms of the concepts of stimulus, response, and reinforcement, then a new avenue is opened for approaching the fundamental problem of interpersonal comparison of utility. This problem not only plagues the foundations of game theory, but for many years has been the focus of much discussion in economics. What is termed the "new welfare economics" is based on a denial of the possibility of obtaining a positive solution.

Since the papers of Bergson [1938], Hotelling [1938], Hicks [1939], and others, the dominant tendency has been toward formulating a theory for economic and social policy-making without interpersonal comparisons of utility.[1] Yet the negative results of Arrow [1951] and the weakness of principles like that of Pareto optimality, which is in essence a version of the sure-thing principle, argue for the desirability of a new approach. It is not appropriate to attempt to work out any details here, but the directions of development may be indicated. (Some additional comments are made in Sec. 11. 7.) Arrow showed, roughly speaking, that if all possible individual preference orderings are admitted, there does not exist a social-decision procedure that satisfies intuitively expected properties.

Applying the theory developed in this book would suggest going beyond the individual preference orderings to the environmental and constitutional conditions that produced them. It has never been very realistic to assume individual orderings as the given data from which a theory of social welfare is to be developed, for the interaction between group decisions and individual preferences is apparent in most social processes. The fundamental theory of learning we have been applying provides dynamic mechanisms for analyzing this interaction. The discussion of social conformity in the next chapter provides a detailed example.

The second major reason for conceptual interest in the derivation of utility functions is more closely connected with the theme of this book. In Sec. 1.6 we remarked that the static character of the utility concept is disturbing to the psychologist, who seeks rather to discover the factors governing the dynamic process by which a particular set of values or utilities is acquired. The derivation of utility functions is one of the first steps in providing a thoroughgoing behavioristic approach to the policy-making sciences. In this chapter we focus on this step. The derivations of Sec. 11.2 proceed from the basic learning axioms of Sec. 1.2, slightly modified to accommodate some new conditioning possibilities. The results of a simple experiment to test the model of Sec. 11.2 are then considered in Secs. 11.3–5. An alternative model that also yields utility functions is developed in Secs. 11.6. The experimental utility functions derived from both models are presented in Secs. 11.7; Luce's multiplication rule is also discussed in this section. Two additional models, in which the same utility derivations are not possible, are considered in the final two sections, 11.8 and 11.9. The observing-response model of Sec. 11.9 challenges some of the assumptions common to many of the experiments on measuring utility.

The experimental situation involves a sequence of discrete trials. There are four responses, denoted by A_1, A_2, A_3, and A_4; and on each trial two of these responses are made available to the subject. The subject must select between

[1] The idea actually originated with Pareto many years earlier. (See particularly his *Manuel d'économie politique.*)

the two. If A_i is available and chosen, then with probability ξ_i the subject wins one cent and with probability $1-\xi_i$ no payoff occurs. The probability ξ_i is independent of the other response available. The subjects are not told the values of ξ_i, and the theoretical problem is to predict their choice behavior over trials as they acquire information. The payoff values used in this study are as follows: $\xi_1 = .2$, $\xi_2 = .4$, $\xi_3 = .6$, and $\xi_4 = .8$. Thus, if a subject is to maximize his probability of a win, he should (1) choose A_4 when (A_1A_4), (A_2A_4), or (A_3A_4) is made available; (2) choose A_3 when (A_1A_3) or (A_2A_3) is made available; and (3) choose A_2 when (A_1A_2) is made available.

11.2. Model

On each trial the subject is allowed to choose between the two responses presented in one of the following six response pairs: (A_1A_2), (A_1A_3), (A_1A_4), (A_2A_3), (A_2A_4), and (A_3A_4). We shall assume that a set of stimulus elements S_{ij} is associated with the presentation of the response pair (A_iA_j). For simplicity the six sets of elements in this experiment are assumed to be pairwise disjoint.

When (A_iA_j) is presented, the subject is required to choose A_i or A_j; therefore, elements in S_{ij} are to be conditioned to either A_i or A_j. Following the axioms of Sec. 1.2, when (A_iA_j) is presented, the subject samples a single element from S_{ij} and makes the response to which the element is conditioned.

In the present analysis one modification will be made. Axioms C2 and C3 are to be replaced by the following:

Axiom C2″. *If a stimulus element is sampled and an A_i response is made and followed by a payoff, then with probability c the element remains conditioned to A_i, and with probability $1-c$ it becomes conditioned to the other available response; if A_i is made and no payoff occurs, then with probability c' the element remains conditioned to A_i, and with probability $1-c'$ it becomes conditioned to the other response.*

Of course, in general one would expect c to be greater than c'.

In analyzing the model, we shall use the following notation:

$A_{i,n}^{(ij)}$ = occurrence of an A_i response on the nth presentation of (A_iA_j) [note that reference is not to the nth trial of the experiment but to the nth presentation of (A_iA_j)];

$w_n^{(ij)}$ = payoff on the nth presentation of (A_iA_j);

$\overline{w}_n^{(ij)}$ = no payoff on the nth presentation of (A_iA_j).

Employing this notation, we have $\xi_i = P(w_n^{(ij)} \mid A_{i,n}^{(ij)})$ for any j and n.

We now examine the model for the case in which there is exactly one element in each set S_{ij}. (The extension to several elements in each set S_{ij} is straightforward; we consider it later.) One of the two trees for the single element in S_{ij} is given in Fig. 11.1. The other tree (for $A_{j,n}^{(ij)}$) is identical except that the i's and j's in their subscripts are reversed.

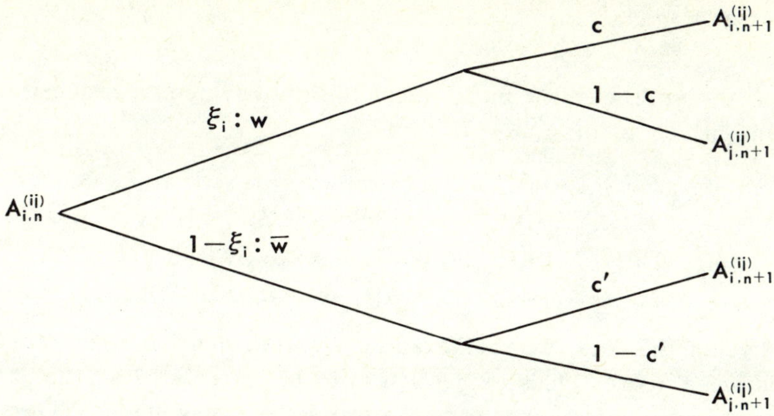

FIGURE 11.1

Thus, the subsequence of trials on which (A_iA_j) is presented can be described by the following transition matrix:

	$A^{(ij)}_{i,n+1}$	$A^{(ij)}_{j,n+1}$
$A^{(ij)}_{i,n}$	$\xi_i c + (1-\xi_i)c'$	$\xi_i(1-c) + (1-\xi_i)(1-c')$
$A^{(ij)}_{j,n}$	$\xi_j(1-c) + (1-\xi_j)(1-c')$	$\xi_j c + (1-\xi_j)c'$

Let $\delta_k = \xi_k(1-c) + (1-\xi_k)(1-c')$. Then, following the development in (1.3.34–35)

$$P(A^{(ij)}_{i,n}) = \frac{\delta_j}{\delta_i+\delta_j} - \left[\frac{\delta_j}{\delta_i+\delta_j} - P(A^{(ij)}_{i,1})\right](1-\delta_i-\delta_j)^{n-1}, \tag{11.2.1}$$

$$P(A^{(ij)}_i) = \lim_{n\to\infty} P(A^{(ij)}_{i,n}) = \frac{\delta_j}{\delta_i+\delta_j}. \tag{11.2.2}$$

Note that the expression for $P(A^{(ij)}_{i,n})$ does not depend on the assumption of one element in each set S_{ij}. To illustrate, assume that there are N_{ij} elements in S_{ij}, and let $\alpha_n = P(A^{(ij)}_{i,n})$. Using the methods of Sec. 3.7, we have

$$\begin{aligned}\alpha_{n+1} &= \alpha_n\left\{\xi_i\left[c\alpha_n + (1-c)\left(\alpha_n - \frac{1}{N_{ij}}\right)\right] + (1-\xi_i)\left[c'\alpha_n + (1-c')\left(\alpha_n - \frac{1}{N_{ij}}\right)\right]\right\}\\ &\quad + (1-\alpha_n)\left\{\xi_j\left[c\alpha_n + (1-c)\left(\alpha_n + \frac{1}{N_{ij}}\right)\right] + (1-\xi_j)\left[c'\alpha_n + (1-c')\left(\alpha_n + \frac{1}{N_{ij}}\right)\right]\right\}\\ &= \alpha_n(1-\delta^{(ij)}_i - \delta^{(ij)}_j) + \delta^{(ij)}_j,\end{aligned}$$

where

$$\delta^{(ij)}_k = \xi_k\frac{(1-c)}{N_{ij}} + (1-\xi_k)\frac{(1-c')}{N_{ij}},$$

and $k = i, j$. Then

$$P(A_{i,n}^{(ij)}) = \frac{\delta_j^{(ij)}}{\delta_i^{(ij)} + \delta_j^{(ij)}} - \left[\frac{\delta_j^{(ij)}}{\delta_i^{(ij)} + \delta_j^{(ij)}} - P(A_{i,1}^{(ij)})\right](1 - \delta_i^{(ij)} - \delta_j^{(ij)})^{n-1}.$$

Also, for the multi-element model the following asymptotic sequential results are given without proof:

$$\begin{aligned}
P_\infty(A_{i,n+1}^{(ij)} \mid A_{i,n}^{(ij)} w_n^{(ij)}) &= [c + \alpha_\infty(N_{ij} - 1)]/N_{ij}, \\
P_\infty(A_{i,n+1}^{(ij)} \mid A_{i,n}^{(ij)} \overline{w}_n^{(ij)}) &= [c' + \alpha_\infty(N_{ij} - 1)]/N_{ij}, \\
P_\infty(A_{i,n+1}^{(ij)} \mid A_{j,n+1}^{(ij)} w_n^{(ij)}) &= [1 - c + \alpha_\infty(N_{ij} - 1)]/N_{ij}, \\
P_\infty(A_{i,n+1}^{(ij)} \mid A_{j,n+1}^{(ij)} \overline{w}_n^{(ij)}) &= [1 - c' + \alpha_\infty(N_{ij} - 1)]/N_{ij}.
\end{aligned}$$

In analyzing the data, we shall examine $P(A_i)$ in addition to the other quantities mentioned above. This expression is to be interpreted as the probability of an A_i response on any trial once the process is at asymptote. Theoretically,

$$\begin{aligned}
P(A_1) &= \tfrac{1}{6}[P(A_1^{(12)}) + P(A_1^{(13)}) + P(A_1^{(14)})], \\
P(A_2) &= \tfrac{1}{6}[P(A_2^{(12)}) + P(A_2^{(23)}) + P(A_2^{(24)})], \\
P(A_3) &= \tfrac{1}{6}[P(A_3^{(13)}) + P(A_3^{(23)}) + P(A_3^{(34)})], \\
P(A_4) &= \tfrac{1}{6}[P(A_4^{(14)}) + P(A_4^{(24)}) + P(A_4^{(34)})].
\end{aligned} \tag{11.2.3}$$

In the present experiment all response pairs (A_iA_j) occur with equal likelihood; hence the probability $\frac{1}{6}$ in the above expressions. Note that if the subject maximizes his probability of success by selecting the response that has the highest likelihood of payoff for any given response pair (A_iA_j), then

$$P(A_1) = 0, \qquad P(A_2) = \tfrac{1}{6}, \qquad P(A_3) = \tfrac{1}{3}, \qquad P(A_4) = \tfrac{1}{2}. \tag{11.2.4}$$

That is, on half the trials either (A_1A_4), (A_2A_4), or (A_3A_4) will be presented, and if A_4 is selected on each occasion, then $P(A_4) = \frac{1}{2}$; and so on.

The derivation of utility functions follows the argument in Suppes [1959*b*]. We seek a function U such that for all responses g, h, i, and j

$$U_g - U_h \leqq U_i - U_j \quad \text{if and only if} \quad P(A_g^{(gh)}) \leqq P(A_i^{(ij)}). \tag{11.2.5}$$

Such a function we call a *stochastic* utility function. In classical economic theory the choice between two alternatives, i and j, is treated deterministically, i. e., one is always preferred and therefore always chosen.

As connections between utility theory and the psychometric literature have been developed, particularly by Luce [1959] (see also Davidson and Marschak [1959]), it has seemed more and more reasonable to assume that i will be chosen with a certain probability rather than deterministically. Condition (11.2.5) reflects this new emphasis. From (11.2.2), we have that

$$U_g - U_h \leqq U_i - U_j \quad \text{if and only if} \quad \frac{\delta_h}{\delta_g + \delta_h} \leqq \frac{\delta_j}{\delta_i + \delta_j}.$$

But by simple arithmetic the right-hand side holds if and only if $\delta_h/\delta_g \leqq \delta_j/\delta_i$; the latter, in turn, holds if and only if

$$\log \frac{1}{\delta_h} - \log \frac{1}{\delta_g} \leqq \log \frac{1}{\delta_j} - \log \frac{1}{\delta_i} .$$

Thus if we define for any response i

$$U_i = \log \frac{1}{\delta_i} = \log \frac{1}{\xi_i(1-c) + (1-\xi_i)(1-c')} , \tag{11.2.6}$$

the function U satisfies (11.2.5). Under appropriate continuity and monotonicity conditions this function may be shown to be unique up to a linear transformation. It is typical of the equilibrium character of the utility concept that the function defined by (11.2.6) is adequate only at asymptote, as may be easily seen from (11.2.1).

11.3. Experimental Method

Subjects. The subjects were 24 undergraduates obtained from introductory courses at Stanford University.

Apparatus and Procedure. The subjects were run individually. Each subject was seated at a table opposite the experimenter. On the table between the subject and the experimenter was a vertical panel 14 inches high and 36 inches long; this panel prevented the subject from watching the experimenter record data. Four vertical bars, numbered 1, 2, 3, and 4, were painted on the panel. Mounted at the base of the panel was a small open box labeled "payoff."

The subject was read the following instructions:

"This experiment is part of a series of studies on the behavior of people in simple game situations. The board in front of you represents the levers of four slot machines. For each play of the game I shall call out two numbers corresponding to two of the levers. You must choose between them by naming one of the numbers. If you lose, I will say 'lose,' and then we will go on to the next play of the game. If you win, I will say 'win' and deposit a poker chip, worth one cent, into the payoff box; we will then go on to the next play. On the other hand, we charge you one dollar to play for the whole session. If you are careful and watch how the various levers pay off, you should win well over a dollar. Make your choices as quickly as you can. Do you have any questions?"

Questions were answered by paraphrasing the appropriate part of the instructions.

A total of 360 trials were run. On each trial one of the pairs (1, 2), (1, 3), (1, 4), (2, 3), (2, 4), and (3,4) was given verbally by the experimenter, and the subject responded with one of the numbers. The sequence of pairs for each subject was generated randomly, with the restriction that each pair occurred equally often in successive 60-trial blocks. Further, there are 24 possible combinations of the responses 1, 2, 3, and 4 with the four ξ values, and each combination was used exactly once.

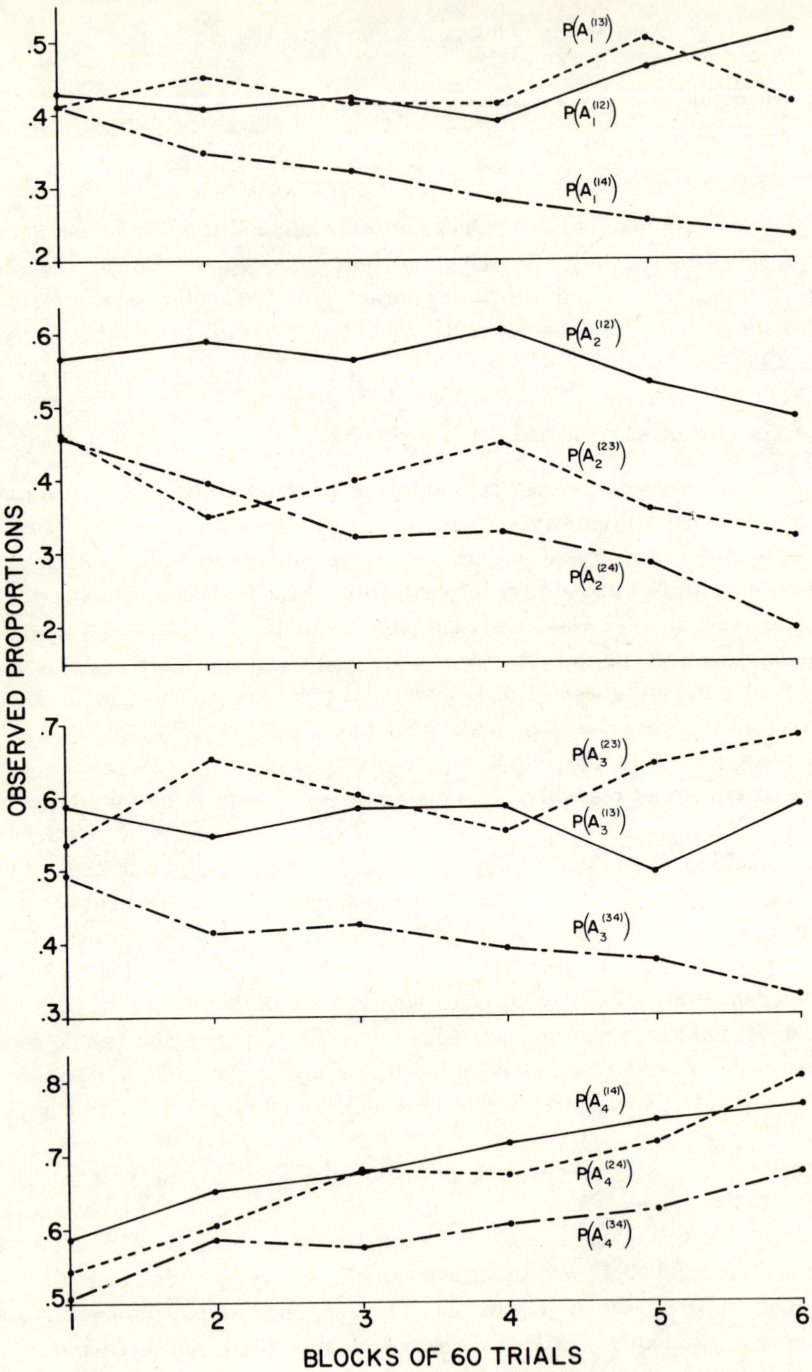

FIGURE 11.2. Observed values of $P(A_{i,n}^{(ij)})$ in successive 60-trial blocks.

11.4. Mean Learning Curves and Transition Data

For each subject the proportion of A_i responses given an (A_iA_j) was computed for successive 60-trial blocks. As indicated earlier, there were ten presentations of (A_iA_j) in successive 60-trial blocks; hence, each proportion was based on exactly ten observations. Figure 11.2 presents the observed means for $P(A_{i,n}^{(ij)})$ in successive blocks of 60 trials. In terms of the ξ values ($\xi_1 = .2, \xi_2 = .4, \xi_3 = .6$, and $\xi_4 = .8$), one would intuitively expect the following order over the trials:

$$\begin{aligned}
&\tfrac{1}{2} > P(A_{1,n}^{(12)}) > P(A_{1,n}^{(13)}) > P(A_{1,n}^{(14)}) \,, \\
&P(A_{2,n}^{(12)}) > \tfrac{1}{2} > P(A_{2,n}^{(23)}) > P(A_{2,n}^{(24)}) \,, \\
&P(A_{3,n}^{(13)}) > P(A_{3,n}^{(23)}) > \tfrac{1}{2} > P(A_{3,n}^{(34)}) \,, \\
&P(A_{4,n}^{(14)}) > P(A_{4,n}^{(24)}) > P(A_{4,n}^{(34)}) > \tfrac{1}{2} \,.
\end{aligned}$$

In general, this ordering holds up fairly well, as can be seen from Fig. 11.2.

Figure 11.3 presents the observed proportions for A_1, A_2, A_3, and A_4 responses in 60-trial blocks. The curves, of course, sum to unity, and the maximum value for any given curve would be $\frac{1}{2}$ (see 11.2.4). From the two figures it would appear that the subjects were not yet completely at asymptote, but in the subsequent analyses we shall assume that the last three blocks of 60 trials represent an approximation of the asymptote.

The observed transition frequencies for the last 180 trials aggregated over subjects are given in (11.4.1). Each matrix corresponds to the transition probabilities associated with a particular (A_iA_j) pair. For example, an entry would be made in the cell corresponding to $A_{1,n}^{(12)} w_n^{(12)} \rightarrow A_{1,n+1}^{(12)}$ if on an (A_1A_2) trial an A_1 occurred and was paid off, and on the next (A_1A_2) trial an A_1 occurred. For each subject 29 transitions occur over the last 180 trials for a particular (A_iA_j) pair; hence, the sum of the entries in a given matrix is $24 \times 29 = 696$.

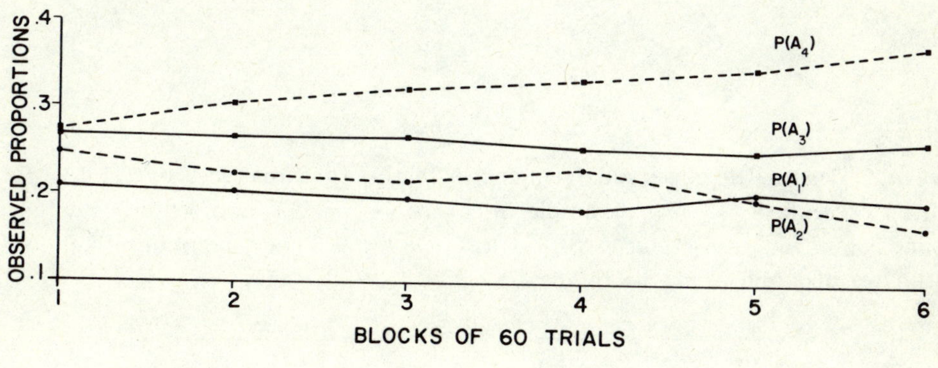

FIGURE 11.3. Observed values of $P(A_{i,n})$ in successive 60-trial blocks.

(11.4.1)

	$A^{(12)}_{1,n+1}$	$A^{(12)}_{2,n+1}$
$A^{(12)}_{1,n}w^{(12)}_n$	36	30
$A^{(12)}_{1,n}\overline{w}^{(12)}_n$	142	107
$A^{(12)}_{2,n}w^{(12)}_n$	48	106
$A^{(12)}_{2,n}\overline{w}^{(12)}_n$	92	135

	$A^{(13)}_{1,n+1}$	$A^{(13)}_{3,n+1}$
$A^{(13)}_{1,n}w^{(13)}_n$	42	21
$A^{(13)}_{1,n}\overline{w}^{(13)}_n$	131	115
$A^{(13)}_{3,n}w^{(13)}_n$	70	155
$A^{(13)}_{3,n}\overline{w}^{(13)}_n$	67	95

	$A^{(14)}_{1,n+1}$	$A^{(14)}_{4,n+1}$
$A^{(14)}_{1,n}w^{(14)}_n$	18	16
$A^{(14)}_{1,n}\overline{w}^{(14)}_n$	61	82
$A^{(14)}_{4,n}w^{(14)}_n$	85	339
$A^{(14)}_{4,n}\overline{w}^{(14)}_n$	24	71

	$A^{(23)}_{2,n+1}$	$A^{(23)}_{3,n+1}$
$A^{(23)}_{2,n}w^{(23)}_n$	50	50
$A^{(23)}_{2,n}\overline{w}^{(23)}_n$	79	84
$A^{(23)}_{3,n}w^{(23)}_n$	79	193
$A^{(23)}_{3,n}\overline{w}^{(23)}_n$	53	108

	$A^{(24)}_{2,n+1}$	$A^{(24)}_{4,n+1}$
$A^{(24)}_{2,n}w^{(24)}_n$	34	36
$A^{(24)}_{2,n}\overline{w}^{(24)}_n$	51	69
$A^{(24)}_{4,n}w^{(24)}_n$	79	319
$A^{(24)}_{4,n}\overline{w}^{(24)}_n$	26	82

	$A^{(34)}_{3,n+1}$	$A^{(34)}_{4,n+1}$
$A^{(34)}_{3,n}w^{(34)}_n$	96	63
$A^{(34)}_{3,n}\overline{w}^{(34)}_n$	58	43
$A^{(34)}_{4,n}w^{(34)}_n$	77	266
$A^{(34)}_{4,n}\overline{w}^{(34)}_n$	25	68

11.5. Maximum-Likelihood Estimates of the Conditioning Parameters and Predicted Asymptotes

In terms of the conditioning assumptions presented in Sec. 11.2 we can write the following transition matrix for the one-element model:

(11.5.1)

	$A^{(ij)}_{i,n+1}$	$A^{(ij)}_{j,n+1}$
$A^{(ij)}_{i,n}w^{(ij)}_n$	c	$1-c$
$A^{(ij)}_{i,n}\overline{w}^{(ij)}_n$	c'	$1-c'$
$A^{(ij)}_{j,n}w^{(ij)}_n$	$1-c$	c
$A^{(ij)}_{j,n}\overline{w}^{(ij)}_n$	$1-c'$	c'

Let $n^{(ij)}_{kl}$ denote the observed frequencies in the 4×2 matrices of (11.4.1) for (A_iA_j) presentations. (For example, $n^{(12)}_{11}=36$, $n^{(24)}_{32}=319$, and so on.) Further, define $n_{kl}=n^{(12)}_{kl}+n^{(13)}_{kl}+n^{(14)}_{kl}+n^{(23)}_{kl}+n^{(24)}_{kl}+n^{(34)}_{kl}$. Then maximum-likelihood estimates of c and c' are as follows:

$$\hat{c}=\frac{n_{11}+n_{32}}{n_{11}+n_{12}+n_{31}+n_{32}}, \qquad \hat{c}\,'=\frac{n_{21}+n_{42}}{n_{21}+n_{22}+n_{41}+n_{42}}.$$

Employing the observed frequencies in (11.4.1), we obtain $\hat{c}=.717$ and $\hat{c}\,'=.579$; as expected, $\hat{c}>\hat{c}\,'$. From these estimates, predictions for $P(A^{(ij)}_i)$ and $P(A_i)$

may be made in terms of (11.2.1). The predicted values and observed proportions over the last 180 trials are presented in Table 11.1. The over-all agreement between the two sets of values is poor. The agreement is still worse when one considers that most of the observed curves appear to be short of a final level; for the curves that are still increasing, the predicted asymptote is already below the observed terminal value; for the curves that are still decreasing, the predicted asymptote is above the observed terminal value.

TABLE 11.1

PREDICTED ASYMPTOTES FOR FOUR MODELS AND OBSERVED PROPORTIONS OVER THE LAST BLOCK OF 180 TRIALS

Asymptotic Value	Predicted Proportion				Observed Proportion
	c,c' Model (Secs. 11.2 and 11.5)	c_i,c_i' Model (Sec. 11.6)	c_{ij},c_{ij}' Model (Sec. 11.8)	Observing-Response Model (Sec. 11.9)	
$P(A_1^{(12)})$	.482	.495	.492	.442	.457
$P(A_1^{(13)})$	.462	.428	.467	.355	.445
$P(A_1^{(14)})$	.400	.318	.398	.228	.270
$P(A_2^{(23)})$	.480	.433	.490	.414	.375
$P(A_2^{(24)})$	.459	.322	.446	.286	.273
$P(A_3^{(34)})$	.478	.384	.488	.372	.368
$P(A_1)$	.224	.207	.226	.171	.195
$P(A_2)$	.243	.210	.241	.210	.199
$P(A_3)$	.256	.254	.255	.267	.258
$P(A_4)$	.277	.329	.278	.352	.348

The poor agreement on mean asymptotes also is reflected in sequential data. For each matrix in (11.4.1) predicted frequencies were generated in terms of the row sums and the estimates of c and c' (see 11.5.1). The respective χ^2's given by (2.4.4) were 10.5, 3.8, 41.1, 34.8, 58.5, and 24.7, which sum to 173.4. Two parameters were estimated from these data, and for each matrix there are four degrees of freedom; hence, we interpret the total χ^2 with $6 \times 4 - 2 = 22$ degrees of freedom. Of course, the result is highly significant.

*11.6. An Alternative Interpretation of Conditioning

We now modify the model presented in Sec. 11.2 by changing our interpretation of the conditioning parameters. Specifically, assume that conditioning parameters c_i and c_i' are associated with each response A_i. If A_i is chosen and followed by payoff, then the sampled element will remain conditioned to A_i with probability c_i, and the element will become conditioned to the other available response with probability $1 - c_i$. When A_i is selected and *not* followed by payoff, then the sampled element remains conditioned to A_i with probability

c_i', and the element becomes conditioned to the other available response with probability $1 - c_i'$.

The intuitive idea underlying these assumptions is that the parameters c_i and c_i' will change over trials as the subject gathers information, and that they will reach some stable level only in the long run; the rate of change and final level are determined in large part by the reinforcement schedule. Before applying the model to all trials of the experiment, one would need, in addition to the above assumptions, a theory describing change in c_i and c_i' over trials; this same comment has been made with regard to the generalized conditioning model of Sec. 2.6. We have some ideas about how to construct such a theory, but they go beyond the scope of the present book (see Sec. 10.5).

In view of these comments concerning a change in the conditioning parameters on early trials, it is clear that the model described in this section should be restricted to choice data for subjects who have achieved a stable level of responding; for only then could we assume that c_i and c_i' were constant. Unfortunately, in the data of this chapter such stability has not yet really been attained. Nevertheless, we apply the model to the results of the last 180 trials to illustrate some of its properties.

Given the above interpretation of the conditioning parameters, the transition matrix for the one-element model is as follows:

(11.6.1)

	$A_{i,n+1}^{(ij)}$	$A_{j,n+1}^{(ij)}$
$A_{i,n}^{(ij)}$	$\xi_i c_i + (1 - \xi_i)c_i'$	$\xi_i(1 - c_i) + (1 - \xi_i)(1 - c_i')$
$A_{j,n}^{(ij)}$	$\xi_j(1 - c_j) + (1 - \xi_j)(1 - c_j')$	$\xi_j c_j + (1 - \xi_j)c_j'$

Let $\varepsilon_k = \xi_k(1 - c_k) + (1 - \xi_k)(1 - c_k')$; then

$$P(A_i^{(ij)}) = \lim_{n \to \infty} P(A_{i,n}^{(ij)}) = \frac{\varepsilon_j}{\varepsilon_i + \varepsilon_j}. \tag{11.6.2}$$

It can be shown that the maximum-likelihood estimates of c_i and c_i' are as follows:

$$\begin{aligned}
\hat{c}_1 &= \frac{n_{11}^{(12)} + n_{11}^{(13)} + n_{11}^{(14)}}{n_{11}^{(12)} + n_{11}^{(13)} + n_{11}^{(14)} + n_{12}^{(12)} + n_{12}^{(13)} + n_{12}^{(14)}}, \\
\hat{c}_2 &= \frac{n_{32}^{(12)} + n_{11}^{(23)} + n_{11}^{(24)}}{n_{32}^{(12)} + n_{11}^{(23)} + n_{11}^{(24)} + n_{31}^{(12)} + n_{12}^{(23)} + n_{12}^{(24)}}, \\
\hat{c}_3 &= \frac{n_{32}^{(13)} + n_{32}^{(23)} + n_{11}^{(34)}}{n_{32}^{(13)} + n_{32}^{(23)} + n_{11}^{(34)} + n_{31}^{(13)} + n_{31}^{(23)} + n_{12}^{(34)}}, \\
\hat{c}_4 &= \frac{n_{32}^{(14)} + n_{32}^{(24)} + n_{32}^{(34)}}{n_{32}^{(14)} + n_{32}^{(24)} + n_{32}^{(34)} + n_{31}^{(14)} + n_{31}^{(24)} + n_{31}^{(34)}}.
\end{aligned} \tag{11.6.3}$$

To obtain the corresponding estimates for c_i', simply add 1 to the first digit of the subscripts in the expression for c_i. Thus

$$\hat{c}_1' = \frac{n_{21}^{(12)} + n_{21}^{(13)} + n_{21}^{(14)}}{n_{21}^{(12)} + n_{21}^{(13)} + n_{21}^{(14)} + n_{22}^{(12)} + n_{22}^{(13)} + n_{22}^{(14)}},$$

and so on.

Using the observed frequencies in (11.4.1), we obtain the following estimates:

$$\hat{c}_1 = .589, \quad \hat{c}_3 = .677, \qquad \hat{c}_1' = .524, \quad \hat{c}_3' = .616,$$
$$\hat{c}_2 = .586, \quad \hat{c}_4 = .793; \qquad \hat{c}_2' = .520, \quad \hat{c}_4' = .747.$$

The nice feature of these estimates is that they are almost perfectly ordered. The estimate of c_i in all cases is larger than c_i', indicating greater perseveration on the A_i response after payoff than after no payoff. Also, $\hat{c}_4 > \hat{c}_3 > \hat{c}_2$ and $\hat{c}_4' > \hat{c}_3' > \hat{c}_2'$, suggesting that frequency of payoff for a given response has a cumulative effect that is reflected in the conditioning parameters. The fact that $\hat{c}_1 \approx \hat{c}_2$ and $\hat{c}_1' \approx \hat{c}_2'$ suggests that subjects do not discriminate carefully between the low probabilities of payoff.

From the estimated values of c_i and c_i', predictions for $P(A_i^{(ij)})$ and $P(A_i)$ may be obtained in terms of (11.7.2) and (11.2.4), respectively. The predicted values and the observed proportions over the last 180 trials are presented in columns 3 and 6 of Table 11.1. The over-all agreement between predicted and observed values is better than that reported in Sec. 11.5, but is still not very good. Again the same comment is in order. The observed curves appear to be short of a final asymptote; for the curves that are still rising, the predicted value generally is already less than the observed terminal value; whereas for the curves that are still decreasing, the predicted asymptote generally is above the observed terminal value.

As in Sec. 11.5 the goodness-of-fit χ^2 test given by (2.4.4) was applied. The respective χ^2's for the matrices in (11.4.1) were 14.4, 2.4, 6.1, 7.2, 7.5, and 5.4, which sum to 42.9. Eight parameters were estimated from these data; hence, we interpret the total χ^2 with $6 \times 4 - 8 = 16$ degrees of freedom. Although these results, like those of Sec. 11.5, are highly significant, the actual level of significance is much better here (approximately 10^{-3}, compared with a level beyond 10^{-8} for Sec. 11.5).

Corresponding to the derivation at the end of Sec. 11.2, the following definition of utility is appropriate for the model of this section:

$$U_i = \log\frac{1}{\varepsilon_i} = \frac{1}{\xi_i(1-c_i) + (1-\xi_i)(1-c_i')}. \tag{11.6.4}$$

11.7. Utility Functions; Luce's Multiplication Rule

Figure 11.4 presents the observed utility functions for the models of Secs. 11.2 and 11.6; these are the two functions defined by (11.2.6) and (11.6.4), respectively. We shall refer to the model of Sec. 11.2 as Model I and that of Sec. 11.6 as Model II. In the figure the utility function derived from Model II has

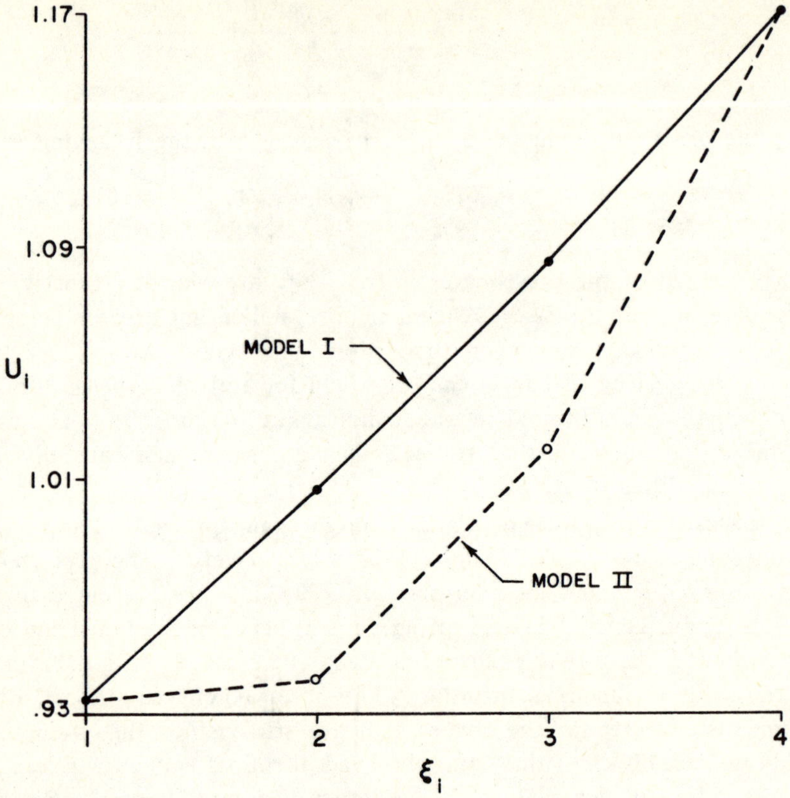

FIGURE 11.4. Observed utility functions for the models of Secs. 11.2 and 11.6.

been normalized by a linear transformation so that $U_1^{(II)} = U_1^{(I)}$, and $U_4^{(II)} - U_1^{(II)} = U_4^{(I)} - U_1^{(I)}$. Prior to this transformation

$$\begin{aligned} &U_1^{(I)} = .934, \quad U_3^{(I)} = 1.084, \qquad U_1^{(II)} = .770, \quad U_3^{(II)} = 1.056, \\ &U_2^{(I)} = 1.006, \quad U_4^{(I)} = 1.169; \qquad U_2^{(II)} = .789, \quad U_4^{(II)} = 1.531. \end{aligned} \tag{11.7.1}$$

Both curves are concave from above, although the curve generated by Model II is decidedly more so. The interpretation is that the value of the expected payoff is not linear in the probability. Further, when parameters are estimated separately for each response, as in Model II, the non-linear result is very clear. In this connection it should be recalled that Model II's over-all goodness-of-fit is considerably better than Model I's. The slight concavity of the curve for Model I is a mathematical result for $c > c'$, independent of the particular values of these parameters. On the other hand, the greater concavity of Model II's curve arises directly from the empirically estimated parameters c_i and c_i'. The greater change in utility from response 3 to response 4 in comparison with the change from 1 to 2 may be represented by

$$\frac{U_4^{(II)} - U_3^{(II)}}{U_2^{(II)} - U_1^{(II)}} \approx 25.$$

This ratio is independent of the arbitrarily chosen unit and zero point of utility.

Perhaps the feature of most theoretical interest is that the exact shape of the utility function depends upon the role of the conditioning parameters and their estimated values. As the theory of these parameters becomes empirically adequate, more definite predictions can be made.

A possible caveat for an economist is that both (11.7.1) and Fig. 11.4. present only mean data for the group of subjects. The classical literature of utility theory is dedicated to the concept of individual utility functions. The presentation of a group utility function might well be regarded as a crude and unsuccessful attempt to solve at one stroke the problem of interpersonal comparison of utility.

The analysis of data in earlier chapters has shown quite clearly that our basic theory is in no sense restricted to group mean data. It would be straightforward to present individual data for Models I and II and the accompanying individual utility functions. We have not done so only because of practical limitations on space and computational effort. It would be of more conceptual interest first to give the derivation of the asymptotic distribution of the conditioning parameters for each model, and then to test the hypothesis that the sample of parameter values estimated from individual data is homogeneous. Recent work by Chernoff and others has made available the statistical tools for such an analysis; applying them, however, is quite a tedious job.[2]

In asking for the existence of a utility function, we asked for the existence of a function that would satisfy (11.2.5). Additional conditions, related to but distinct from utility considerations, have been enunciated or derived from more general postulates by various authors. We select for brief discussion Luce's multiplication rule (Luce [1959], pp. 16–18). From his basic choice axiom he derives the following result, expressed in our notation:

$$P(A_i^{(ij)})P(A_j^{(jk)})P(A_k^{(ik)}) = P(A_i^{(ik)})P(A_k^{(jk)})P(A_j^{(ij)}) .$$

Define

$$\omega_{ij} = \frac{P(A_i^{(ij)})}{P(A_j^{(ij)})} . \tag{11.7.2}$$

Then the rule may be written more simply as

$$\omega_{ij}\omega_{jk} = \omega_{ik} . \tag{11.7.3}$$

[2] In later publications we hope to present some results in this direction. Their theoretical interest for the problem of interpersonal comparison of utility is apparent. Even though the individual utility functions differ, if there are no significant differences in individual conditioning parameters for a given population, a good case can be made for taking the group mean utility function as the expression of the "general will." Moreover, our theory of conditioning provides at least the outlines of a method for making scientific sense of the notion of the general will—a notoriously vague concept in the history of political theory.

That (11.7.3) holds asymptotically for Models I and II is easily seen from the fact that in Model I $\omega_{ij} = \delta_j/\delta_i$, and in Model II $\omega_{ij} = \varepsilon_j/\varepsilon_i$.

Empirical data on the multiplication rule are presented below. In terms of the four responses, there are two predictions for each ω_{ij}. For example, ω_{12} can be estimated directly from the data and can also be predicted by means of the products $\hat{\omega}_{13}\hat{\omega}_{32}$ and $\hat{\omega}_{14}\hat{\omega}_{42}$. Both predictions for each $\hat{\omega}_{ij}$ are given below:

$$(11.7.4)\qquad \begin{array}{lll} \hat{\omega}_{12} = .84\,, & \hat{\omega}_{13}\hat{\omega}_{32} = 1.34\,, & \hat{\omega}_{14}\hat{\omega}_{42} = .82\,, \\ \hat{\omega}_{13} = .80\,, & \hat{\omega}_{12}\hat{\omega}_{23} = .51\,, & \hat{\omega}_{14}\hat{\omega}_{43} = .64\,, \\ \hat{\omega}_{14} = .37\,, & \hat{\omega}_{12}\hat{\omega}_{24} = .32\,, & \hat{\omega}_{13}\hat{\omega}_{34} = .47\,, \\ \hat{\omega}_{23} = .60\,, & \hat{\omega}_{24}\hat{\omega}_{43} = .65\,, & \hat{\omega}_{21}\hat{\omega}_{13} = .95\,, \\ \hat{\omega}_{24} = .38\,, & \hat{\omega}_{21}\hat{\omega}_{14} = .44\,, & \hat{\omega}_{23}\hat{\omega}_{34} = .35\,, \\ \hat{\omega}_{34} = .58\,, & \hat{\omega}_{31}\hat{\omega}_{14} = .46\,, & \hat{\omega}_{32}\hat{\omega}_{24} = .63\,. \end{array}$$

Since the multiplication rule was not well substantiated by our data, a statistical analysis has not been made. For statistical tests of the multiplication rule the reader is referred to the works of Bradley and Terry [1952] and Bradley [1954*a*], [1954*b*], and [1955]. However, in this connection it may be mentioned that the probability $P(A_i^{(ij)})$ is the asymptotic expectation of the response random variable for the pair (A_iA_j). Luce, as well as Bradley and Terry, treats the probability of choice not as a random variable but as a given number.

*11.8. A Third Interpretation of Conditioning

We examine still another interpretation of the conditioning parameters by assuming that values of c and c' may differ over the stimulus sets S_{ij}. The comments made at the begining of Sec. 11.7 concerning changes in the conditioning parameters over trials and the related difficulties also apply to this section.

Assume that parameters c_{ij} and c'_{ij} are associated with each stimulus set S_{ij}. If an element from S_{ij} is sampled and the response made leads to payoff, then that element remains conditioned to the elicited response with probability c_{ij}, or it becomes conditioned to the other response with probability $1 - c_{ij}$. The quantity c'_{ij} is defined similarly if there is no payoff.

The transition matrix for the one-element model is as follows:

$$\begin{array}{c|cc} & A_{i,n+1}^{(ij)} & A_{j,n+1}^{(ij)} \\ \hline A_{i,n}^{(ij)} & \xi_i c_{ij} + (1-\xi_i)c'_{ij} & \xi_i(1-c_{ij}) + (1-\xi_i)(1-c'_{ij}) \\ A_{j,n}^{(ij)} & \xi_j(1-c_{ij}) + (1-\xi_j)(1-c'_{ij}) & \xi_j c_{ij} + (1-\xi_j)c'_{ij} \end{array}$$

Let $\gamma_k^{(ij)} = \xi_k(1-c_{ij}) + (1-\xi_k)(1-c'_{ij})$, where $k = i, j$. Then

$$(11.8.1)\qquad P(A_i^{(ij)}) = \frac{\gamma_j^{(ij)}}{\gamma_j^{(ij)} + \gamma_i^{(ij)}}\,.$$

Using this model, we obtain maximum-likelihood estimates of c_{ij} and c'_{ij} in terms of the matrices given in (11.4.1) as follows:

$$\hat{c}_{ij} = \frac{n_{11}^{(ij)} + n_{32}^{(ij)}}{n_{11}^{(ij)} + n_{12}^{(ij)} + n_{31}^{(ij)} + n_{32}^{(ij)}},$$

(11.8.2)

$$\hat{c}'_{ij} = \frac{n_{21}^{(ij)} + n_{42}^{(ij)}}{n_{21}^{(ij)} + n_{22}^{(ij)} + n_{41}^{(ij)} + n_{42}^{(ij)}}.$$

Using the observed frequencies in (11.4.1), we obtain the following estimates:

$$\begin{array}{llll} \hat{c}_{12} = .646, & \hat{c}_{23} = .653, & \hat{c}'_{12} = .582, & \hat{c}'_{23} = .577, \\ \hat{c}_{13} = .684, & \hat{c}_{24} = .754, & \hat{c}'_{13} = .554, & \hat{c}'_{24} = .583, \\ \hat{c}_{14} = .780, & \hat{c}_{34} = .721; & \hat{c}'_{14} = .555, & \hat{c}'_{34} = .650. \end{array}$$

From the estimated values of c_{ij} and c'_{ij}, predictions for $P(A_i^{(ij)})$ and $P(A_i)$ may be obtained in terms of (11.8.1) and (11.2.4). The predicted values and observed proportions over the last 180 trials are given in columns 4 and 6 of Table 11.1. As in the case of Model I, the agreement between predicted and observed values is poor. This poor correspondence is further supported by the results of the χ^2 goodness-of-fit test. The respective χ^2's for the matrices in (11.4.1) were 4.4, 1.3, 37.2, 25.7, 58.1, and 21.2 which sum to 147.9. Twelve parameters have been estimated from these data; hence, the total χ^2 is interpreted with $6 \times 4 - 12 = 12$ degrees of freedom. The χ^2 result here, like that of Sec. 11.5, has a significance level beyond 10^{-8}. Of the three models considered thus far, all of which are variants of a generalized conditioning model that assumes no overlap between the stimulus sets S_{ij}, the model of Sec. 11.6 is clearly the best.

Finally, we remark that since the conditioning parameters c_{ij} depend upon both i and j, neither the utility derivation of Sec. 11.2 nor the multiplication rule of Sec. 11.7 holds for this model. Additional comments on this point are made at the end of the next section.

11.9. An Observing-Response Model for Choice Behavior

In this section a somewhat different theoretical approach is taken toward the class of experimental problems introduced in the present chapter. The major change is to postulate an additional implicit response by the organism prior to making the observed response A_i. This implicit response—actually observable in some experimental situations—has been referred to in the literature as an *observing, orienting,* or *approach-avoidance* response, depending on the theoretical orientation of the author.

We assume (1) that associated with each response A_i there is a tendency to approach or avoid the response, and (2) that the response finally selected on a trial depends on the subject's observing or orienting behavior in the pre-decision period of the trial. The basic notions underlying the model are similar to those presented by Bower [1959], Estes [1960], and Audley [1960]. The Estes formulation is quite general, encompassing a broad range of choice experiments. In special cases the Estes model reduces to the Bower and Audley models, as well as to the linear model described in Sec. 1.5. Unfortunately, the Estes model is too complex to be applied to our present experiment. We introduce some major simplifications to obtain a mathematically tractable process.

The assumptions for the model will be presented for an experimental situation that is more general than the one used in this chapter. We assume that there are K responses and that on any trial the experimenter presents (i.e., makes available) to the subject any subset of these responses; for the experiment reported in this chapter only two responses were presented on each trial. The subject is required to select a response from the available subset, and the trial terminates either with payoff or with no payoff. The assumptions are as follows:

1. On every trial each response has an approach value of 1 or 0.
2. At the start of each trial the subject randomly observes one of the available responses.[3]
3. If the approach value for a particular response is 1 and the response is observed, then that response will be made. If the approach value is 0, and the response is observed, then the subject will randomly re-orient and observe one of the other available response alternatives.
4. If all available responses have been observed on a trial and no response has been selected (i.e., if all available responses have values of 0), then the subject terminates the trial by randomly selecting one of the available responses.
5. If a response is selected on a trial and followed by a win, then with probability ρ' its approach value becomes 1 and with probability $1-\rho'$ its value remains unchanged. If a response is selected and followed by a loss, then with probability ρ'' its value becomes 0 and with probability $1-\rho''$ its value remains unchanged. The reinforcement parameters ρ' and ρ'' are independent of the trial number and the preceding pattern of events.
6. The approach value associated with a response not selected on a given trial does not change on that trial.

In terms of these assumptions we can define a 16-state Markov chain for our particular experiment. Denote the states of the process as $(ijkl)$, where i, j, k, l are 1 or 0. The first entry denotes the approach value assigned to A_1, the second the value assigned to A_2, and so on. We may use the above assumptions to derive a representative tree for the Markov chain. Only two responses are available on each trial; let $a, b, c, d, e,$ and f denote the probabilities of presenting (A_1A_2), (A_1A_3), (A_1A_4), (A_2A_3), (A_2A_4), and (A_3A_4), respectively. (These probabilities are all $\frac{1}{6}$ in the present experiment.) The tree starting in state (1001) is given in Fig. 11.5. As indicated on the top branch, when (A_1A_2) or (A_1A_3) is

[3] In our experiment the observing response may be thought of as the subject's focusing attention on one of the two response numbers presented verbally to him. Direct observation of this response is not possible. This situation is to be contrasted with that of an animal in a T-maze. At the choice point the animal orients toward the left or right arm of the maze, and this overt orienting response may be recorded by the experimenter.

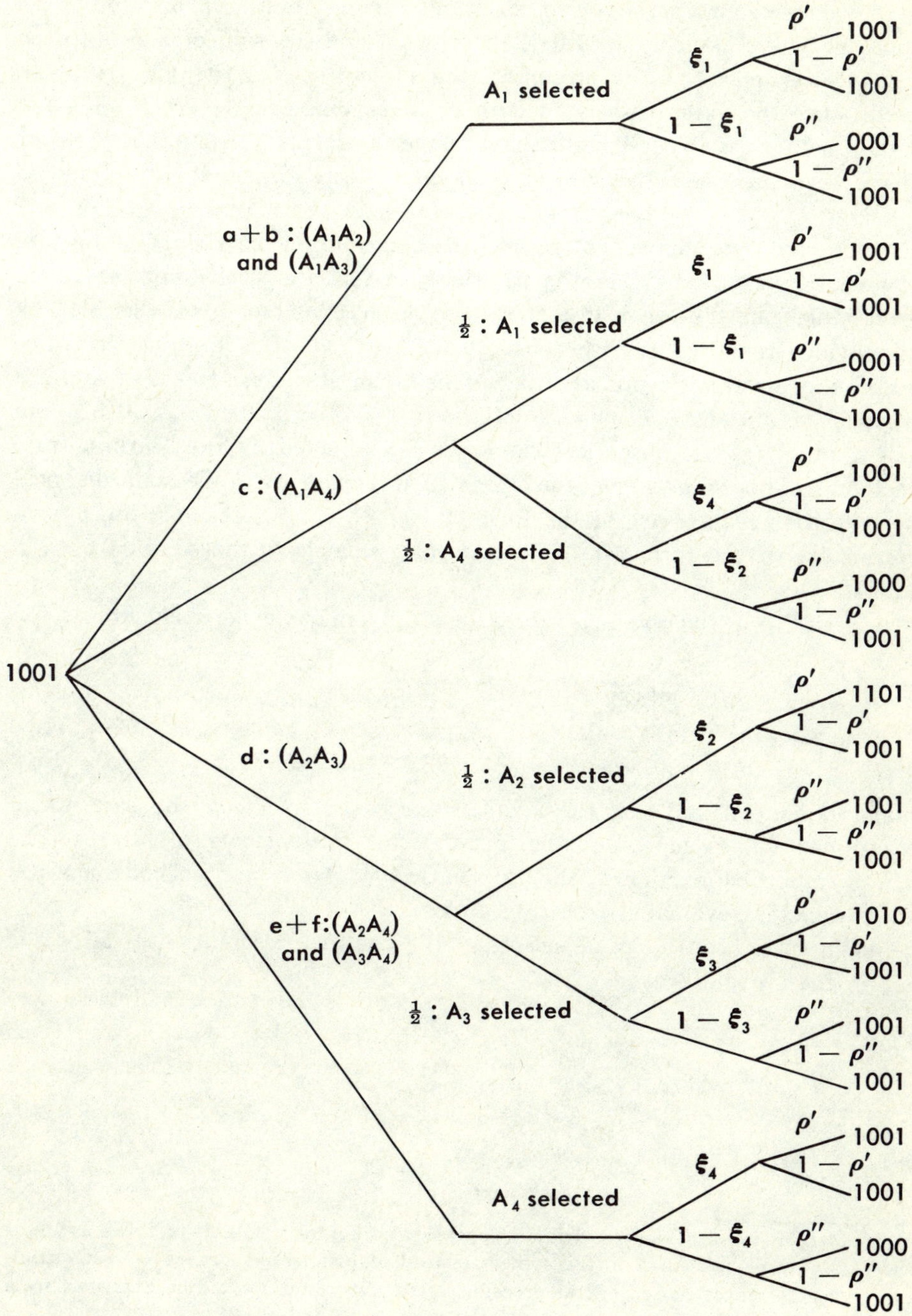
1001
a+b : (A₁A₂) and (A₁A₃)
c : (A₁A₄)
d : (A₂A₃)
e+f:(A₂A₄) and (A₃A₄)
A₁ selected
½ : A₁ selected
½ : A₄ selected
½ : A₂ selected
½ : A₃ selected
A₄ selected
ξ₁
1 − ξ₁
ξ₄
1 − ξ₂
ξ₂
1 − ξ₂
ξ₃
1 − ξ₃
ξ₄
1 − ξ₄
ρ′
1 − ρ′
ρ″
1 − ρ″
1001
1001
0001
1001
1001
1001
0001
1001
1001
1001
1000
1001
1101
1001
1001
1001
1010
1001
1001
1001
1001
1001
1000
1001

FIGURE 11.5

presented, the subject will select A_1, since the value assigned to A_1 is 1 and the value of both A_2 and A_3 is 0. If payoff terminates the trial, no change in the approach values occurs; however, if no payoff occurs, then with probability ρ'' the approach value associated with A_1 becomes 0 and the subject moves to the state (0001). When (A_1A_4) is presented, the subject selects the first response observed, since the value associated with both responses is 1. When (A_2A_3) is presented, both available responses have value 0; and by Assumption 4, after both have been observed, the subject randomly selects either A_2 or A_3; and so on.[4]

Construction of the analogous trees yields a transition matrix for a 16-state Markov chain. The state (1111) is transient (in fact the probability of re-entering it is 0), but the other 15 states form an irreducible, aperiodic Markov chain, provided the ξ values are neither 0 nor 1.[5] Thus the limiting quantities u_{ijkl}—i.e., the asymptotic probabilities of being in states $(ijkl)$—exist and are independent of initial conditions. Further, if $\rho' = \rho'' = \rho$, then u_{ijkl} also is independent of ρ; i.e., asymptotic behavior is not affected by the learning rate parameter ρ. This follows from the observation that terms on the main diagonal of the transition matrix are of the form $(1-\rho) + \rho \cdot f_{ii}$, whereas all other non-zero terms are of the form $\rho \cdot f_{ij}$. Therefore, ρ cancels in the solution of u_{ijkl} by (1.3.17).

Given $\rho' = \rho'' = \rho$, the obtained values for u_{ijkl} from our experiment are as follows:

$$u_{1110} = \tfrac{1}{10} u_{1000} \approx .0018, \qquad u_{1101} = u_{1100} = \tfrac{1}{10} u_{0100} \approx .0049,$$
$$u_{1011} = u_{1010} = \tfrac{1}{10} u_{0010} \approx .0111, \qquad u_{1001} = u_{0111} = u_{0110} = \tfrac{1}{10} u_{0001} \approx .0295,$$
$$u_{0101} \approx .0788, \qquad u_{0011} \approx .1772, \qquad u_{0000} \approx .1477.$$

It should be emphasized that these values are based solely on the experimentally assigned values of ξ_i and the fact that the presentation of all response pairs was equiprobable. Thus, they are entirely *a priori* and do not make use of any parameters evaluated from the data.

The asymptotic probability of selecting A_i given the presentation of (A_iA_j) is defined in the obvious way:

(11.9.1)

$$\begin{aligned} P(A_1^{(12)}) &= u_{1011} + u_{1010} + u_{1001} + u_{1000} \\ &\quad + \tfrac{1}{2}(u_{1110} + u_{1101} + u_{1100} + u_{0011} + u_{0010} + u_{0001} + u_{0000}), \\ P(A_1^{(13)}) &= u_{1101} + u_{1100} + u_{1001} + u_{1000} \\ &\quad + \tfrac{1}{2}(u_{1110} + u_{1011} + u_{1010} + u_{0101} + u_{0100} + u_{0001} + u_{0000}), \end{aligned}$$

---

[4] It should be noted that the tree would be somewhat different if subsets of three and four responses also were presented to the subject instead of just two responses on each trial. However, even for these more general procedures the experiment would be described by a 16-state Markov chain.

[5] A state of a Markov chain is transient if the probability of ever returning to it is less than 1.

$$P(A_1^{(14)}) = u_{1110} + u_{1100} + u_{1010} + u_{1000} + \tfrac{1}{2}(u_{1101} + u_{1011} + u_{1001} + u_{0110} + u_{0100} + u_{0010} + u_{0000}),$$

$$P(A_2^{(23)}) = u_{1101} + u_{1100} + u_{0101} + u_{0100} + \tfrac{1}{2}(u_{1110} + u_{1001} + u_{1000} + u_{0111} + u_{0110} + u_{0001} + u_{0000}),$$

$$P(A_2^{(24)}) = u_{1110} + u_{1100} + u_{0110} + u_{0100} + \tfrac{1}{2}(u_{1101} + u_{1010} + u_{1000} + u_{0111} + u_{0101} + u_{0010} + u_{0000}),$$

$$P(A_3^{(34)}) = u_{1110} + u_{1010} + u_{0110} + u_{0010} + \tfrac{1}{2}(u_{1100} + u_{1011} + u_{1000} + u_{0111} + u_{0100} + u_{0011} + u_{0000}).$$

Of course, the quantity $P(A_i)$ still is to be computed by (11.2.3).

The predicted asymptotic values and the observed proportions over the last block of 180 trials are given in columns 5 and 6 of Table 11.1. The agreement between predicted and observed values is fairly good. Further, as noted earlier, several of the observed curves in Figs. 11.2 and 11.3 are still rising or falling over the terminal trials of the experiment; when these observed trends are taken into account, the correspondence between asymptotic predictions and observed proportions is even better.

On the whole, the model gives a satisfactory account of the asymptotic mean data when predictions are based solely on experimentally determined parameter values. The correspondence between theory and data undoubtedly could be improved if ρ' and ρ'' were estimated from the data and used in generating predictions.

Sequential predictions can be obtained by the same techniques used elsewhere in this book. However, we need some new notation. Let $A_{i,n}$ denote response A_i on trial n of the experiment, W_n denote payoff on trial n of the experiment, $D_n^{(ij)}$ denote the presentation of the response pair (A_iA_j), and $C_{i,n}$ denote conditioning state i on trial n of the experiment. Then, as a typical sequential prediction, we have

$$\lim_{n\to\infty} P(A_{1,n+1}D_{n+1}^{(12)}w_nA_{1,n}D_n^{(12)}) = \tfrac{1}{36}\xi_1\{u_{1011} + u_{1010} + u_{1001} + u_{1000} + \tfrac{1}{4}(u_{1110} + u_{1101} + u_{1100}) + [\tfrac{1}{2}\rho' + \tfrac{1}{4}(1-\rho')](u_{0010} + u_{0001} + u_{000})\}.$$

The proof is as follows:

$$P(A_{1,n+1}D_{n+1}^{(12)}w_nA_{1,n}D_n^{(12)}) = \sum_{i,j} P(A_{1,n+1}D_{n+1}^{(12)}C_{i,n+1}W_nA_{1,n}D_n^{(12)}C_{j,n}).$$

In terms of the axioms we may rewrite the sum as

$$\sum_{i,j} P(A_{1,n+1} \mid D_{n+1}^{(12)}C_{i,n+1})P(D_{n+1}^{(12)})P(C_{i,n+1} \mid W_nA_{1,n}D_n^{(12)}C_{j,n}) \cdot\xi_1 P(A_{1,n} \mid D_n^{(12)}C_{j,n})P(D_n^{(12)})P(C_{j,n}).$$

Applying the axioms again, carrying out the summation, and then taking the limit yields the desired result. It should be noted that this model does not yield deterministic predictions of the sort generated by the one-element stimulus-sampling model [e.g., see (1.3.5)].

We have neither tested the sequential predictions of the model nor attempted a goodness-of-fit test because of the computational difficulties in working with a 16-state process. However, the model seems promising enough to warrant more detailed investigation in the future, including its application to presentation sets of more than just two response alternatives.

The good asymptotic predictions of this observing-response model have disturbing implications for the kind of utility analysis considered in this chapter. Clearly, the choice of one of two available responses depends not only on conditioning parameters that are a function of these two responses but also on previous trials in which either of these responses occurred. For example, the actual state of the Markov chain occupied on trial n, and thus the probability of choosing A_1 rather than A_2 on trial n, is affected by the choice of A_1 instead of A_3, say, on trial $n-1$. To take account of this, the utility analysis would have to be in terms of the utility of a *sequence* of responses; in principle there may be no difficulty with such an analysis, but in actual practice this makes the application of utility concepts to choice behavior unfeasible.

Admission of this kind of sequential dependency implies that it is not appropriate to look at the subsequence of trials on which a specific set of possible responses or choices was presented. The sequential analyses of Secs. 11.5 and 11.6 must be replaced by the analysis of sequential data in terms of all trials in order to make an adequate goodness-of-fit test of the observing-response model. It is worth noting that this same difficulty arises in the models of Secs. 11.2 and 11.6 when we remove the restriction that the sets S_{ij} of stimuli are pairwise disjoint.

11.10. Brief Account of a Second Experiment

Since the above sections were written we have completed a more elaborate experiment, the results of which we briefly describe here. The total number of responses was reduced to three, and on each trial a choice was made by the subject from (A_1A_2), (A_1A_3), (A_2A_3), or $(A_1A_2A_3)$. There were 96 subjects and 400 trials. The four response sets just enumerated were presented with equal probability, restricted to 25 of each, for 100 trials.

The main reason for presenting the results, even if in cursory form, is that a very stable asymptotic level of responding was reached in the last 150 trials, in contrast to the first experiment. This result is consistent with the first experiment, for by trial 251 of the second experiment the subjects had already been exposed to each response set on more trials than in the whole of the first experiment, in which each response set occurred on a total of 60 trials.

The inclusion of the triple $(A_1A_2A_3)$ is straightforward in the observing-response model, as we have already remarked in the previous section. For application of the model of Sec. 11.2, it is necessary to postulate another stimulus element (or set of stimulus elements for the multi-element case) associated with the presentation of $(A_1A_2A_3)$. When this element is sampled and a response is chosen and not reinforced, then the probability $1-c'$ of the element's becoming conditioned to some other response is uniformly distributed over the remaining two available responses. Also, for simplicity we take $c=1$ and thereby have a model close to the one-element θ model of earlier chapters, with $\theta = 1 - c'$. Under these assumptions it can be shown (Suppes [1959*b*]) that the asymptotic probability of choosing A_i, given the presentation of $(A_1A_2A_3)$, is

$$P(A_i^{(123)}) = \frac{\Pi_{j\neq i}(1-\xi_j)}{(1-\xi_1)(1-\xi_2)+(1-\xi_1)(1-\xi_3)+(1-\xi_2)(1-\xi_3)}\,.$$

And, by analogy with (11.2.2),

$$P(A_i^{(ij)}) = \frac{1-\xi_j}{1-\xi_i+1-\xi_j}\,.$$

In the present experiment $\xi_1 = \frac{2}{3}$, $\xi_2 = \frac{4}{10}$, and $\xi_3 = \frac{2}{10}$. The 96 subjects were split into two groups; subjects in one group won or lost one cent on each trial, and those in the other group won or lost five cents on each trial. The observed proportions of A_1, A_2, and A_3 responses in successive 40-trial blocks are given in Table 11.2. It is clear from these data that the mean learning curves are quite stable over the last three or four blocks of trials.

TABLE 11.2

OBSERVED PROPORTIONS OF A_i RESPONSES IN SUCCESSIVE 40-TRIAL BLOCKS

Block of 40 Trials	One-Cent Group			Five-Cent Group		
	$P(A_1)$	$P(A_2)$	$P(A_3)$	$P(A_1)$	$P(A_2)$	$P(A_3)$
1	.360	.329	.311	.375	.309	.316
2	.372	.320	.308	.401	.311	.288
3	.394	.310	.296	.418	.304	.278
4	.401	.317	.282	.433	.302	.265
5	.423	.309	.268	.478	.274	.248
6	.424	.307	.269	.456	.294	.250
7	.448	.293	.259	.474	.294	.232
8	.454	.307	.239	.494	.273	.233
9	.482	.298	.220	.496	.267	.237
10	.468	.289	.243	.484	.284	.232

The predicted asymptotes for the model of Sec. 11.2 (as modified above) and for the observing-response model of Sec. 11.9 are given in Table 11.3. Also, the observed proportions of A_1, A_2, and A_3 responses on the last 150 trials are given in the table. Both models provide quite good fits to the observed asymptotic proportions; the data for the one-cent group are particularly well fit

TABLE 11.3

PREDICTED ASYMPTOTES AND OBSERVED PROPORTIONS OVER THE LAST BLOCK OF 150 TRIALS

Asymptotic Value	Observed Proportion		Predicted Proportion	
	One-Cent group	Five-Cent group	c' Model (Sec. 11.2)	Observing-Response Model (Sec. 11.9)
$P(A_1)$	.465	.489	.464	.457
$P(A_2)$	.296	.279	.299	.311
$P(A_3)$	.239	.232	.237	.232

by the simple one-element model. It is to be emphasized that all of these predictions are based solely on the experimentally specified parameter values. In terms of the comparisons presented in Table 11.3 it is not possible to select between the two models or between related multi-element models and linear models. Currently a detailed analysis of the sequential data is being made to provide a basis for selection.

11.11. Summary

In this chapter we derive a utility function for simple choice behavior in terms of stimulus-sampling theory. There are at least two reasons why such a derivation is of interest. First, if an empirically successful theory can be given for deriving utility functions in terms of learning theory, then a new approach is available to the fundamental problem of interpersonal comparisons of utility. Second, the derivation of such a function is a first step toward providing a thoroughgoing behavioristic approach to the policy-making sciences.

The basic theory is presented in Sec. 11.2; the derivation proceeds from the axioms of Sec. 1.2, with a slight modification of the conditioning assumptions. In Sec. 11.3 we describe an experiment designed to test the model. The experimental situation involved a series of 360 trials. There were four responses (A_1, A_2, A_3, A_4), and on each trial two of these responses were made available to the subject. The subject chose one of them; if A_i was chosen, the subject won one cent with probability ξ_i, and no payoff occurred with probability $1-\xi_i$.

In Secs. 11.4 and 11.5 theoretical predictions are compared with the observed data; the correspondence between theory and data is not very good. In Sec. 11.6 an alternative model is examined that also yields a utility function. The experimental utility functions derived from the two models are presented in Sec. 11.7. Both models yield functions that are concave from above. The feature of most interest is that the exact shape of the utility function depends on the role of the conditioning parameters and their estimated values.

Two additional models, for which the same utility derivation is not possible, are presented in Secs. 11.8 and 11.9. The observing-response model of Sec. 11.9 provides a particularly good fit between predicted asymptotic choices and observed proportions. Further, these predictions are completely *a priori*, since they depend only on the experimentally specified parameters ξ_i and do not make use of parameters estimated from the data. Finally in Sec. 11.10 some additional data from a second experiment are reported. Both the simple stimulus-sampling model of Sec. 11.2 and the observing-response model provide very good fits to asymptotic response proportions. More detailed analysis of the sequential data is necessary to discriminate between the models.

12

Extensions and Conclusions

12.1. Introduction

The experiments reported in earlier chapters have centered around interaction situations resembling fairly simple games. To indicate that our basic theory has a broader range of application, we will devote the next three sections of this final chapter to sketching different extensions. The first extension is to the kind of social conformity situations made prominent by Asch [1956], the second to economic oligopolies, and the third to two-person games involving a continuum of responses rather than simply two. The three extensions are independent of one another and can be read in any order.

No experimental data are presented in connection with these extensions, but the theory is set forth in sufficient detail to make the discussion self-contained.

The final section of this chapter offers an over-all review of our results and conclusions.

12.2. Social Conformity

A major thesis of this book is that social behavior may be analyzed without remainder in terms of the concepts of stimulus, response, and reinforcement and the processes of stimulus sampling and conditioning. We choose the Asch-type experiment [1956] on social conformity to illustrate how the verbal responses of one subject may serve as stimuli and reinforcements for other subjects. Asch's technique was to set up groups consisting of one naïve subject and various numbers of other subjects who were actually in collaboration with the experimenter. These collaborators (or stooges, as we shall call them) were instructed beforehand to give unanimously incorrect judgments at certain points during a series of visual judgment tests. The subjects were asked to judge which one of three line segments (projected on a screen before the group) was equal in length to a standard. The naïve subject was the last of the group to give his judgment. The purpose of the experiments was to determine whether his response would be influenced by the responses of the stooges, i.e., whether he would conform to the group's response and thus give an answer contrary to his own judgment. Asch found the effect of group pressure to be an important factor even when the difference between the correct line and the line chosen by the stooges was so large that under normal conditions subjects rarely, if ever, made a mistake. Further, Asch showed that the number of subjects acting as stooges was a significant factor in determining the amount of conformity.

In this section we consider two variants of the Asch experiment. We analyze first the situation in which the stooge or stooges respond after the subject on each trial; and second, the situation in which each trial is initiated by the stooge's response and terminated by the subject's response.

The analysis of the first situation will be similar to that of Suppes and Krasne [1959]. On each of a sequence of trials the subject is asked to make a perceptual judgment—for instance, to indicate by his response which of two line segments is longer or which of two lights flashed first. We restrict ourselves to the comparison of two objects, labeled 1 and 2. Subjects respond A_1 or A_2 to indicate which object has property X as defined by the experimenter. (Property X may be the property of being longer, of occurring first, of having a longer duration, etc.)

After the subject makes his response, a second person, employed as a stooge, makes his response; we treat the stooge's response as a social reinforcer.[1] The experimenter can manipulate the effectiveness of this social reinforcer in terms of both the stooge's authority status and his particular response schedule.

We define the following events for $i = 1, 2$:

$S_{i,n}$ = the event of object i's having property X on trial n;
$A_{i,n}$ = the subject's statement (i.e., response) that object i has property X;
$G_{i,n}$ = social reinforcement of the A_i response; i.e., the event of the stooge's saying that object i has property X;
$I_{i,n}$ = self-reinforcement of the A_i response; i.e., the subject's perception that object i has property X.

For simplicity of notation,

$$\rho_i = P(G_{i,n} \mid S_{i,n}) , \qquad \gamma = P(S_{1,n}) .$$

Further, we postulate two stimulus elements, s_1 and s_2; the element s_i is sampled when the subject perceives object i as having property X. For our first analysis we assume that s_i is sampled with probability 1 on $S_{i,n}$ trials; later, when we consider more difficult discriminations between S_1 and S_2, this restriction is removed. At the start of each trial, s_1 and s_2 are each conditioned to either A_1 or A_2. One of the elements is sampled, and the conditioned response occurs. Then the reinforcing events $I_{i,n}$ and $G_{i,n}$ occur, giving rise to a possible change in the state of conditioning at the start of the next trial. Four states

[1] It is not necessary to assume only one stooge. The same analysis applies for more than one stooge when we require that they all make the same response. The size of the stooge group, of course, would be reflected in the parameter values of the model.

of conditioning are possible; as in (1.3.8), they are denoted simply by the *set* of stimulus elements conditioned to A_1, i.e., the states $\{s_1, s_2\}$, $\{s_1\}$, $\{s_2\}$, and 0.

We assume that on every trial exactly one of the following outcomes occurs, independent of the response actually made by the subject: (1) the social reinforcement G_i is effective; (2) the self-reinforcement I_i is effective; (3) neither is effective. This type of assumption leads to the following general reformulation of the conditioning axiom C2:

AXIOM C2‴. *Let there be m reinforcing events $E^{(1)}, \cdots, E^{(m)}$, of which at most one is effective on a given trial. The probability of effectiveness of $E^{(i)}$ is θ_i; if a stimulus element is sampled on a trial, it becomes conditioned with probability θ_i to the response that is reinforced by $E^{(i)}$.*

It follows from the axiom that $\sum\theta_i \leq 1$. In the present case $m = 2$. A still more general formulation is possible by introducing the generalized conditioning parameters c_{ij} instead of the parameters θ_i, but we shall not explore this alternative. With this reformulation of Axiom C2 a minor change is necessary in

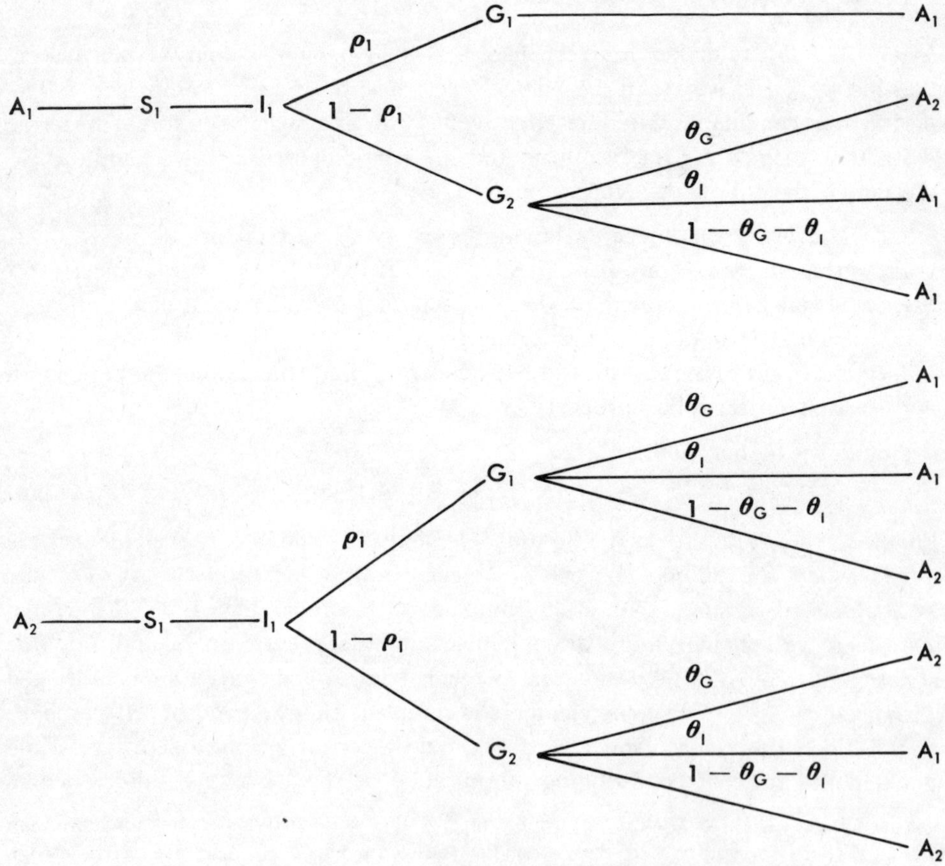

FIGURE 12.1

the independence-of-path axiom C5, but the change is obvious and we shall not explicitly formulate it here.

Rather than use θ_1 and θ_2, we write θ_G and θ_I for the probabilities of the effectiveness of social reinforcement and self-reinforcement, respectively. With the other concepts and the above notations, the construction of the trees for the four states of conditioning is straightforward. However, as in Chapter 6, a considerable simplification can be made by treating the S_1 and S_2 trials separately. The justfication for this procedure rests primarily on the assumption that the conditioning of a stimulus can change only when it is sampled, and in the present formulation s_i is sampled with probability 1 on an S_i trial. When stimuli are always sampled singly, as we are assuming here, the subsequence of trials on which a given stimulus element is sampled forms an independent learning process.

For S_1 trials we then have the two trees of Fig. 12.1. The A_i state represents the conditioning of s_1 to response A_i. From these two trees we obtain the following transition matrix for S_1 trials:

$$\text{(12.2.1)} \qquad \begin{array}{c|cc} & A_1 & A_2 \\ \hline A_1 & 1-\theta_G(1-\rho_1) & \theta_G(1-\rho_1) \\ A_2 & \theta_I+\theta_G\rho_1 & 1-\theta_I-\theta_G\rho_1 \end{array}$$

In similar fashion we may derive the following transition matrix for S_2 trials:

$$\text{(12.2.2)} \qquad \begin{array}{c|cc} & A_1 & A_2 \\ \hline A_1 & 1-\theta_I-\theta_G\rho_2 & \theta_I+\theta_G\rho_2 \\ A_2 & \theta_G(1-\rho_2) & 1-\theta_G(1-\rho_2) \end{array}$$

Solving these two transition matrices, we have for the mean asymptotic conditional probabilities of response

$$\text{(12.2.3)} \qquad P_\infty(A_1 \mid S_1) = \frac{\rho_1 + (\theta_I/\theta_G)}{1+(\theta_I/\theta_G)}, \qquad P_\infty(A_2 \mid S_2) = \frac{\rho_2 + (\theta_I/\theta_G)}{1+(\theta_I/\theta_G)}.$$

As θ_I approaches 1, $P_\infty(A_i \mid S_i)$ also approaches 1; that is, as the self-reinforcer becomes more important, social conformity ceases to be effective. Similarly, as θ_G approaches 1, $P_\infty(A_i \mid S_i)$ approaches ρ_1; in other words, as the social reinforcer becomes more important, social conformity reaches a maximum. It is to be noted that the asymptotes depend only on the ratio θ_I/θ_G of the conditioning parameters. Several psychologically interesting hypotheses can be formulated in terms of this ratio. For example, it is reasonable to suppose that the value of θ_G increases monotonically as the authority status of the stooge (or the number of stooges) increases, and thus the ratio should become smaller with increased authority, indicating the increasing importance of social reinforcement. Further, Crutchfield [1955] has shown that there are large individual differences in the tendency to conform, and that these differences correlate with personality variables such as intelligence, originality, anxiety, and social attitude.

Such inter-subject variability would of course be reflected in the estimates of θ_G and θ_I for individual subjects. Some hypothetical relations between psychological variables and the conditioning parameters θ_G and θ_I, as well as methods of estimation, are presented in Suppes and Krasne [1959] and will not be pursued here.

If the discrimination between S_1 and S_2 is made difficult, it is a natural next step to introduce the psychophysical parameter δ of correctly judging property X.

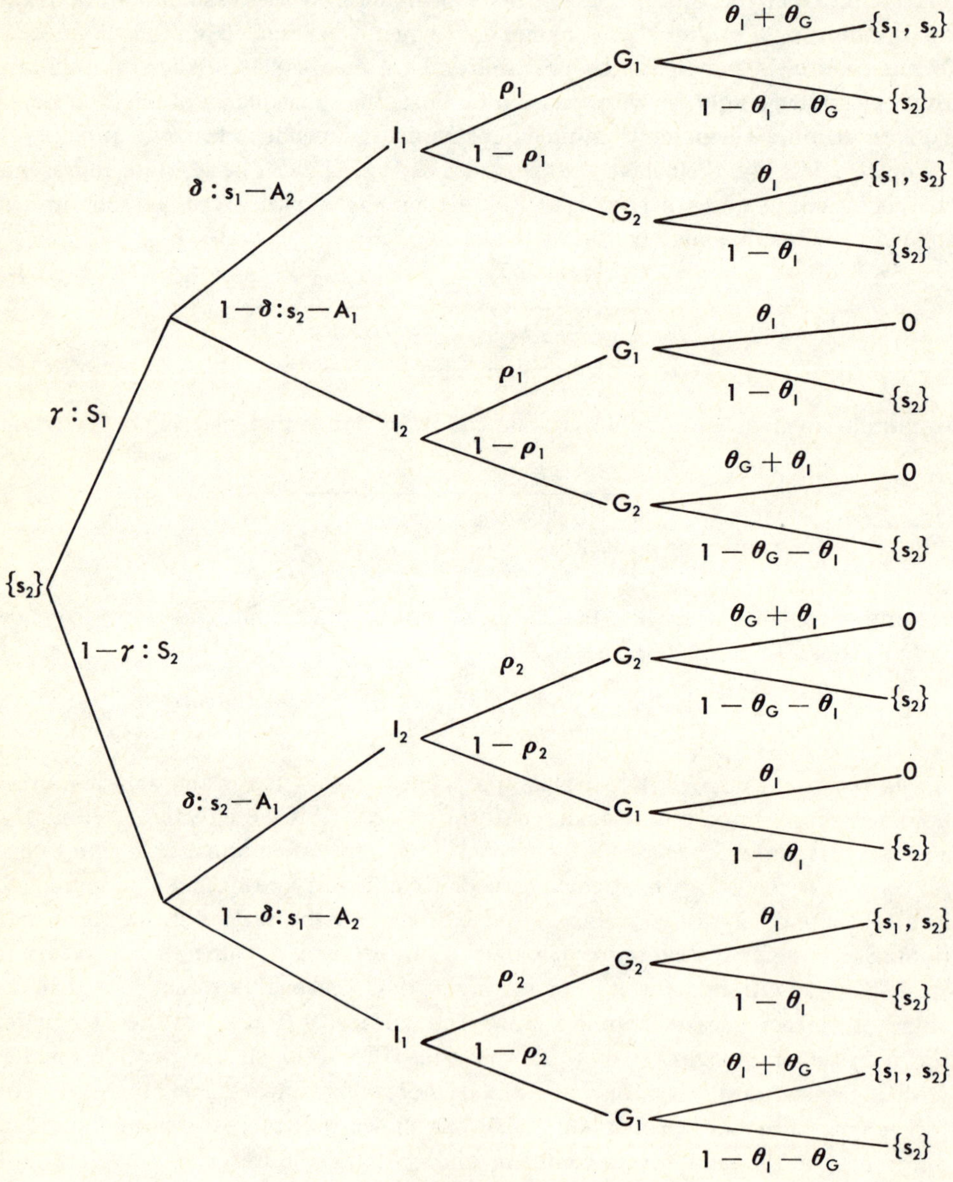

FIGURE 12.2

For example, both Asch [1956] and Flament [1959] have shown that as the psychophysical problem is made more difficult, the amount of conformity increases. Let δ be the probability that a subject makes a correct discrimination when no social pressure is applied and no information is given him about the the correctness of his response.[2] On the assumption that $\delta < 1$, it is not reasonable to separate S_1 and S_2 trials, since the subject on some trials actually confuses them in his perceptual discriminations, independent of questions of social conformity. In other words, when an S_1 trial occurs, the element s_1 is sampled with probability δ, and the element s_2 with probability $1-\delta$; thus when $\delta<1$, it is necessary to revert to the full four-state process. The states of conditioning $\{s_1, s_2\}$, $\{s_1\}$, $\{s_2\}$, and 0 are defined as above—the state is indicated by the set of stimuli conditioned to the A_1 response. The tree for the state $\{s_2\}$ is given in Fig. 12.2. Note that the perceived stimulus s_i determines the self-reinforcing event $I_{i,n}$ and also is the only stimulus whose conditioning can be changed. For example, consider the case in which the stooge makes an A_1 response on an S_1 trial, and assume that the stimulus s_2 is sampled (with probability $1-\delta$). The stooge's response specifies a G_1 event, and the sampling of s_2 is an I_2 event even though we have an S_1 trial. With probability θ_G, s_2 is conditioned to A_1; with probability θ_I it is conditioned to A_2; and with probability $1-\theta_G-\theta_I$ no change in conditioning occurs.

The three analogous trees are similar in form. The resulting transition probabilities are as follows:

(12.2.4)
$$
\begin{aligned}
P(\{s_1, s_2\}|\{s_1, s_2\}) &= 1-\theta_I[\gamma(1-\delta)+(1-\gamma)\delta]-\theta_G[\gamma(1-\rho_1)+(1-\gamma)\rho_2],\\
P(\{s_1\}|\{s_1, s_2\}) &= \gamma(1-\delta)[\theta_I+\theta_G(1-\rho_1)]+(1-\gamma)\delta(\theta_I+\theta_G\rho_2),\\
P(\{s_2\}|\{s_1, s_2\}) &= \gamma\delta(1-\rho_1)\theta_G+(1-\gamma)(1-\delta)\rho_2\theta_G,\\
P(0|\{s_1, s_2\}) &= 0;\\
P(\{s_1, s_2\}|\{s_1\}) &= \gamma(1-\delta)\rho_1\theta_G+(1-\gamma)\delta(1-\rho_2)\theta_G,\\
P(\{s_1\}|\{s_1\}) &= 1-\theta_G\gamma[(1-\delta)\rho_1+\delta(1-\rho_1)]-\theta_G(1-\gamma)[\delta(1-\rho_2)+(1-\delta)\rho_2],\\
P(\{s_2\}|\{s_1\}) &= 0,\\
P(0|\{s_1\}) &= \gamma\delta(1-\rho_1)\theta_G+(1-\gamma)(1-\delta)\rho_2\theta_G;\\
P(\{s_1, s_2\}|\{s_2\}) &= \gamma\delta(\theta_I+\theta_G\rho_1)+(1-\gamma)(1-\delta)[\theta_I+\theta_G(1-\rho_2)],\\
P(\{s_1\}|\{s_2\}) &= 0,\\
P(\{s_2\}|\{s_2\}) &= 1-\theta_I-\theta_G\gamma[\delta\rho_1+(1-\delta)(1-\rho_1)]-\theta_G(1-\gamma)[\delta\rho_2\\
&\quad +(1-\delta)(1-\rho_2)],\\
P(0|\{s_2\}) &= \gamma(1-\delta)[\theta_I+\theta_G(1-\rho_1)]+(1-\gamma)\delta[\theta_I+\theta_G\rho_2];\\
P(\{s_1, s_2\}|0) &= 0,\\
P(\{s_1\}|0) &= \gamma\delta(\theta_I+\theta_G\rho_1)+(1-\gamma)(1-\delta)[\theta_I+\theta_G(1-\rho_2)],\\
P(\{s_2\}|0\} &= \gamma(1-\delta)\rho_1\theta_G+(1-\gamma)\delta(1-\rho_2)\theta_G,\\
P(0|0) &= 1-\theta_I[\gamma\delta+(1-\gamma)(1-\delta)]-\theta_G[\gamma\rho_1+(1-\gamma)(1-\rho_2)].
\end{aligned}
$$

[2] Empirically δ will not be constant but, as an approximation, such an assumption seems reasonable. For related experimental evidence on variability in δ see Verplanck, Collier, and Cotton [1952].

The expressions for the asymptotic probabilities u_i [defined by (1.3.17)] are given in terms of (1.4.4). From an inspection of the states of the chain it is clear that

$$
\begin{aligned}
P_\infty(A_1 \mid S_1) &= u_{\{s_1, s_2\}} + u_{\{s_1\}}\delta + u_{\{s_2\}}(1-\delta)\,, \\
P_\infty(A_2 \mid S_2) &= u_0 + u_{\{s_2\}}\delta + u_{\{s_1\}}(1-\delta)\,.
\end{aligned} \tag{12.2.5}
$$

It can easily be shown that u_i and $P_\infty(A_i \mid S_j)$ are dependent on the subject-determined parameters θ_G, θ_I, and δ as well as on the experimentally specified parameters γ, ρ_1, and ρ_2. More detailed theoretical results can be obtained, but they are not presented, since we do not here apply the model to data. Parenthetically, an interesting psychophysical case arises when $\rho_1 = \rho_2 = 1$; under these conditions the stooge is providing the subject with correct information at the end of each trial. If the experimenter takes the role of the stooge, then the situation is equivalent to a standard psychophysical experiment in which the subject is informed of the correct response after each judgment. The predicted value for $P_\infty(A_i \mid S_i)$ would be given by (12.2.5) for $\rho_1 = \rho_2 = 1$, and it should be noted that the prediction is not necessarily δ but actually a function of $\theta_G, \theta_I, \gamma$, and δ. This result is not contrary to expectation, for recall that δ is obtained when the subject is *not* informed of the correct response after each judgment; a difference between $P_\infty(A_i \mid S_i)$ and δ is theoretically possible and is not contrary to empirical findings.[3]

Our analysis of social conformity is somewhat different when the stooge (or group of stooges) responds first. In this situation, which corresponds closely to Asch's original experiment and to the procedure used by Cohen [1958], it is natural to treat the stooge's response as a stimulus to which the subject makes a discrimination response. This type of approach corresponds to the analyses given in Chapter 6.

Let σ_1 and σ_2 be the stimuli representing the possible responses of the stooge; specifically, if the stooge makes the A_i response at the onset of a trial, then σ_i is available for sampling. The stimuli s_1 and s_2 are defined similarly; if the subject perceives object i as having property X, then s_i is available for sampling. The psychophysical parameter δ plays essentially the same role as before; i.e., when an S_1 trial occurs, s_1 is available for sampling with probability δ and s_2 with probabitity $1-\delta$.

The four stimuli σ_1, σ_2, s_1, and s_2 could each be conditioned to A_1 or A_2, but to simplify the analysis we shall assume that s_1 and s_2 are always conditioned to A_1 and A_2, respectively. Then the states of the process are the four possible states of conditioning of s_i and σ_j. Specifically, the states are denoted $\{s_1, \sigma_1, \sigma_2\}$,

[3] The case where $\rho_1 = \rho_2 = 1$ clearly illustrates one feature of the model that should be made explicit. Assume s_2 is conditioned to A_1. Then when s_2 is sampled, A_1 occurs; but of course the sampling of s_2 implies that the subject "perceives" object 2 (and not object 1) as having property X. At first glance, it may seem unrealistic that a subject should perceive object 2 as having property X and yet respond that object 1 has property X. However, more careful consideration indicates that such a conditioning result is reasonable and in fact not counter to intuition.

$\{s_1, \sigma_1\}$, $\{s_1, \sigma_2\}$ and $\{s_1\}$; they indicate the *set* of stimuli conditioned to A_1. At the start of each trial, σ_i $(i = 1, 2)$ and s_j $(j = 1, 2)$ are available for sampling. We assume that the available σ_i is sampled with probalility ω, and the available s_j with probability $1 - \omega$. The parameter ω is completely subject-controlled and must be estimated from the data; social conformity is characterized by a large value of ω.

The conditioning of s_j is fixed throughout the experiment, and a change in conditioning of σ_i can occur only when σ_i is sampled. We make the following three assumptions: (1) with probability θ_G the social reinforcing event $G_{i,n}$ is effective, and if σ is sampled, it becomes conditioned to A_i; (2) with probability θ_I the self-reinforcing event $I_{j,n}$ is effective, and if σ is sampled, it becomes conditioned to A_j; and (3) with probability $1-\theta_G-\theta_I$ the conditioning of σ remains unchanged.

The tree based on these assumptions for the state $\{s_1, \sigma_2\}$ is shown in Fig. 12.3. The analogous trees can readily be constructed and the appropriate transition matrix given. Again, for convenience we number the states as follows: $1 = \{s_1, \sigma_1, \sigma_2)$, $2 = \{s_1, \sigma_1\}$, $3 = \{s_1, \sigma_2\}$, $4 = \{s_1\}$. It is obvious that the anti-diagonal of the transition matrix for this process is identically 0, and hence the asymptotic probabilities u_i for the conditioning states are given by (1.4.4). Inspection of the states of the chain reveals that

$$P_\infty(A_1 \mid S_1) = (1 - \omega)\delta + \omega[u_1 + \rho_1 u_2 + (1 - \rho_1)u_3] \, ,$$
$$P_\infty(A_2 \mid S_2) = (1 - \omega)\delta + \omega[u_4 + (1 - \rho_2)u_3 + \rho_2 u_2] \, .$$

It is easy to show that u_i and $P_\infty(A_i \mid S_j)$ depend on the subject-determined parameters θ_G, θ_I, ω, and δ as well as on the experimentally specified parameters γ, ρ_1, and ρ_2. The social conformist in this situation is characterized by high values of ω and θ_G. In fact, if $\omega = 1$ and $\theta_G = 1$, the subject exactly imitates the stooge's response pattern. It can also be shown that for fixed values of all parameters except δ, social conformity will increase as δ approaches $\frac{1}{2}$; in other words, as the psychophysical problem becomes more difficult, conformity to the stooge's response tends to increase.

The particular discrimination process we have outlined for analyzing this situation is not the only possible one. Extensive experimentation will be required before we can choose between various formulations. The important point is the demonstration that a detailed analysis of social conformity leading to quantitative predictions is feasible.

It should now be clear how the theory may be extended to include the presence of several stooges, experimental manipulation of the authority variable, or still more complex conditions. In our opinion the two main difficulties hindering rapid development of the theory are not conceptual but computational and experimental. The computational problem is generated by the staggering number of states (and often parameters) needed as the experimental situation becomes more complex; the coalition structures mentioned in Chapter 7 are an example.

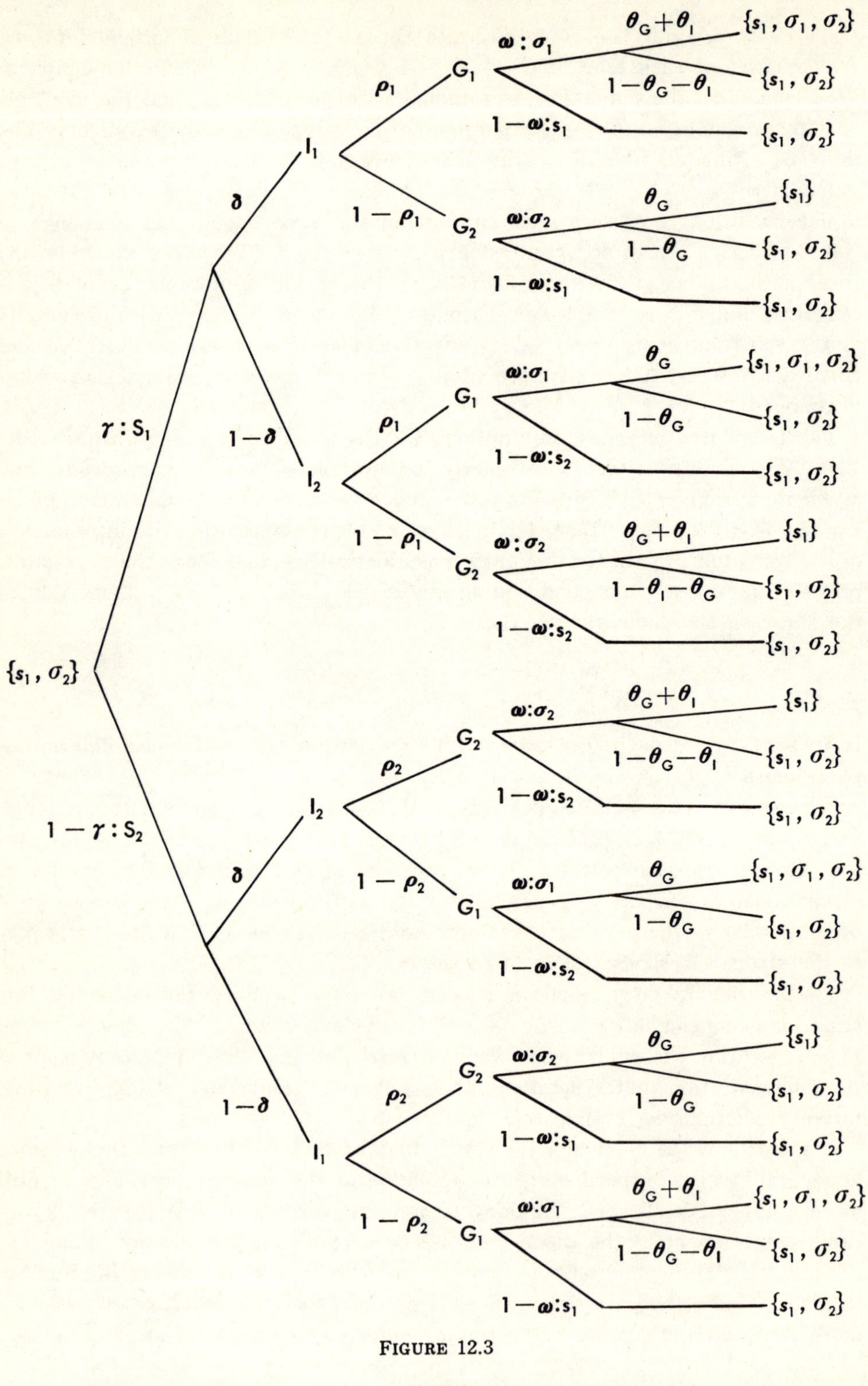
{s1, σ2}
γ : S1
1 − γ : S2
δ
1−δ
δ
1−δ
I1
I2
I2
I1
ρ1
1 − ρ1
ρ1
1 − ρ1
ρ2
1 − ρ2
ρ2
1 − ρ2
G1
G2
G1
G2
G2
G1
G2
G1
ω : σ1
1−ω:s1
ω:σ2
1−ω:s1
ω:σ1
1−ω:s2
ω : σ2
1−ω:s2
ω:σ2
1−ω:s2
ω:σ1
1−ω:s2
ω:σ2
1−ω:s1
ω:σ1
1−ω:s1
θG+θ1
1−θG−θ1
θG
1−θG
θG
1−θG
θG+θ1
1−θ1−θG
θG+θ1
1−θG−θ1
θG
1−θG
θG
1−θG
θG+θ1
1−θG−θ1
{s1, σ1, σ2}
{s1, σ2}
{s1, σ2}
{s1}
{s1, σ2}
{s1, σ2}
{s1, σ1, σ2}
{s1, σ2}
{s1, σ2}
{s1}
{s1, σ2}
{s1, σ2}
{s1}
{s1, σ2}
{s1, σ2}
{s1, σ1, σ2}
{s1, σ2}
{s1, σ2}
{s1}
{s1, σ2}
{s1, σ2}
{s1, σ1, σ2}
{s1, σ2}
{s1, σ2}

FIGURE 12.3

The experimental problem involves exact and clear identification of the stimuli. At present, we may not always be able to identify the relevant stimuli experimentally, for many of the nuances of social communication and activity depend on a long history of verbal usage and interaction. Nevertheless, we are convinced that ultimately such an analysis can be provided.

12.3. Economic Oligopoly

The highly structured, clearly motivated character of simple economic behavior makes it a natural topic for analysis within the framework of this book. It would be desirable to develop the utility theory begun in Chapter 11 into a new theory of consumer demand, but our aims in this section must be more modest. To illustrate briefly the application of our basic theory we have selected a simplified oligopoly situation.

Before describing a particular oligopoly set up, we want to indicate some of the general limitations of our analysis. The most severe one is the unrealistically small number of variables considered. We take no account of technological change, the introduction of competing products, the role of inventories, the cost of entry into the market, the effects of advertising, the liquidity of assets, secular changes in general business conditions, and the like. For this reason it is preferable perhaps to think of our theory as a theory for the experimental simulation of economic markets. In such simulation studies variables not included in the present conceptual framework of the theory can be controlled, as they cannot be in real economic data.

A second limitation is that in the model presented we restrict the firms to offering their products at one of two prices, high (H) and low (L). Relaxation of this restriction is discussed in Sec. 12.4.

The salient feature of our approach to oligopoly is to provide a specific psychological theory of behavior for the decision-makers of the firms that form the oligopoly. The classical economic theories of Cournot, Edgeworth, and more recently Chamberlain, Mrs. Robinson, and others provide no such detailed theory of behavior. Or rather they offer a highly doubtful theory. Mrs. Robinson [1933] p. 6, puts the matter as follows: "If individuals act in an erratic way only statistical methods will serve to discover the laws of economics, and if individuals act in a predictable way, but from a large number of complicated motives, the economist must resign his task to the psychologist. It is the assumption that any individual, in his economic life, will never undertake an action that adds more to his losses than to his gains, and will always undertake an action which adds more to his gains than to his losses, which makes the analysis of value possible."

This theory of rational economic behavior suffers from at least two major weaknesses. First, neither the formulation nor the method of application to situations involving uncertainty and risk is clear. Second, the theory is not dynamic in character; no provision is made for mechanisms of change or adjustment.

The modern approaches to oligopoly via game theory, as exemplified by Shubik [1959], do yield specific dynamic models, but their basic postulates are normative rather than empirical in orientation. So far as we know, there have been no serious attempts to test their empirical adequacy. In fact, the experimental psychologist, accustomed to demanding rigorous empirical tests of a proposed theory of behavior, might be surprised at the economist's apparent lack of concern. However, it is undoubtedly true that economists have a much more difficult task of data collection than do psychologists. In an experimental simulation of an oligopoly it is a trivial matter to run a 200-trial experiment with each trial corresponding to a quarterly decision on prices and production quotas. In contrast, it is very unlikely that economic data exist for a duopoly of fifty years' duration; and even if such data do exist, gross perturbations like major depressions or wars would seriously affect their interpretation.

The difficulty of empirical confirmation suggests that it is not possible to join directly the issue between rationalistic and psychological theories of economic behavior. Without doubt economists can easily offer evidence against our theory of stimulus sampling and conditioning as an empirically adequate theory of oligopoly. But the same objections would hold for the economic theories of oligopoly. They suffer equally from oversimplification and elimination of important variables. The results of an experimental simulation of a symmetric duopoly by Suppes and Carlsmith [1960] indicate that the learning-theory approach to economic behavior may not be as unrealistic as it first seems. We emphasize that the model of oligopoly presented in this section is in certain respects highly artificial, but none of the artificialities is fundamental. The power of the theory should not be judged in terms of the obvious limitations of this particular model.

Turning now to systematic considerations, our procedure is first to describe a one-commodity market with k consumers and N firms or producers, and then to consider the special case of duopoly in some detail. From the theoretical standpoint there is nothing unique about $N=2$. The same behavioral assumptions hold for $N=3$, $N=4$, etc., but the computations rapidly become more tedious.

The market is simply defined in probabilistic terms. We do not begin by postulating an underlying demand function, although our assumptions come to essentially the same thing.[4] Let $\rho_1, \cdots, \rho_N$ be the prices (H or L) set by the N firms. Then for each firm h, the probability $p_h(\rho_1, \cdots, \rho_N)$ is the probability that an arbitrarily selected consumer will purchase one unit of the commodity from firm h when the N firms have set prices $\rho_1, \cdots, \rho_N$. The probability $p_0(\rho_1, \cdots, \rho_N)$ is the probability that the consumer will make no purchase.

[4] The market assumptions made here are somewhat similar to those made by Shubik [1959], p. 81. If we had begun with individual demand schedules, we would have needed a concept similar to that of the contingent demand function which he uses. Actually our market description is objective and avoids any explicit mention of demand functions because our theory does not require them.

Various intuitive relations between the probabilities p_h should hold. We shall not enumerate them here; the important point is that the exact specification of the p_h would vary from market to market. For firms setting the same price, inequalitities in the probabilities p_h reflect the relative power or dominance of the firms in the market, and possibly differences in product quality.

As indicated already, the quarterly decisions about prices correspond to trials. We define certain events in terms of our earlier notation:

$A_{i,n}(h) =$ the event of firm h's setting price i on trial n, where i is H or L
$E_{j,n}(h) =$ the sale of j commodities by firm h on trial n.

We then define conditioning parameters in much the same fashion as for the generalized conditioning model:

(12.3.1) $$c_i(j, h) = P[A_{i,n+1}(h) \mid E_{j,n}(h) A_{i,n}(h)] \, .$$

In words, $c_i(j, h)$ is the probability that when price i has been set and j commodities have been sold by firm h, price i will be set by firm h at the beginning of the next quarter. In psychological language, $c_i(j, h)$ is the probability that when response i is made and reinforcement j received, the single stimulus remains conditioned to i.

Given (12.3.1) and the other assumptions of the generalized conditioning model, it is possible to draw trees for the states $(\rho_1, \cdots, \rho_N)$, where ρ_h is H or L (for $h = 1, \cdots, N$). However, each tree has an enormous number of branches on it and we shall not draw any figures. The transition probabilities are given in terms of the multinomial distribution defined by our market assumptions. For example, the probability that every firm will bid L, having bid L the last quarter (or trial), is

(12.3.2) $$P(L, \cdots, L \mid L, \cdots, L)$$
$$= \sum_{\substack{j_0, \cdots, j_N \\ j_0 + j_1 + \cdots + j_N = k}} c_L(j, 1) \cdots c_L(j, N) \frac{k!}{j_0!\, j_1! \cdots j_N!} \prod_{m=0}^{N} p_m(L, \cdots, L)^{j_m}$$

The number j_0 is, of course, the number of consumers who make no purchases; j_1 the number of consumers who purchase from firm 1 and so on.

Without more specific assumptions about the conditioning parameters $c_i(j, h)$, it is difficult to proceed much further. In general it is reasonable to suppose that the probability $c_i(j, h)$ is monotonically increasing in j, and for mathematical simplicity we assume it is linear in j. (Experimental evidence in Suppes and Carlsmith [1960] supports this linearity assumption.) We express the linearity assumption as follows:

(12.3.3) $$c_L(j, h) = a_h + b_h j \, , \qquad c_H(j, h) = \alpha_h + \beta_h j \, ,$$

where $a_h, \alpha_h \geqq 0$, $b_h, \beta_h > 0$, and $a_h + b_h k$, $\alpha_h + \beta_h k \leqq 1$. The effects of different gross profits for the same gross sale j at the two different prices are absorbed in the coefficients b_h and β_h.

To avoid cumbersome notation we now limit ourselves to the duopoly market

$(N=2)$. For compactness of notation, we define

$$p_h = p_h(L, L)\,, \qquad q_h = p_h(L, H)\,, \qquad r_h = p_h(H, L)\,, \qquad s_h = p_h(H, H)\,.$$

As the specialization of (12.3.2) we have

$$\begin{aligned} P(L, L \mid L, L) &= \sum_{\substack{j_0, j_1, j_2 \\ j_0+j_1+j_2=k}} (a_1 + b_1 j_1)(a_2 + b_2 j_2) \frac{k!}{j_0!\, j_1!\, j_2!} p_0^{j_0} p_1^{j_1} p_2^{j_2}\,, \\ &= a_1 a_2 + a_1 b_2 E(j_2) + a_2 b_1 E(j_1) + b_1 b_2 E(j_1 j_2)\,, \end{aligned}$$

where the expectations are with respect to the given multinomial distribution. Now $E(j_1) = kp_1$, $E(j_2) = kp_2$, and $E(j_1 j_2) = k(k-1)p_1 p_2$. Whence

$$P(L, L \mid L, L) = a_1 a_2 + a_1 b_2 k p_2 + a_2 b_1 k p_1 + b_1 b_2 k(k-1) p_1 p_2\,. \tag{12.3.4}$$

The other fifteen transition probabilities are as follows:

(12.3.5)

$$\begin{aligned}
P(L, H \mid L, L) &= (1-a_2)a_1 - a_1 b_2 k p_2 + (1-a_2) b_1 k p_1 - b_1 b_2 k(k-1) p_1 p_2\,, \\
P(H, L \mid L, L) &= (1-a_1)a_2 + (1-a_1) b_2 k p_2 - a_2 b_1 k p_1 - b_1 b_2 k(k-1) p_1 p_2\,, \\
P(H, H \mid L, L) &= (1-a_1)(1-a_2) - (1-a_1) b_2 k p_2 - (1-a_2) b_1 k p_1 + b_1 b_2 k(k-1) p_1 p_2\,; \\
P(L, L \mid L, H) &= (1-\alpha_2)a_1 - a_1 \beta_2 k q_2 + (1-\alpha_2) b_1 k q_1 - b_1 \beta_2 k(k-1) q_1 q_2\,, \\
P(L, H \mid L, H) &= a_1 \alpha_2 + a_1 \beta_2 k q_2 + \alpha_2 b_1 k q_1 + b_1 \beta_2 k(k-1) q_1 q_2\,, \\
P(H, L \mid L, H) &= (1-a_1)(1-\alpha_2) - (1-a_1) \beta_2 k q_2 - (1-\alpha_2) b_1 k q_1 + b_1 \beta_2 k(k-1) q_1 q_2\,, \\
P(H, H \mid L, H) &= (1-a_1)\alpha_2 + (1-a_1) \beta_2 k q_2 - \alpha_2 b_1 k q_1 - b_1 \beta_2 k(k-1) q_1 q_2\,; \\
P(L, L \mid H, L) &= (1-\alpha_1)a_2 + (1-\alpha_1) b_2 k r_2 - a_2 \beta_1 k r_1 - \beta_1 b_2 k(k-1) r_1 r_2\,, \\
P(L, H \mid H, L) &= (1-\alpha_1)(1-a_2) - (1-\alpha_1) b_2 k r_2 - (1-a_2) \beta_1 k r_1 + \beta_1 b_2 k(k-1) r_1 r_2\,, \\
P(H, L \mid H, L) &= \alpha_1 a_2 + \alpha_1 b_2 k r_2 + a_2 \beta_1 k r_1 + \beta_1 b_2 k(k-1) r_1 r_2\,, \\
P(H, H \mid H, L) &= (1-a_2)\alpha_1 - \alpha_1 b_2 k r_2 + (1-a_2) \beta_1 k r_1 - \beta_1 b_2 k(k-1) r_1 r_2\,; \\
P(L, L \mid H, H) &= (1-\alpha_1)(1-\alpha_2) - (1-\alpha_1) \beta_2 k s_2 - (1-\alpha_2) \beta_1 k s_1 + \beta_1 \beta_2 k(k-1) s_1 s_2\,, \\
P(L, H \mid H, H) &= (1-\alpha_1)\alpha_2 + (1-\alpha_1) \beta_2 k s_2 - \alpha_2 \beta_1 k s_1 - \beta_1 \beta_2 k(k-1) s_1 s_2\,, \\
P(H, L \mid H, H) &= (1-\alpha_2)\alpha_1 - \alpha_1 \beta_2 k s_2 + (1-\alpha_2) \beta_1 k s_1 - \beta_1 \beta_2 k(k-1) s_1 s_2\,, \\
P(H, H \mid H, H) &= \alpha_1 \alpha_2 + \alpha_1 \beta_2 k s_2 + \alpha_2 \beta_1 k s_1 + \beta_1 \beta_2 k(k-1) s_1 s_2\,.
\end{aligned}$$

To gain some insight into this process and the character of the theory of oligopoly proposed here, it is useful to consider some special cases. For the present, we set the coefficients of (12.3.3) equal for both firms, $a = \alpha = 0$, $\beta = 2b = 1/k$. The simplest interpretation of $\beta = 2b$ is that the actual price called H is twice as much as price L.

Simple monopoly. Here we may simply assume that $p_2 = q_2 = r_2 = s_2 = 0$ to represent the fact that firm 2 has no appreciable portion of the market. Then the transition matrix obtained from (12.3.4) and (12.3.5) has the simple form

(12.3.6)

	(L, L)	(L, H)	(H, L)	(H, H)
(L, L)	0	$\frac{1}{2}p_1$	0	$1-\frac{1}{2}p_1$
(L, H)	$\frac{1}{2}q_1$	0	$1-\frac{1}{2}q_1$	0
(H, L)	0	$1-r_1$	0	r_1
(H, H)	$1-s_1$	0	s_1	0

Moreover, if the monopolist (firm 1) always sells to all k consumers regardless of the price set, i.e., if $p_1 = q_1 = r_1 = s_1 = 1$, then the matrix is

$$\left|\begin{array}{cccc} 0 & \frac{1}{2} & 0 & \frac{1}{2} \\ \frac{1}{2} & 0 & \frac{1}{2} & 0 \\ 0 & 0 & 0 & 1 \\ 0 & 0 & 1 & 0 \end{array}\right.$$

which has the proper subset $\{(H, L), (H, H)\}$ as an absorbing set. The appropriate result is then that the monopolist always sets the high price. The other firm alternates between H and L. This latter result, which is not too realistic, stems from the simplifying assumptions made about the values of the conditioning parameters.

On the other hand, suppose the monopoly is such that no customers will make purchases when the high price is set, i.e., $r_1 = s_1 = 0$, but some will at the low price, i.e., $p_1, q_1 \neq 0$. Putting these assumptions into (12.3.6) and solving for the asymptotic responses of the monopolist (firm 1), we find that the probability of his setting the low price is $(p_1 + q_1)/(2p_1 + 2q_1 - p_1q_1)$, which is greater than $\frac{1}{2}$. Thus, owing to the absence of profit in quarters when the high price was in effect, the monopolist (firm 1) would in each subsequent quarter revert to the low price; further, over the long run he would tend to set the low price more frequently.

Symmetric duopoly. We assume the firms are of equal strength. Also for simplicity, let us assume that if at least one firm sets the low price, all k consumers will buy. Thus $p_1 = p_2 = \frac{1}{2}$, and on the basis of symmetry $q_1 = r_2$, and $s_1 = s_2$. To eliminate subscripts, let $p = p_1$, $q = q_1 = r_2$, and $s = s_1$. The transition matrix obtained from (12.3.4) and (12.3.5) may be simplified by taking $k \approx k - 1$ and combining the symmetric states (H, L) and (L, H) to obtain the following 3×3 transition matrix, where state i $(i = 0, 1, 2)$ corresponds to the number of firms setting the high price:

$$\text{(12.3.7)} \qquad \begin{array}{c|ccc} & 0 & 1 & 2 \\ \hline 0 & 1/16 & 6/16 & 9/16 \\ 1 & \frac{1}{2}q(1-q) & 1 - \frac{3}{2}q + q^2 & q(1 - \frac{1}{2}q) \\ 2 & (1-s)^2 & 2s(1-s) & s^2 \end{array}$$

The asymptotic probabilities of the three states may be expressed as follows:

$$u_0 = \frac{32q(1-s)[1 - q + \frac{1}{2}(1-s)]}{87q - 56q^2 + 36s - 48s^2 - 64sq + 16qs^2 + 32q^2s + 12},$$

$$u_1 = \frac{16s(1-s) + [3 - 16s(1-s)]u_0}{12q - 8q^2 + 16s(1-s)}, \qquad u_2 = 1 - u_0 - u_1 .$$

The first thing we notice about u_0, u_1, and u_2 is that whatever the values of q and s (given that $0 \leqq q \leqq 1$ and $0 \leqq s \leqq \frac{1}{2}$), the firms do not, in this model, adopt the monopoly price of H or the competitive price of L with probability 1. The probabilistic equilibrium has $0 < u_0, u_1, u_2 < 1$ for all q and s. This result may be compared with the game-theory analysis in terms of equilibrium points. Setting the low price as 1 and the high price as 2, we obtain the following expected payoff matrix:

	L	H
L	$(\frac{1}{2}k, \frac{1}{2}k)$	$(qk, 2(1-q)k)$
H	$(2(1-q)k, qk)$	$(2sk, 2sk)$

It is completely reasonable to impose the restraint $q > \frac{1}{2} > s$; this means simply that if one firm sets the low price and the other the high price, the low-price firm's share q of the market is greater than $\frac{1}{2}$, whereas when both firms set the high price each firm's share s of the market is less than $\frac{1}{2}$. But this constraint is not sufficient to fix the equilibrium points of the payoff matrix. If $2(1-q) < \frac{1}{2}$, the pair of strategies (L, L) is a pure equilibrium point. If $2s > q$, the pair (H, H) is a pure equilibrium point. The significant thing is that for many intuitively reasonable values of q and s, the game-theory solution yields a non-probabilistic pure strategy, whereas this is not the case for the learning theory. The same result was encountered in Chapters 4 and 5.

The kind of asymptotic probabilistic predictions made by (12.3.7) may be better understood by taking a numerical example. Let $q = 1$ and $s = .1$. This means that the firm which sets the high price when the other firm sets the low price sells no products, and when both firms set the high price each sells to .1 of the consumers. In this case, $u_0 \approx .30$, $u_1 \approx .35$, and $u_2 \approx .35$, whence the probability of either firm's setting the low price is $u_0 + \frac{1}{2}u_1 = .475$. From an economic standpoint this probability may seem low. The expected payoff matrix is

	L	H
L	$(.5k, .5k)$	$(k, 0)$
H	$(0, k)$	$(.2k, .2k)$

which has (L, L) as its unique equilibrium point. We emphasize that the exact quantitative value .475 for $P_\infty(L)$ is a direct result of our rather artificial assumptions about the linear relationships (12.3.3) between sales and conditioning probabilities. The experimental results of Suppes and Carlsmith [1960] suggest that the hypothesis of flatter slopes for (12.3.3) with $a_i, \alpha_i \neq 0$ is empirically more sound. In their study $P_\infty(L)$ is approximately .87 under economic assumptions that are more favorable to setting the high price than the present ones.

If we use values for a, b, α, and β of the order of magnitude suggested by the Suppes and Carlsmith study, the results are quite different. We may interpret these new values as defining a new market. Let $a = .7$, $b = .2/k$, $\alpha = .4$, and $\beta = .5/k$. The inequality $a > \alpha$ is in line with our earlier results for the generalized conditioning model: if the average return from setting the low price is better than from setting the high price, the effect of zero return will not affect the conditioning of the low bid as much.

Taking again $q=1$, $s=.1$, and $k \approx k-1$, we obtain the following 3×3 transition matrix, where [as in (12.3.7)] state i indicates the number of firms setting the high price:

	0	1	2
0	.64	.32	.04
1	.54	.42	.04
2	.30	.50	.20

The asymptotic probabilities of the three states are $u_0 \approx .59$, $u_1 \approx .36$, $u_2 \approx .05$, whence the probability of either firm's setting the low price is .77.

Economists have frequently discussed a duopoly model in which each duopolist acts as if the other's price were going to remain the same as it was in the previous period. Information about the other firm's price can be taken account of in our model in the manner of the discrimination model given in Chapter 7: i.e., the information would be treated as a stimulus to which a differential response is made. In our model of only two prices the Markov process would have 64 states, as is clear from the analysis in Chapter 7. The important point is that our conditioning model does not restrict the use of this information to the simplified assumption of price repetition. The specific dynamics of our discrimination model are described in Chapters 6 and 7. Exactly how this price information would be used in our model by a firm depends on details of the particular market and the behavior of the other firm. In the Suppes-Carlsmith experimental simulation of a duopoly it was found that giving subjects this additional information about their opponent's prices did not significantly affect their behavior. However, this might well not be the case in a multiprice market or when there is a delay in announcing gross profits.

Unfortunately it does not seem feasible to offer economic data in support of our models. It certainly is true that many more or less duopolistic markets do not seem to show the kind of statistical fluctuation in prices that we predict. Many factors can account for this discrepancy. We make no case for the economic realism of the models introduced in this section; they omit too many significant variables. However, we do feel that they suggest a new line of approach, and that with development and elaboration they may well prove more realistic than the classical economic and game-theory models of oligopoly. The central idea is to replace the concept of rational action with a specific psychological theory of behavior.

Methods for removing the severe restriction to two possible prices are suggested in the next section.

12.4. Continuous Games

Up to this point our applications of learning theory to one-person and multiperson studies have been restricted to problems involving two or four possible responses. Nevertheless, it has been clear from the start that the theory can be readily extended to situations involving any finite number of responses, although the computations rapidly become tedious. Matters take a different turn, however, when a continuum of responses is considered. A new concept is needed to yield a psychologically sound and mathematically tractable theory. The concept (introduced in Suppes [1959*a*], [1960]) is that of a *smearing* distribution for the conditioning of a stimulus. The discontinuous assumption that each stimulus is conditioned to exactly one response is replaced with the postulate that the conditioning of each stimulus is smeared over a certain interval of responses. This smearing is represented by a probability distribution.

The following axioms are very similar to those given in Sec. 1.2. In stating the axioms we use x for the response variable and z for the parameter of the smearing distribution $K_s(x; z)$ of any stimulus s. It is assumed that the smearing distributions are a one-parameter family of distributions, with the parameter z of each distribution being its mode.[5]

Conditioning Axioms

CC1. *For each stimulus s there is on every trial a unique smearing distribution $K_s(x; z)$ on the interval $[a, b]$ of possible responses such that* (a) *the distribution $K_s(x; z)$ is determined by its mode z and its variance;* (b) *the variance is constant over trials for a fixed stimulating situation;* (c) *the distribution $K_s(x; z)$ is continuous and piecewise differentiable in both variables.*

CC2. *If a stimulus is sampled on a trial, the mode of its smearing distribution becomes, with probability θ, the point of the response (if any) which is reinforced on that trial; with probability $1-\theta$ the mode remains unchanged.*

CC3. *If no reinforcement occurs on a trial, there is no change in the smearing distributions of sampled stimuli.*

CC4. *Stimuli that are not sampled on a given trial do not change their smearing distributions on that trial.*

CC5. *The probability θ that the mode of the smearing distribution of a sampled stimulus will become the point of the reinforced response is independent of the trial number and the preceding pattern of events.*

Sampling Axioms

CS1. *Exactly one stimulus is sampled on each trial.*

CS2. *Given the set of stimuli available for sampling on a given trial, the probability of sampling a given element is independent of the trial number and the preceding pattern of events.*

Response Axioms

CR1. *If the sampled stimulus s and the mode z of its smearing distribution*

[5] The axioms are the same as those given in Suppes [1960].

are given, then the probability of a response in the interval $[a_1, a_2]$ *is* $K_s(a_2; z) - K_s(a_1; z)$.

CR2. *This probability of response is independent of the trial number and the preceding pattern of events.*

Comparing these axioms with those of Sec. 1.2, we see that in all essentials CC3=C3, CC4=C4, CC5=C5, CS1=S1, and CS2=S2. The modifications in Axioms CC1, CC2, and CR1 are minimal in order to accommodate the continuum of responses. Axiom CR2 is a new independence-of-path assumption required by the probabilistic character of the first response axiom.

It is shown in Suppes [1960] that for simple reinforcement schedules the sequence of random variables that take as values the modes of the smearing distribution of the stimuli constitutes a continuous-state, discrete-time Markov process. For simplicity of exposition we shall assume in this section that there is only one stimulus for each subject. With this restriction, we introduce the following notation for random variables on trial n:

(i) The *response* random variable $\boldsymbol{X}_n$, with values x_n or simply x, distribution R_n, and density r_n;

(ii) the *reinforcement* random variable $\boldsymbol{Y}_n$, with values y_n or y, distribution F_n, and density f_n;

(iii) the *smearing-parameter* random variable $\boldsymbol{Z}_n$ of the stimulus s, with values z_n or z, distribution G_n, and density g_n;

(iv) the *effectiveness-of-conditioning* random variable $\boldsymbol{F}_n$, with value 1 for effectiveness and 0 for non-effectiveness; the probability of value 1 is θ (see Axiom CC2). The three random variables $\boldsymbol{X}_n$, $\boldsymbol{Y}_n$, and $\boldsymbol{Z}_n$ have continuous distributions, whereas $\boldsymbol{F}_n$ has a discrete distribution. We use J_n for the joint distribution of any finite sequence of these random variables, the last of which occurs on trial n, and j_n for the corresponding density. To distinguish the continuous from the discrete random variables, we use the event notation $F_{i,n}$ for values of the effectiveness random variable rather than a lower-case letter.

For the continuous case we apply the same general techniques of proof used in the finite case: first expand, then conditionalize, then apply the axioms, and finally simplify to obtain the desired result. To illustrate, we first derive the expression for the mean response density $r_n(x)$ when reinforcement is determinate, i.e., when exactly one possible response is reinforced on each trial.

THEOREM.

$$r_n(x) = \int_a^b k(x; z) g_n(z)\, dz\,. \tag{12.4.1}$$

PROOF. First expanding, we have

$$r_n(x) = \int_a^b j_n(x, z)\, dz\,;$$

then conditionalizing, we have

$$r_n(x) = \int_a^b j_n(x \mid z) j_n(z)\, dz \,.$$

Now $j_n(z)$ is simply $g_n(z)$. Further, by virtue of Axioms CC1 and CR1, $j_n(x \mid z) = k(x; z)$; and our theorem immediately follows.

For application to two-person continuous games, we prove one important general theorem—a recursion for the density g_n of the smearing parameter z_n. By using the Dirac delta function, we obtain a more algorithmic proof than the one given in Suppes [1960]. Use of the delta function is important for computational purposes in more complicated contexts. We note that if $f(x)$ is a continuous density defined on the interval (a, b) and point a_0 is in this interval, then

$$\int_a^b \delta(x - a_0) f(x)\, dx = f(a_0) \,.$$

THEOREM.

(12.4.2) $$g_{n+1}(z) = (1 - \theta) g_n(z) + \theta f_n(z) \,.$$

PROOF. Expanding, we have

$$g_{n+1}(z) = \sum_i \int_a^b \int_a^b \int_a^b j_{n+1}(z, F_{i,n}, y_n, x_n, z_n)\, dy_n\, dx_n\, dz_n \,;$$

then conditionalizing, we have

$$g_{n+1}(z) = \sum_i \int_a^b \int_a^b \int_a^b j_{n+1}(z \mid F_{i,n}, y_n, x_n, z_n) j_n(F_{i,n} \mid y_n, x_n, z_n) \cdot j_n(y_n \mid x_n, z_n) j_n(x_n \mid z_n) g_n(z_n)\, dy_n\, dx_n\, dz_n \,.$$

Now by Axioms CC2 and CC5, $j_n(F_{1,n} \mid y_n, x_n, z_n) = \theta$, independent of y_n, x_n, and z_n; and correspondingly, $j_n(F_{0,n} \mid y_n, x_n, z_n) = 1 - \theta$. When $i = 1$, we have $z = y_n$ by Axiom CC2, and thus we replace $j_{n+1}(z \mid F_{1,n}, y_n, x_n, z_n)$ by the delta function $\delta(z - y_n)$. When $i = 0$, we have $z = z_n$ (again by virtue of Axiom CC2), and we replace $j_{n+1}(z \mid F_{0,n}, y_n, x_n, z_n)$ by $\delta(z - z_n)$. We now have

$$g_{n+1}(z) = \theta \int_a^b \int_a^b \int_a^b \delta(z - y_n) j_n(y_n \mid x_n, z_n) j_n(x_n \mid z_n) g_n(z_n)\, dy_n\, dx_n\, dz_n + (1 - \theta) \int_a^b \int_a^b \int_a^b \delta(z - z_n) j_n(y_n \mid x_n, z_n) j_n(x_n \mid z_n) g_n(z_n)\, dy_n\, dx_n\, dz_n \,.$$

Integrating out x_n and z_n in the first term and y_n and x_n in the second term, we obtain

$$g_{n+1}(z) = \theta \int_a^b \delta(z - y_n) f_n(y_n)\, dy_n + (1 - \theta) \int_a^b \delta(z - z_n) g_n(z_n)\, dz_n = \theta f_n(z) + (1 - \theta) g_n(z) \,,$$

which is the desired result.

The payoff function for a zero-sum, two-person continuous game may be taken to be a conditional density $f(y \mid x_1, x_2)$, where y is the payoff to the first player, and x_1 and x_2 are the responses of the two players. If the game is non-zero-sum, a payoff function for each player is needed. To stay within the framework of determinate reinforcement, we formulate a non-zero-sum, two-person game as follows. On each play of the game, i.e., on each trial, the players try to hit targets. The players' responses consist of predicting where the target will be on a continuum of possible positions. The position y of player i's target on trial n is given by the conditional density $f^i(y \mid x_1, x_2)$, where x_j is player j's prediction.

In this target game, then, for each player i

$$f_n^i(y) = \int_a^b \int_a^b f^i(y \mid x_1, x_2) r_n(x_1, x_2)\, dx_1\, dx_2 .$$

And expanding further (in terms of the smearing parameters z_1 and z_2), we have, on the basis of Axiom CR2 and (12.4.1),

$$f_n^i(y) = \int_a^b \int_a^b \int_a^b \int_a^b f^i(y \mid x_1, x_2) k_1(x_1 ; z_1) k_2(x_2 ; z_2) g_n(z_1, z_2)\, dx_1\, dx_2\, dz_1\, dz_2 .$$

If we select $f^i(y \mid x_1, x_2)$ so that it may be represented as

$$f^i(y \mid x_1, x_2) = \alpha_0^i h_0^i(y) + \alpha_1^i h_1^i(y \mid x_1) + \alpha_2^i h_2^i(y \mid x_2) ,$$

where $\alpha_j^i \geqq 0$, $\sum_j \alpha_j^i = 1$, and each function h_j^i is a density on $[a, b]$, then we obtain after some simplification

$$f_n^i(y) = \alpha_0^i h_0^i(y) + \alpha_1^i \int_a^b \int_a^b h_1^i(y \mid x_1) k_1(x_1 ; z_1) g_n^1(z_1)\, dx_1\, dz_1 + \alpha_2^i \int_a^b \int_a^b h_2^i(y \mid x_2) k_2(x_2 ; z_2) g_n^2(z_2)\, dx_2\, dz_2 . \tag{12.4.3}$$

Substituting (12.4.3) into (12.4.2) and letting $n \to \infty$, we obtain two simultaneous integral equations for $g_1(z)$ and $g_2(z)$, where now the subscripts on $g(z)$ refer to the player and not the trial number:

$$\begin{aligned} g_1(z) &= \alpha_0 h_0^1(z) + \alpha_1 \int_a^b \int_a^b h_1^1(z \mid x_1) k_1(x_1 ; z_1) g_1(z_1)\, dx_1\, dz_1 \\ &\quad + \alpha_2 \int_a^b \int_a^b h_2^1(z \mid x_2) k_2(x_2 ; z_2) g_2(z_2)\, dx_2\, dz_2 , \\ g_2(z) &= \beta_0 h_0^2(z) + \beta_1 \int_a^b \int_a^b h_1^2(z \mid x_1) k_1(x_1 ; z_1) g_1(z_1)\, dx_1 dz_1 \\ &\quad + \beta_2 \int_a^b \int_a^b h_2^2(z \mid x_2) k_2(x_2 ; z_2) g_2(z_2)\, dx_2\, dz_2 , \end{aligned} \tag{12.4.4}$$

where $\alpha_j = \alpha_j^1$ and $\beta_j = \alpha_j^2$ (for $j = 0, 1, 2$).

What can be done toward solving (12.4.4) for $g_1(z)$ and $g_2(z)$ depends upon the character of the smearing distribution and the reinforcement distributions. To illustrate the possibilities, we solve a particular example.

Two subjects are asked to predict the position of a target on the circumference of a circular screen. Subjects can be told that they are aiming at a common target. The evasion tactics adopted by the target pilot will be a complicated function of both predictions.

We take as the smearing distribution for both subjects

$$k(x; z) = \frac{1}{\pi}\cos^2\left(\frac{x-z}{2}\right), \tag{12.4.5}$$

which is a proper density on the interval $(-\pi, \pi)$ and periodic with period 2π. For the two reinforcement distributions, we take

$$\begin{aligned} f^1(y \mid x_1, x_2) &= \frac{\alpha}{\pi}\cos^2\frac{y}{2} + \frac{1-\alpha}{\pi}\cos^2\left(\frac{x_2-y}{2}\right), \\ f^2(y \mid x_1, x_2) &= \frac{\beta}{\pi}\cos^2\left(\frac{y-\gamma}{2}\right) + \frac{1-\beta}{\pi}\cos^2\left(\frac{x_1-y}{2}\right). \end{aligned} \tag{12.4.6}$$

Each player's reinforcement distribution has a noncontingent component and a contingent component, depending on the response of the other player. In terms of the coefficients of (12.4.4) we have let $\alpha_1 = \beta_2 = 0$, but only to reduce the length of the computations. We have also rotated by γ radians the mean of the second player's noncontingent component in order to avoid a symmetric situation, as may also be done in the final selection of α and β.

The smearing distribution

$$\frac{8}{5\pi}\cos^6\left(\frac{x-z}{2}\right)$$

has more experimental justification (see Suppes and Frankmann [1960]), but (12.4.5) is simpler to work with and will suffice to indicate the method of solution.

Substituting (12.4.5) and (12.4.6) into the two integral equations (12.4.4) and taking $(-\pi, \pi)$ as the response continuum, we obtain

$$\begin{aligned} g_1(z) &= \frac{\alpha}{\pi}\cos^2\frac{z}{2} + \frac{1-\alpha}{\pi^2}\int_{-\pi}^{\pi}\int_{-\pi}^{\pi}\cos^2\left(\frac{x-z}{2}\right)\cos^2\left(\frac{x-z'}{2}\right)g_2(z')\,dx\,dz', \\ g_2(z) &= \frac{\beta}{\pi}\cos^2\left(\frac{z-\gamma}{2}\right) + \frac{1-\beta}{\pi^2}\int_{-\pi}^{\pi}\int_{-\pi}^{\pi}\cos^2\left(\frac{x-z}{2}\right)\cos^2\left(\frac{x-z'}{2}\right)g_1(z')\,dx\,dz'. \end{aligned} \tag{12.4.7}$$

The kernel

$$\cos^2\left(\frac{x-z}{2}\right)\cos^2\left(\frac{x-z'}{2}\right)$$

of the two integral equations is degenerate, i.e., it may be expressed as a finite sum $\sum_n a_n(z)b_n(z')$. Consequently, finding a solution of the equations essentially reduces to an algebraic problem of solving a set of simultaneous linear equations (see Mikhlin [1957], pp. 19–22), as will become clear in the sequel.

Our first step is to rewrite the kernel as

$$\cos^2\left(\frac{x-z}{2}\right)\cos^2\left(\frac{x-z'}{2}\right) = \tfrac{1}{4}(1+\cos x\cos z+\sin x\sin z)(1+\cos x\cos z'+\sin x\sin z').$$

We next integrate with respect to x over the interval $(-\pi, \pi)$:

$$\int_{-\pi}^{\pi} \cos^2\left(\frac{x-z}{2}\right)\cos^2\left(\frac{x-z'}{2}\right)dx = \frac{\pi}{2} + \frac{\pi}{4}\cos z\cos z' + \frac{\pi}{4}\sin z\sin z' .$$

Putting this result in (12.4.7) and simplifying, we obtain

$$\begin{aligned} g_1(z) &= \frac{1}{2\pi} + A_1\cos z + B_1\sin z , \\ g_2(z) &= \frac{1}{2\pi} + A_2\cos z + B_2\sin z , \end{aligned} \tag{12.4.8}$$

where the coefficients A_1, A_2, B_1, and B_2 include the value of the definite integrals containing $g_1(z')$ or $g_2(z')$ in their integrands. Substituting these expressions for $g_1(z)$ and $g_2(z)$ into (12.4.7), we obtain

$$g_1(z) = \frac{1}{2\pi} + \frac{\alpha}{2\pi}\cos z + \frac{1-\alpha}{4\pi}\int_{-\pi}^{\pi}(\cos z\cos z' + \sin z\sin z')\left(\frac{1}{2\pi} + A_2\cos z' + B_2\sin z'\right)dz',$$

$$g_2(z) = \frac{1}{2\pi} + \frac{\beta}{2\pi}\cos\gamma\cos z + \frac{\beta}{2\pi}\sin\gamma\sin z$$
$$+ \frac{1-\beta}{4\pi}\int_{-\pi}^{\pi}(\cos z\cos z' + \sin z\sin z')\left(\frac{1}{2\pi} + A_1\cos z' + B_1\sin z'\right)dz' .$$

Integrating both these equations with respect to z' yields

$$\begin{aligned} g_1(z) &= \frac{1}{2\pi} + \left[\frac{\alpha}{2\pi} + \frac{(1-\alpha)A_2}{4}\right]\cos z + \frac{(1-\alpha)B_2}{4}\sin z , \\ g_2(z) &= \frac{1}{2\pi} + \left[\frac{\beta}{2\pi}\cos\gamma + \frac{(1-\beta)A_1}{4}\right]\cos z + \left[\frac{\beta}{2\pi}\sin\gamma + \frac{(1-\beta)B_1}{4}\right]\sin z . \end{aligned} \tag{12.4.9}$$

Since (12.4.8) and (12.4.9) must hold for all z in the interval $(-\pi, \pi)$, we may equate the two coefficients of $\cos z$ and the two of $\sin z$ in the expressions for $g_1(z)$ and $g_2(z)$. This yields four linear equations in A_1, A_2, B_1, and B_2. From the two equations for $g_1(z)$ we have

$$A_1 = \frac{\alpha}{2\pi} + \frac{(1-\alpha)A_2}{4} , \qquad B_1 = \frac{(1-\alpha)B_2}{4} ;$$

and from the two equations for $g_2(z)$ we obtain

$$A_2 = \frac{\beta}{2\pi}\cos\gamma + \frac{(1-\beta)A_1}{4} , \qquad B_2 = \frac{\beta}{2\pi}\sin\gamma + \frac{(1-\beta)B_1}{4} .$$

Solving these four equations, we find

$$A_1 = \frac{8\alpha + 2(1-\alpha)\beta\cos\gamma}{\pi[16 - (1-\alpha)(1-\beta)]} , \qquad B_1 = \frac{2(1-\alpha)\beta\sin\gamma}{\pi[16 - (1-\alpha)(1-\beta)]}$$

$$A_2 = \frac{8\beta\cos\gamma + 2\alpha(1-\beta)}{\pi[16 - (1-\alpha)(1-\beta)]} , \qquad B_2 = \frac{8\beta\sin\gamma}{\pi[16 - (1-\alpha)(1-\beta)]} .$$

The determination of these coefficients completes the solution of the integral equations for $g_1(z)$ and $g_2(z)$. Since (12.4.1) holds as $n \to \infty$, we may use it finally to determine the asymptotic mean response densities $r_1(x)$ and $r_2(x)$:

$$\begin{aligned} r_1(x) &= \int_{-\pi}^{\pi} k(x; z) g_1(z) dz \\ &= \frac{1}{\pi} \int_{-\pi}^{\pi} \cos^2\left(\frac{x-z}{2}\right)\left(\frac{1}{2\pi} + A_1 \cos z + B_1 \sin z\right) dz \\ &= \frac{1}{2\pi} + \frac{A_1}{2} \cos x + \frac{B_1}{2} \sin x \,. \end{aligned}$$

Similarly,

$$r_2(x) = \frac{1}{2\pi} + \frac{A_2}{2} \cos x + \frac{B_2}{2} \sin x \,.$$

The asymptotic response densities are determined by the numbers α, β, and γ, and the smearing density $k(x; z)$. The numbers α, β, and γ are determined by the experimenter. The smearing density, empirically estimated from other experiments using simple noncontingent reinforcement schedules, can be used to make predictions of $r_1(x)$ and $r_2(x)$ prior to the running of any multiperson experiments. Prediction of sequential dependencies will require the estimation of θ, although again estimates of θ from other experiments on the same apparatus can be used.

The identical expressions for $r_1(x)$ and $r_2(x)$ also hold for the linear model defined in Suppes [1959*a*]. This linear model is an extension of the one defined in Sec. 1.5, just as the model of this section is an extension of the axioms of Sec. 1.2 to the continuous response problem. (These matters are treated in more detail in Suppes and Zinnes [1960].)

For comparison of $r_1(x)$ and $r_2(x)$ with strategies derived from the normative theory of continuous games, some further considerations are necessary. The conditional reinforcement densities $f^i(y \mid x_1, x_2)$ are not payoff functions in the sense of game theory. What is needed is a utility or loss function of the distance between the predicted and actual positions of the target. In the literature of statistical decision theory, the loss function is the negative of the utility function. A loss function commonly assumed by statisticians is a quadratic function of the distance between the predicted and actual observations. The loss functions resulting from the reinforcement densities are, in our particular example, not too interesting from the standpoint of game theory, because the loss function of one player is a function of the response of the other player only. We shall not enter here into detailed consideration of the possible equilibrium points in this non-zero-sum continuous game.

12.5. Conclusions

Our purpose in this book has been to present and test a quantitative theory of learning. We have tried to show how learning theory can be applied in a rigorous manner to social behavior. Emphasis has been placed on testing the consequences of a particular theory, namely, stimulus-sampling theory as described by the axioms of Sec. 1.2. However, for most of the empirical data we have also tested predictions resulting from various modifications of these basic axioms.

We shall review briefly the results for the different models, as well as the general outcome of the many statistical tests for Markov properties.

One-element models. A model with one stimulus element and conditioning parameter θ does not yield realistic predictions concerning the fine structure of the data. However, it does a surprisingly good job of predicting mean response probabilities, which are usually independent of (or only slightly dependent on) the number of stimuli assumed. Thus the one-element model may be viewed as a simple device for computing the grosser predictions of the general theory.

For exploratory work in complex situations, then, we recommend using the one-element model because of the great difficulty of computing almost all quantities for the multi-element models. In advocating this approach we are taking a methodological position with which some scientists do not agree. Our position is in contrast to one which asserts that a model should be discarded once it is clear that certain of its predictions are in error, but this position postulates a closer connection between learning models and the actual behavior of organisms than we are willing to grant exists. Naturally, it is always better to choose from the available models the one that best fits the data, but in the present state of psychological knowledge no single model is clearly superior to all others in every facet of analysis. The one-element model, despite some of its erroneous features, is in our judgment a valuable instrument for the rapid exploration of a wide variety of experimental situations.

The prediction that is most seriously in error for the one-element model is the prediction that $P(A_{i,n+1} \mid E_{i,n}A_{i,n}) = 1$, i.e., that a response which is reinforced will be repeated on the next trial with probability 1. Perhaps the most striking instance of this is to be found in Chapter 10 for Group *Z*. The predicted transition matrix for the sequence of response random variables fits the asymptotic data extraordinarily well. However, when we consider the more detailed process consisting of the sequence of response-reinforcement random variables, the fit is very poor.

One good result for the one-element model is the high degree of stability of the maximum-likelihood estimates of θ. For many of the experiments, $\theta = .6$ is a reasonably good estimate. A review of the estimates is given in Table 12.1.

TABLE 12.1

ESTIMATES OF θ FOR THE ONE-ELEMENT MODEL

Experimental Group		$\hat{\theta}$	Experimental Group		$\hat{\theta}$
Chapter 3:	Group *M*	.64	Chapter 6:	Group *M*	.61
	Group *P*	.59		Group *S*	.59
	Group *S*	.61	Chapter 8:	Group I	.48
Chapter 4:	Group *M*	.72		Group II	.49
	Group *S*	.91		Group III	.52
Chapter 5:	Single Group	.59		Group IV	.53
			Chapter 10:	Group *Z*	.89

The tabulated estimates assume that the two or three interacting subjects have the same θ value; also they are based on the aggregated group data.

Estimates of individual θ values for the one-element model were given in Tables 3.4 and 6.3. In many respects it would have been desirable to carry out more analysis of the data for individual subjects, not only for the one-element model but also for our other models. Unfortunately, the analysis of each subject's protocol often is extremely expensive and time-consuming, and thus a practical balance must be struck between considerations of individual and aggregate data. The point to be emphasized is that the models we have examined are in no sense restricted in applicability to group data.

Multi-element models. In Chapter 3 and various chapters thereafter we have analyzed multi-element models with particular reference to those aspects of the data that are inadequately accounted for by the one-element model. The most serious weakness of the one-element model, i.e., the prediction that $P(A_{i,n+1} \mid E_{i,n}A_{i,n}) = 1$, is not shared by the N-element models for $N \geqq 2$.

Two other desirable features of multi-element models have been emphasized in our discussions. In the first place, the sequence of observable response random variables (or the sequence of response-reinforcement random variables) is not a first-order Markov chain but a chain of infinite order. This means that sequential effects of some length can be adequately reflected in the theory. The sequential analyses of data in Chapter 10 provide evidence for this claim. Second, the transition probabilities are functions of the trial number, which provides a means of accounting for the observed non-stationarity in most of the transition data on the early trials of our experiments.

Unfortunately, in the majority of our experiments detailed evaluation of the multi-element models has been ruled out by the enormous number of algebraic and numerical computations involved. Only in Chapters 3 and 10 have substantial applications been made, and in both cases the results are encouraging. This is particularly true of the results in Sec. 10.6, where the procedure given in Sec. 2.3 for jointly estimating N and θ is applied.

Generalized conditioning models. As far as we know, generalized conditioning models receive their first systematic treatment in this book. They represent a direct generalization of the one-element models with the single conditioning parameter θ. The most impressive accomplishment of the generalized conditioning models is with regard to the results obtained on goodness-of-fit tests. In Table 12.2 we have presented the χ^2 values, the degrees of freedom (after subtraction for the number of estimated parameters), and the approximate significance levels. The experimental groups are independent, and consequently we can combine them for an over-all test of significance. The total χ^2 is 228.8 with 188 degrees of freedom, which has a P value of approximately .043.

These goodness-of-fit results are far better than those obtained for any of our other models. Their better fit obviously results from the larger number of parameters to be estimated from the data and from the removal of various constraints in the one-element θ model. But the goodness of fit is not an arti-

TABLE 12.2
GOODNESS OF FIT OF GENERALIZED CONDITIONING MODELS

Experimental Group		χ^2	d.f.	P
Chapter 3:	Group *M*	13.8	13	.4
	Group *P*	19.3	13	.1
	Group *S*	16.3	13	.25
Chapter 4:	Group *M*	19.1	19	.5
	Group *S*	33.7	28	.1
Chapter 5:	Single Group	24.3	22	.3
Chapter 9:	Group *C*	15.0	16	.5
	Group *E*	16.7	16	.4
Chapter 10:	Group *Z*	28.4	16	.03
	Group *F*	15.0	16	.5
	Group *T*	23.2	16	.1
Total		228.8	188	

fact due to the addition of more parameters, for the degrees of freedom used in determining the significance level take account of the number of estimated parameters.

The estimates of the conditioning parameters c_{ij} for the various applications of the generalized conditioning model clearly indicate that the c_{ij}'s are highly dependent on the particular reinforcement schedule and the amount of reinforcement. We have not yet been able to formulate a satisfactory theory of the elementary process of conditioning that will reflect this dependency between the parameters c_{ij} and reinforcement variables. Perhaps the most promising direction is indicated in Chapter 10; namely, a theory that accounts for the effects of reinforcement in terms of changes in the values of the conditioning parameters over time. To work out such a theory for the possibilities even now envisaged is beyond the scope of this book. We do feel that the goodness-of-fit results recorded in Table 12.2 justify further study of the generalized conditioning model.

General Markov properties. Because of their importance for future theoretical work, we shall briefly review the results of the χ^2 tests of stationarity and order. These tests were made for the experimental groups of Chapters 3, 4, 5, 8, 9, and 10. The stationarity tests for the sequence of response random variables indicated that the first-order transition probabilities were not stationary on the early trials of our experiments but attained constant values over the last 100 or so trials. This result is confirmed by all our tests and supports the class of models that are like the multi-element ones; i.e., models for which the transition probabilities in the observable responses are a function of the trial number and are constant only at asymptote.

Tests of the hypothesis that the sequence of response variables is a chain of zero order against the alternative hypothesis that it is a first-order chain were in all cases highly significant. The results for first-order versus second-order are less consistent. For the Mixed Group of Chapter 3, the single group of Chapter 5, Group IV of Chapter 8, both groups of Chapter 9, and Groups *F* and *T* of Chapter 10 they are significant, indicating at least a two-stage dependency in the response transitions. However, for the Pure and Sure Groups of Chapter 3, both groups of Chapter 4, Groups I, II, and III of Chapter 8, and Group *Z* of Chapter 10 the results of the first-order versus second-order tests were not significant. The conclusion to be drawn is that for some experimental problems a first-order chain in the observable responses may provide a reasonably good approximation.

The results of a first-order versus second-order test in the sequence of response-reinforcement random variables is of special importance for the evaluation on generalized conditioning models. If neither this result nor the result of the goodness-of-fit test is significant, the experimental data are in a statistical sense being fully accounted for by the model. It is of course important to realize that the order test may have varying results for different sequences of random variables; e.g., for response variables as opposed to response-reinforcement variables. As remarked in earlier chapters, the order test for the sequence of response-reinforcement variables was not made, since the number of observations was insufficient with regard to the large number of categories involved. Detailed results in this direction are highly desirable in future investigations.

Game-theory comparisons. We shall not review the analyses given in earlier chapters of the applicability of game-theory notions to the behavior of subjects in the various experimental groups. The results are fairly decisive; there is little if any evidence that game theory is an effective descriptive theory of actual behavior for any of the experiments we performed.

A more stringent comparison of learning-theory and game-theory concepts would be afforded by further analysis of economic markets in terms of the class of models exemplified in Sec. 12.3. It is too soon to decide how favorably the conditioning mechanisms characteristic of learning theory will compare with the maximizing concepts of game theory in their ability to account for the complexities of actual market behavior.

Role of mathematical models in psychology. Perhaps one of the best justifications for the use of models in experimental psychology is that they provide a framework within which the sequential aspects of behavior can be analyzed. An experiment designed only to establish the existence of a relation between two variables, such as response magnitude and amount of reward, ignores the many sequential properties of psychological phenomena. Examination of these properties is an important step forward in studying the fine structure of behavioral data. In our judgment, theories stated only in qualitative terms cannot provide an adequate detailed account of such sequential effects.

The use of mathematical models is virtually synonymous with the construction of a quantitative theory of behavior. From a mathematical standpoint it is logically possible to have a theory of behavior that leads only to qualitative predictions—"There will be a tendency to make this response more often than that one," "This reward will usually be effective," "After many trials the pattern of behavior will become stable," etc. However it is difficult to find in the history of science, let alone in the history of psychology, theories of this sort that have had sustained empirical significance. From the systematic standpoint a theory or model based only on qualitative distinctions leads to a small number of testable predictions. Aristotle's physics and Lewin's topological field theory [1951] are good examples. The absence of precise systematization leads usually to pseudo-derivations from the theory. By "pseudo-derivation" we mean the derivation of a prediction that requires many additional assumptions that are not part of the original theory.

Because psychology is a branch of empirical science, and not of mathematics, some psychologists are not concerned whether the predictions tested by an experiment follow in a logical sense from the basic postulates about behavior. They maintain that the essential thing is the making of the prediction, not its derivation from fundamental theory. The reply to this argument appears obvious. The inability to yield significant predictions without additional *ad hoc* assumptions is an indication that the theory does not provide an objective and fundamental analysis of behavior.

An important function of mathematical models is to clarify this aspect of a theory. As we have seen in previous chapters, many different models can stem from the same fundamental theory. Even when we restrict ourselves to considering a single stimulus element, the model with conditioning parameter θ is different from the generalized conditioning model; and given the parameter θ, the model defined by the sequence of response random variables is, strictly speaking, different from the model defined by the sequence of response-reinforcement random variables. The important factor is whether the theory will yield at least one well-defined model in a non-arbitrary manner. The attempt to specify a model will in itself require an exact characterization of the theory and will often reveal unstated assumptions.

Of course, we do not believe that the use of models will in itself solve the significant problems of experimental psychology. A quantitative model is defined by any numerical equation. Classical psychophysics is replete with numerical relationships, yet its limitation as a scientific discipline has been apparent for many years. What it lacks is a theory of the elementary process by which the organism makes psychophysical judgments. Fortunately, in the recent work of Luce [1959] and others the elements of such a theory are being constructed.

The importance of constructing a theory of the elementary process that leads to choice behavior was emphasized in Chapter 1. The fact that stimulus-sampling theory provides more details of this process than the linear model of Sec. 1.5 was a primary reason for giving a secondary role to linear models in this book.

The physiologically oriented psychologist will be quick to point out that stimulus-sampling theory as we have formulated it is a highly schematized version of the elementary process of organism activity in choice behavior. The difficulties of introducing neurophysiological variables in behavior theory were mentioned early in the book. But there are additional behavioral variables with direct physiological significance that we have not considered. The most important is probably that of latency. In all the experiments we have reported, important information was lost because we did not measure response times. Future theoretical developments in this direction are much to be desired.

A second variable of the same nature is that of an observing or orienting response. The single application of this concept in the final section of Chapter 11 yielded good empirical predictions in comparison with the structurally simpler stimulus-sampling models derived from the axioms of Sec. 1.2. All indications point to fruitful applications of the observing-response concept in multiperson interactions. Once again, at present the major difficulty in such applications is computational rather than theoretical.

The models considered in this book are thoroughly probabilistic. This is a property they share with most of the mathematical models now being investigated in the behavioral sciences. As our understanding of neurophysiology increases, it is possible that more deterministic models will be constructed, but probabilistic models seem likely to play a major role for a long time to come.

Finally, we remark that we have avoided a technical dissussion of the general nature of models and theories. A statement of our views on these matters is to be found in the final chapter of Suppes [1957]. The words "model" and "mathematical model" are used in a variety of related senses by behavioral scientists and mathematicians. Often the most reasonable interpretation is that the model is the set of mathematically formulated postulates that express in precise form the intuitive notions of the relevant psychological theory. Despite the dominance of this usage in the literature of the behavioral sciences, we prefer to use the more precise concept of model that has been adopted by mathematical logicians. Roughly speaking, a model is an abstract object that satisfies a theory. A theory given in axiomatic form can, if one chooses, be identified with its set of axioms. It is not to the point to enter into technical details here. What is important in our opinion is that models have a role to play whenever theory is constructed. From our standpoint, if a theory has systematic content and is not simply a vague collection of heuristic ideas, then there exist models that satisfy the theory, and it is up to the experimentalist to determine whether these models provide an adequate analysis of behavioral phenomena. We believe that the role of models in psychology is not really separate from the role of systematic theorizing. We have intended this book to contribute to systematic theory in psychology, as well is to the development of mathematical models.

Bibliography

ANDERSON, N. H. 1960. Effect of first-order conditional probability in a two-choice learning situation. *J. Exptl. Psychol.*, **59**, 73–93.

——— and GRANT, D. A. 1957. A test of a statistical learning theory model for two-choice behavior with double stimulus events. *J. Exptl. Psychol.*, **54**, 305–16.

——— and HOVLAND, C. I. 1957. The representation of order effects in communication research. Appendix in *The Order of Presentation in Persuasion*, C. I. Hovland, *et al.*, eds. New Haven, Conn.: Yale Univ. Press.

ANDERSON, T. W., and GOODMAN, L. A. 1957. Statistical inference about Markov chains. *Ann. Math. Stat.*, **28**, 89–110.

ARROW, K. J. 1951. *Social Choice and Individual Values.* (Cowles Commission Monograph 12.) New York: Wiley.

ASCH, S. E. 1956. Studies of independence and conformity: I. A minority of one against a unanimous majority. *Psychological Monograph*, No. 9.

ATKINSON, R. C. 1956. An analysis of the effect of non-reinforced trials in terms of statistical learning theory. *J. Exptl. Psychol.*, **52**, 28–32.

———. 1957. A stochastic model for rote serial learning. *Psychometrika*, **22**, 87–96.

———. 1958. A Markov model for discrimination learning. *Psychometrika*, **23**, 309–22.

———. 1959. The Observing Response in Discrimination Learning. Technical Report No. 4, Contract Nonr 233(58), Department of Psychology, University of California, Los Angeles, (To be published in *Journal of Experimental Psychology.*)

———. 1960*a*. A theory of stimulus discrimination learning. Chap. 15 in *Mathematical Methods in the Social Sciences*, K. J. Arrow, S. Karlin, and P. Suppes, eds. Stanford, Calif.: Stanford Univ. Press.

———. 1960*b*. A Generalization of Stimulus Sampling Theory. Technical Report No. 29, Contract Nonr 225(17), Institute for Mathematical Studies in the Social Sciences, Applied Mathematics and Statistics Laboratories, Stanford University.

———, BOGARTZ, W., and TURNER, R. 1959. Discrimination learning with probabilistic reinforcement schedules. *J. Exptl. Psychol.*, **57**, 349–50.

——— and SOMMER, G. 1960. Decision making by children as a function of amount of reinforcement. *Psychol. Rep.*, **6**, 299–306.

——— and SUPPES, P. 1958. An analysis of two-person game situations in terms of statistical learning theory. *J. Exptl. Psychol.*, **55**, 369–78.

——— and ——— 1959. Applications of a Markov model to two-person noncooperative games. Chap. 3 in *Studies in Mathematical Learning Theory*, R. R. Bush and W. K. Estes, eds. Stanford, Calif.: Stanford Univ. Press.

AUDLEY, R. J. 1957. A stochastic description of the learning behavior of an individual subject. *Quart. J. Exptl. Psychol.*, **9**, 12–20.

———. 1958. The inclusion of response times within a stochastic description of the learning behaviour of individual subjects. *Psychometrika*, **23**, 25–31.

———. 1960. A stochastic model for individual choice behavior. *Psychol. Rev.*, **67**, 1–15.

BALES, R. F. 1950. A set of categories for the analysis of small group interaction. *Amer. Sociol. Rev.*, **15**, 257–63.

———, STRODTBECK, F. L., MILLS, T. M., and ROSEBOROUGH, M. E. 1951. Channels of communication in small groups. *Amer. Sociol. Rev.*, **16**, 461–68.

BARTLETT, M. S. 1951. The frequency goodness of fit test for probability chains. *Proc. Cambridge Phil. Soc.*, **47**, 86–95.

———. 1955. *Introduction to Stochastic Processes.* Cambridge: Cambridge Univ. Press.

BAVELAS, A. 1950. Communication patterns in task oriented groups. *J. Acoust. Soc. Amer.*, **22**, 725–30.

BERGSON, A. 1938. A reformulation of certain aspects of welfare economics. *Quart. J. Econ.*, **52**, 310–34.

BLACKWELL, D., and GIRSHICK, M. A. 1954. *Theory of Games and Statistical Decisions.* New York: Wiley.

BOWER, G. H. 1959. Choice-point behavior. Chap. 6 in *Studies in Mathematical Learning Theory*, R. R. Bush and W. K. Estes, eds. Stanford, Calif.: Stanford Univ. Press.

BRADLEY, R. A. 1954*a*. Rank analysis of incomplete block designs, II. Additional tables for the method of paired comparisons. *Biometrika*, **41**, 502–37.

———. 1954*b*. Incomplete block rank analysis: On the appropriateness of the model for a method of paired comparisons. *Biometrics*, **10**, 375–90.

———. 1955. Rank analysis of incomplete block designs, III. Some large-sample results on estimation and power for a method of paired comparisons. *Biometrika*, **42**, 450–70.

——— and TERRY, M. E. 1952. Rank analysis of incomplete block designs, I. The method of paired comparisons. *Biometrika*, **39**, 324–45.

BRAND, H., WOOD, P. J., and SAKODA, J. M. 1956. Anticipation of reward as a function of partial reinforcement. *J. Exptl. Psychol.*, **52**, 18–22.

BRODY, A. L. 1957. Statistical learning theory applied to an instrumental avoidance situation. *J. Exptl. Psychol.*, **54**, 240–45.

BRUNER, J. S., GOODNOW, J. J., and AUSTIN, G. A. 1956. *A Study of Thinking.* New York: Wiley.

BURKE, C. J. 1959. Applications of a linear model to two-person interactions. Chap. 9 in *Studies in Mathematical Learning Theory*, R. R. Bush and W. K. Estes, eds. Stanford, Calif.: Stanford Univ. Press.

———. 1960. Some two-person interactions. Chap. 16 in *Mathematical Methods in the Social Sciences*, K. J. Arrow, S. Karlin, and P. Suppes, eds. Stanford, Calif.: Stanford Univ. Press.

——— and ESTES, W. K. 1957. A component model for stimulus variables in discrimination learning. *Psychometrika*, **22**, 133–45.

———, ———, and HELLYER, S. 1954. Rate of verbal conditioning in relation to stimulus variability. *J. Exptl. Psychol.*, **48**, 153–61.

BUSH, R. R., and MOSTELLER, F. 1951. A model for stimulus generalization and discrimination. *Psychol. Rev.*, **58**, 413–23.

——— and ———. 1955. *Stochastic Models for Learning.* New York: Wiley.

CERVIN, V. B. 1957. Relationship of ascendant submissive behavior in dyadic groups of human subjects to their emotional responsiveness. *J. Abnor. and Soc. Psychol.*, **54**, 241–49.

———. 1958. Preliminary Test of a Mathematical Model Conceptualizing the Persuasion Process Taking Place in Debates Between Two Initially Disagreeing Subjects. (Paper presented at the meeting of the American Psychological Association in Washington, D.C., August 28, 1958.)

CERVIN, V. B. 1959. Preliminary test of a mathematical model conceptualizing the persuasion process in debates between two initially disagreeing subjects. *Psychol. Rep.*, **5**, 404.

CHRISTIE, L. S., LUCE, R. D., and MACY, J., JR. 1952. Communication and Learning in Task Oriented Groups. Massachusetts Institute of Technology RLE Report.

COHEN, B. P. 1958. A probability model for conformity. *Sociometry*, **21**, 69–81.

COLEMAN, J. S. 1954. A Survey of Mathematical Models of Small Group Behavior. Technical Report No. 10, Contract No. NR 042–115, Bureau of Applied Social Research, New York.

CRUTCHFIELD, R. S. 1955. Conformity and character. *Amer. Psychol.*, **10**, 191–98.

DAVIDSON, D., and MARSCHAK, J. 1959. Experimental tests of a stochastic decision theory. Chap. 13 in *Measurement: Definition and Theories*, C. W. Churchman and Ph. Ratoosh, eds. New York: Wiley.

DETAMBEL, M. H. 1955. A test of a model for multiple-choice behavior. *J. Exptl. Psychol.*, **49**, 97–104.

ESTES, W. K. 1950. Toward a statistical theory of learning. *Psychol. Rev.*, **57**, 94–107.

———. 1954. Individual behavior in uncertain situations: An interpretation in terms of statistical association theory. Chap. 9 in *Decision Processes*, R. M. Thrall, C. H. Coombs, and R. L. Davis, eds. New York: Wiley.

———. 1957*a*. Theory of learning with constant, variable, or contingent probabilities of reinforcement. *Psychometrika*, **22**, 113–32.

———. 1957*b*. Of models and men. *Amer. Psychol.*, **12**, 609–17.

———. 1959. Component and pattern models with Markovian interpretations. Chap. 1 in *Studies in Mathematical Learning Theory*, R. R. Bush and W. K. Estes, eds. Stanford, Calif.: Stanford Univ. Press.

———. 1960. A random-walk model for choice behavior. Chap. 18 in *Mathematical Methods in the Social Sciences*, K. J. Arrow, S. Karlin, and P. Suppes, eds. Stanford, Calif.: Stanford, Univ. Press.

——— and BURKE, C. J. 1953. A theory of stimulus variability in learning. *Psychol. Rev.*, **60**, 276–86.

——— and ———. 1955. Application of a statistical model to simple discrimination learning in human subjects. *J. Exptl. Psychol.*, **50**, 81–88.

———, ———, ATKINSON, R. C., and FRANKMANN, J. P. 1957. Probabilistic discrimination learning. *J. Exptl. Psychol.*, **54**, 233–39.

——— and STRAUGHAN, J. H. 1954. Analysis of a verbal conditioning situation in terms of statistical learning theory. *J. Exptl. Psychol.*, **47**, 225–34.

——— and SUPPES, P. 1959. Foundations of linear models. Chap. 8 in *Studies in Mathematical Learning Theory*, R. R. Bush and W. K. Estes, eds. Stanford, Calif.: Stanford Univ. Press.

——— and ———. 1959. Foundations of Statistical Learning Theory, II. The Stimulus Sampling Model for Simple Learning. Technical Report No. 26, Contract Nonr 225(17), Institute for Mathematical Studies in the Social Sciences, Applied Mathematics and Statistics Laboratories, Stanford University.

FELLER, W. 1957. *An Introduction to Probability and Its Applications*. 2d ed. New York: Wiley.

FESTINGER, L. 1950. Informal social communication. *Psychol. Rev.*, **57**, 271–82.

FLAMENT, C. 1958. Influence sociale et perception. *L'Année Psychol.*, **58**, 377–400.

———. 1959. Modèle stratégique des processus d'influence sociale sur les jugements perceptifs. *Psychol. française*, IV, 91–101.

FLOOD, M. M. 1954*a*. A stochastic model for social interaction. *Trans. N.Y. Acad. Sci.*, **16**, 202–5.

———. 1954*b*. Game-learning theory and some decision-making experiments. Chap. 10 in *Decision Processes*, R. M. Thrall, C. H. Coombs, and R. L. Davis, eds. New York: Wiley.

FOA, U. G., and ZACKS, S. 1959. A Stochastic Facet Theory of Social Interaction in the Dyad. Tech. Note No. 1, Contract No. AF 61 (052)–121, The Israel Inst. Appl. Soc. Res., Jerusalem, Israel.

FRÉCHET, M. 1938. *Recherches théoriques modernes sur le calcul des probabilités.* Vol. 2, Paris: Gauthier-Villars.

GALANTER, E. H., and SMITH, W. A. S. 1958. Some experiments on a simple thought-problem. *Amer. J. Psychol.*, **71**, 359–66.

GOODNOW, J. J 1955. Response-sequences in a pair of two-choice probability situations. *Amer. J. Psychol.*, **68**, 624–30.

——— and PETTIGREW, T. F. 1955. Effect of prior patterns of experience upon strategies and learning sets. *J. Exptl. Psychol.*, **49**, 381–89.

———, RUBENSTEIN, I., and LUBIN, A. 1957. Response to Changing Patterns of Events. Memorandum 4: Problem-solving project. Walter Reed Army Institute of Research, Washington, D.C.

GRANT, D. A., HAKE, H. W., and HORNSETH, J. P. 1951. Acquisition and extinction of a verbal conditioned response with differing percentages of reinforcement. *J. Exptl. Psychol.*, **42**, 1–5.

GREEN, E. J. 1958. A simplified model for stimulus discrimination. *Psychol. Rev.*, **65**, 56–63.

GUTHRIE, E. R. 1952. *The Psychology of Learning*. Rev. ed. New York: Harper.

HALL, R. L., and ROSENBERG, S. 1957. An S-R analysis of social interaction. *Amer. Psychol.*, **12**, 370 (abstr.).

HAYS, D. G., and BUSH, R. R. 1954. A study of group action. *Amer. Sociol. Rev.*, **19**, 693–701.

HICKS, J. R. 1939. The foundations of welfare economics. *Econ. J.*, **49**, 696–712.

HOEL, P. G. 1954. A test for Markoff chains. *Biometrika*, **41**, 430–33.

HOMANS, G. C. 1950. *The Human Group*. New York: Harcourt, Brace.

HOTELLING, H. 1938. The general welfare in relation to problems of taxation and of railway and utility rates. *Econometrica*, **6**, 242–69.

HULL, C. L. 1943. *Principles of Behavior: An Introduction to Behavior Theory*. New York: Appleton-Century-Crofts.

HUMPHREYS, L. G. 1939. Acquisition and extinction of verbal expectations in a situation analogous to conditioning. *J. Exptl. Psychol.*, **25**, 294–301.

JARVIK, M. E. 1951. Probability learning and a negative recency effect in the serial anticipation of alternative symbols. *J. Exptl. Psychol.*, **41**, 291–97.

KALISCH, C., MILNOR, J. W., NASH, J., and NERING, E. D. 1954. Some experimental *n*-person games. Chap. 19 in *Decision Processes*, R. M. Thrall, C. H. Coombs, and R. L. Davis, eds. New York: Wiley.

KANAREFF, F. T., and LANZETTA, J. T. 1958. The acquisition of imitative and opposition responses under two conditions of instruction-induced set. *J. Exptl. Psychol.*, **56**, 516–28.

KEMENY, J. G., and SNELL, J. L. 1957. Markov processes in learning theory. *Psychometrika*, **22**, 221–30.

——— and ———. 1960. *Finite Markov Chains*. Princeton, N.J.: Van Nostrand.

LABERGE, DAVID. 1959*a*. A model with neutral elements. Chap. 2 in *Studies in Mathematical Learning Theory*, R. R. Bush and W. K. Estes, eds. Stanford, Calif.: Stanford Univ. Press.

———. 1959*b*. The effect of preliminary trials on rate of conditioning in a simple prediction situation. *J. Exptl. Psychol.*, **57**, 20–24.

LAMPERTI, J., and SUPPES, P. 1959. Chains of infinite order and their applications to learning theory. *Pacific J. Math.*, **9**, 739–54.

LAZARSFELD, P. F., ed. 1954. *Mathematical Thinking in the Social Sciences*. Glencoe, Ill.: The Free Press.

LEWIN, K. 1936. *Principles of Topological Psychology.* New York: McGraw-Hill.

LUCE, R. D. 1959. *Individual Choice Behavior: A Theoretical Analysis.* New York: Wiley.

——— and RAIFFA, H. 1957. *Games and Decisions.* New York: Wiley.

MCKINSEY, J. C. C. 1952. *Introduction to the Theory of Games.* New York: McGraw-Hill.

MIKHLIN, S. G. 1957. *Integral Equations and Their Applications to Certain Problems in Mechanics, Mathematical Physics and Technology,* A. H. Armstrong, trans. New York: Pergamon Press.

MILLER, G. A. 1952. Finite Markov processes in psychology. *Psychometrika,* **17**, 149–68.

MILLER, N. E., and DOLLARD, J. 1941. *Social Learning and Imitation.* New Haven, Conn.: Yale Univ. Press.

MUELLER, C. G. 1950. Theoretical relationships among some measures of conditioning. *Proc. Nat. Acad. Sci.,* **36**, 123–30.

NASH, J. F. 1950. The bargaining problem. *Econometrica,* **18**, 155–62.

———. 1951. Non-cooperative games. *Ann. Math.,* **54**, 286–95.

NEIMARK, E. D. 1956. Effects of type of non-reinforcement and number of alternative responses in two verbal conditioning situations. *J. Exptl. Psychol.,* **52**, 209–20.

——— and ROSENBERG, S. 1959. The effect of "social" discriminative cues on probability learning. *J. Exptl. Psychol.,* **58**, 302–11.

NEWELL, A., SHAW, J. C., and SIMON, H. A. 1958*a*. The Processes of Creative Thinking. (Paper presented at a Symposium on Creative Thinking, University of Colorado, Boulder, Colorado, May 16, 1958.)

———, ———, and ———. 1958*b*. Elements of a theory of human problem solving. *Psychol. Rev.,* **65**, 151–66.

PILLAI, K. C. S., and RAMACHANDRAN, K. V. 1954. On the distribution of the ratio of the *i*th observation in an ordered sample from a normal population to an independent estimate of the standard deviation. *Ann. Math. Stat.,* **25**, 565–72.

POPPER, J., and ATKINSON, R. C. 1958. Discrimination learning in a verbal conditioning situation. *J. Exptl. Psychol.,* **56**, 21–26.

RASHEVSKY, N. 1951. *Mathematical Biology of Social Behavior.* Chicago: Univ. Chicago Press.

RATOOSH, PH. 1958. Decision Making under Conditions of Risk. (Paper presented at fall meetings of the Ohio Psychological Association, 1958.)

RESTLE, F. 1955*a*. A theory of discrimination learning. *Psychol. Rev.,* **62**, 11–19.

———. 1955*b*. Axioms of a theory of discrimination learning. *Psychometrika,* **20**, 201–8.

———. 1957. A theory of selective learning with probable reinforcements. *Psychol. Rev.,* **64**, 182–91.

ROBINSON, JOAN. 1933. *The Economics of Imperfect Competition.* London: Macmillan.

ROSS, B. M., and LEVY, N. 1958. Patterned predictions of chance events by children and adults. *Psychol. Rep.,* **4**, 87–124.

SAVAGE, L. J. 1957. The theory of statistical decision. *J. Amer. Stat. Assoc.,* **48**, 238–48.

SCODEL, A., RATOOSH, PH., and MINAS, J. S. 1959. Some personality correlates of decision making under conditions of risk. *Behavioral Sciences,* **4**, 19–28.

SHUBIK, M. 1959. *Strategy and Market Structure.* New York: Wiley.

SHWARTZ, N. 1953. An Experimental Study of Imitation: The Effects of Reward and Age. Senior honors thesis, Radcliffe College.

SIDOWSKI, J. B. 1957. Reward and punishment in a minimal social situation. *J. Exptl. Psychol.,* **54**, 318–26.

———. 1959. Reinforcement and Punishment in a Social Situation. (Paper presented at spring meetings of Western Psychological Association, 1959.)

———, WYCKOFF, L. B., and TABORY, L. 1956. The influence of reinforcement and punishment in a minimal social situation. *J. Abnor. and Soc. Psychol.,* **52**, 115–19.

SIEGEL, S. 1959. Theoretical models of choice and strategy behavior: Stable state behavior in the two-choice uncertain outcome situation. *Psychometrika,* **24**, 303–16.

SIMON, H. A. 1952. A formal theory of interaction in social groups. *Amer. Sociol. Rev.*, **17**, 202–11.

———. 1956. A comparison of game theory and learning theory. *Psychometrika*, **21**, 267–72.

———. 1957. *Models of Man.* New York: Wiley.

——— and GUETZKOW, H. 1955. A model of short and long run mechanisms involved in pressures toward unification in groups. *Psychol. Rev.*, **62**, 56–68.

SPENCE, K. W. 1956. *Behavior Theory and Conditioning.* New Haven, Conn.: Yale Univ. Press.

STEPHAN, F., and MISCHLER, E. G. 1952. Distribution of participation in small groups. *Amer. Sociol. Rev.*, **17**, 598–608.

STONE, J. J. 1958. An experiment in bargaining games. *Econometrica*, **26**, 286–96.

SUPPES, P. 1959*a*. A linear model for a continuum of responses. Chap. 19 in *Studies in Mathematical Learning Theory*, R. R. Bush and W. K. Estes, eds. Stanford, Calif. Stanford Univ. Press.

———. 1959*b*. Behavioristic Foundations of Utility. Technical Report No. 23, Contract Nonr 225(17), Institute for Mathematical Studies in the Social Sciences, Applied Mathematics and Statistics Laboratories, Stanford University. (To be published in *Econometrica*.)

———. 1957. *Introduction to Logic.* Princeton, N.J.: Van Nostrand.

———. 1960. Stimulus sampling theory for a continuum of responses, Chap. 23 in *Mathematical Methods in the Social Sciences*, K. J. Arrow, S. Karlin, and P. Suppes, eds. Stanford, Calif.: Stanford Univ. Press.

——— and CARLSMITH, J. M. 1960. Experimental Analysis of a Duopoly Situation from the Standpoint of Mathematical Learning Theory. Technical Report No. 30, Contract Nonr 225(17), Institute for Mathematical Studies in the Social Sciences, Applied Mathematics and Statistics Laboratories, Stanford University.

——— and FRANKMANN, R. W. 1960. Test of stimulus sampling theory for a continuum of responses with unimodal noncontingent determinate reinforcement. (To be published in *Journal of Experimental Psychology*.)

——— and KRASNE, F. 1959. Application of Stimulus Sampling Theory to Situations Involving Social Pressure. Technical Report No. 24, Contract Nonr 225(17), Institute for Mathematical Studies in the Social Sciences, Applied Mathematics and Statistics Laboratories, Stanford University. (To be published in *Psychological Review*.)

——— and ZINNES, J. 1960. Stochastic learning models for a response continuum with nondeterminate reinforcement. Technical Report No. 34, Contract Nonr 225(17), Institute for Mathematical Studies in the Social Sciences, Applied Mathematics and Statistics Laboratories, Stanford University.

THRALL, R. M., COOMBS, C. H., and DAVIS, R. L. 1954. *Decision Processes.* New York: Wiley.

VERPLANCK, W. S., COLLIER, G. H., and COTTON, J. W. 1952. Nonindependence of successive responses in measurements of the visual threshold. *J. Exptl. Psychol.*, **44**, 273–82.

VON NEUMANN, J., and MORGENSTERN, O. 1944. *Theory of Games and Economic Behavior.* Princeton: Princeton Univ. Press (2d ed., 1947).

WYCKOFF, L. B., JR. 1952. The role of observing responses in discrimination behavior. *Psychol. Rev.*, **59**, 431–42.

ZERMELO, E. 1912. Über eine Anwendung der Mengenlehre auf ide Theorie des Schachspiels. *Proceedings of the Fifth International Congress of Mathematicians*, Cambridge II, 501–10.

Author Index

Subject Index